Microwave Heating

Microwave Heating
Fundamentals and Applications

Pawel Blasiak
Kossi Tepe

Kruger Brentt
Publishers
2025

Kruger Brentt Publishers UK. LTD.
Company Number 9728962

Regd. Office: 68 St Margarets Road, Edgware, Middlesex HA8 9UU

ISBN: 978-1-78715-421-6

For information on all our publications visit our website at http://krugerbrentt.com/

PREFACE

Microwave heating has emerged as a powerful and efficient technique for rapid and uniform heating of materials in various industrial, scientific, and domestic applications. With its ability to penetrate materials, generate heat volumetrically, and achieve precise temperature control, microwave heating offers numerous advantages over conventional heating methods such as conduction, convection, and radiation. From food processing and pharmaceutical manufacturing to polymer synthesis and materials processing, microwave heating has revolutionized numerous fields, enabling faster processing times, improved product quality, and energy savings. Understanding the fundamentals and applications of microwave heating is essential for researchers, engineers, and practitioners seeking to harness the full potential of this transformative technology.

This handbook, titled “Microwave Heating: Fundamentals and Applications,” is a comprehensive exploration of the principles, methodologies, and applications of microwave heating. Drawing upon the expertise of physicists, engineers, and researchers, this volume offers theoretical insights, practical guidance, and case studies for understanding and utilizing microwave heating techniques in various applications.

From the fundamentals of electromagnetic theory and microwave interactions with materials to the latest advancements in microwave system design, process optimization, and safety considerations, the “Microwave Heating: Fundamentals and Applications” covers a wide range of topics essential for mastering the field of microwave heating. Through clear explanations, mathematical derivations, and experimental techniques, readers will gain a deep understanding of the mechanisms and parameters that govern microwave heating processes and the factors that influence heating performance.

Moreover, this handbook explores the applications of microwave heating across different industries and disciplines, including food processing, chemical synthesis, pharmaceutical manufacturing, and environmental remediation. By examining case studies, research findings, and technological trends, we highlight the diverse ways in which microwave heating is used to enhance process efficiency, product quality, and sustainability.

As we continue to explore new frontiers in materials science, chemistry, and engineering, the importance of microwave heating in enabling rapid and efficient processing has never been greater. It is our hope that the "Microwave Heating: Fundamentals and Applications" handbook will serve as a valuable resource for researchers, engineers, and practitioners in the field of microwave processing. Whether you are developing new microwave heating processes, optimizing existing applications, or exploring innovative uses of microwave technology, this handbook invites you to explore the capabilities and opportunities of microwave heating and contribute to the advancement of materials processing technologies.

Pawel Blasiak
Kossi Tepe

CONTENTS

Part 3: Health and Environment

Part 4: Technology

Part 5: Synthesis

PART 1
GENERAL

1 Chapter

MICROWAVE SYNTHESIS: A PHYSICAL PRINCIPLE

1. INTRODUCTION: AN OVERVIEW

The phrase 'microwaves' refers to wavelengths ranging from approximately 1 meter to 0.1 centimeters, or frequencies between 300 MHz and 300 GHz. Microwaves are predominantly utilized in point-to-point communication, satellite television broadcasting, and RADAR systems. In addition to these, they are utilized in industrial, medicinal, chemical, and scientific research applications. The recent widespread adoption of microwave technology is mostly for "microwave heating" in the synthesis of diverse chemicals (both organic and inorganic) in research laboratories and industry. Dr. Percy Spencer's serendipitous radar-oriented research in 1946 revealed the heating properties of microwaves, evidenced by a candy bar melting in his pocket. Since then, microwave radiation has facilitated the emergence of a novel technology. The application is no longer restricted to the food business but has demonstrated effects in other sectors as well. Microwaves are increasingly utilized in the industrial sectors of inorganic synthesis, polymer curing, and textile drying. The initial published accounts of employing microwave irradiation for organic chemical transformations by Gedye (1986) and Giguere et al. (1986) have established it as a promising technology in medicinal chemistry, polymer synthesis, material sciences, nanotechnology, and biochemical processes. In the initial phase of microwave synthesis, experiments were conducted in sealed Teflon or glass containers within a household oven, lacking pressure or temperature monitoring, occasionally leading to violent explosions due to rapid, uncontrolled heating of organic solvents in a closed vessel environment. In the 1990s, synthesis commenced with solvent-free reactions, mitigating the risk of explosions and subsequently enhancing device designs. Microwave heating serves as an alternative method while enhancing yield and reproducibility. Furthermore, it may result in innovative products by enabling access to alternative kinetic paths and stabilizing various energy minima inside a reaction vessel. Numerous facets of microwave application, particularly in chemical synthesis, have been documented and analyzed since that time (Rao et al. 1999, Vanetsev and Tretyakov 2007, Tsuji et al. 2005, Yoshikawa 2010, Bilecka & Niederberger 2010). Numerous specific frequency bands have been designated for industrial, scientific, and medicinal use. 24.15 MHz, 2.45 GHz, and 5.80 GHz are globally recognized frequencies. There remain a few countries where varying frequencies are utilized. In the United Kingdom, 896 MHz and 40.6 MHz are prevalent, whereas the Netherlands use 3.39 and 6.78 GHz. Metaxas and Meredith (1983)

2. EXPERIMENTAL SET UP

Significant progress has been made in the synthesis of materials, driven by the acceleration of chemical reactions and the necessity for advancements in experimental equipment during microwave heating. Typically, a household microwave oven functioning in multimode configuration is utilized to conduct the experiment. These microwave ovens possess specific constraints. Various modes are generated within the primary chamber or cavity of the oven, disrupting the power matching upon the introduction of the sample. The utilization of domestic microwave ovens was hindered by potential heterogeneity of the magnetic field, occasionally resulting in inadequate repeatability, non-uniform heating, mixing, and exact measurement of the reaction temperature. The dimensions, configuration, and capacity are the primary limitations that must be enhanced to achieve superior and more efficient outcomes. The implementation of a mono-mode system facilitated the concentration of the microwave beam on the sample. Consequently, it is prudent for the researcher to be acquainted with the various components of the experimental apparatus employed for microwave heating. Figure 1 illustrates the block diagram of the microwave processing unit. The primary chamber and some components of this unit can be adjusted to meet the specific requirements of each researcher. Microwave heating employs a high-power source; thus, caution is necessary about the potential hazards of microwaves. The subsequent sub-sections delineate the construction and operation of various components.

2.1. Microwave source: magnetron

Magnetrons, Klystrons, Gyrotrons, and Traveling Wave Tubes (TWT) are employed to produce microwave power. These can provide high microwave power. Each possesses distinct advantages. Klystron delivers accurate regulation of amplitude, frequency, and phase, whereas Gyrotron gives far greater power output and beam focusing capabilities. TWTs can deliver changeable and regulated microwave energy frequencies. Solid-state devices are utilized in applications requiring low-power microwaves. Magnetrons are extensively utilized in microwave heating owing to their accessibility and affordability. The dimensions and design of a cylindrical magnetron are highly appropriate for microwave ovens and various other microwave heating applications. Magnetron can generate either continuous or pulsed power up to megawatt and frequency between 1-40 GHz. Its power efficiency is around 85% and life time is ~ 5000 hours. The basic structure of a cylindrical magnetron consists of a number of identical cavity resonators arranged in a cylindrical pattern around a cylindrical cathode (Figure 2).

Two large pole pieces of permanent magnets are used to produce a strong magnetic field normal to the plane of cavities. The anode is kept at higher potential relative to cathode. The electrons emitted from the central cathode are accelerated towards the anode but the presence of transverse magnetic field exerts a torque which causes the electrons to move in a curved path in the drift space. Due to initial perturbation or interaction of electrons an r.f. field is induced in the cavities which propagate in the azimuthal direction with a certain phase velocity. For a constant value of magnetic field the potential difference between anode and cathode is applied to obtain synchronism between the radial velocity of electrons and phase velocity of nth spatial harmonics of r.f. field in azimuthal direction. The condition is achieved in which the electrons continuously interact with r.f. field in retarding phase and give up their energy to the field. Waveguides are used to couple the microwave power from the magnetron to the main chamber of the oven. The conducting walls of the guides / resonators confine the electromagnetic fields and couple the power from magnetron to the cavity of the oven. A number of distinct field configuration or modes can exist in wave guides. Inside the wave guide either TE or TM modes can propagate but not TEM. These modes are labelled by two identifying integer subscripts 'n' and 'm'i.e. TE_{nm} or TM_{nm}. The integer n and m shows the number of half variation of the respective fields along the two transverse coordinates. Each mode has associated with it a characteristic cut-off frequency $f_{c,nm}$ below which the mode does not propagate or a characteristic cut-off wavelength $\lambda_{c,nm}$ above which the mode does not propagate. The predominant mode in a specific guide is the mode with the lowest cutoff frequency or the highest cutoff wavelength. The field configurations for various modes in both rectangular and circular waveguides are illustrated in Figures 3 and 4.

2.2. Circulator

A circulator is a multi-port device that facilitates unidirectional sequential transmission of power among its ports (Figure 5). Port 1 connects exclusively to port 2, while port 2 connects solely to port 3, and so forth. An arrow on the device indicates the circulation direction of the circulator. The circulator is fundamentally constructed using a three-port Y junction of rectangular waveguides, which is loaded with ferrite cylinders or discs magnetized perpendicular to the plane of the junctions. The three-port junction is fully matched at all its terminals. In a microwave heating configuration, power is transmitted from port 1 to port 2, while port 3 is terminate d with a fake load that absorbs reflected power from the mismatched load.

2.3. Tuner

A tuner is a passive device utilized to align power from a source to a mismatched circuit load. Tuners offer adjustable equivalent inductive or capacitive loads by altering the depth of the sliding screws within the waveguide. The standard design of the sliding screw tuner is tailored to align with the experimental configuration (figure 6). The primary attribute of a tuner is its matching performance, quantified as the highest magnitude of the load reflection coefficient that can be perfectly matched regardless of phase. The maximum magnitude limit is contingent upon the insertion depth of screws and the generator's power level. As input power exceeds the maximum threshold, power dissipation in the tuning stub escalates, resulting in overheating that can melt and damage the adjacent softer components. Consequently, it is essential to ascertain the safe limits of input power for the tuner.

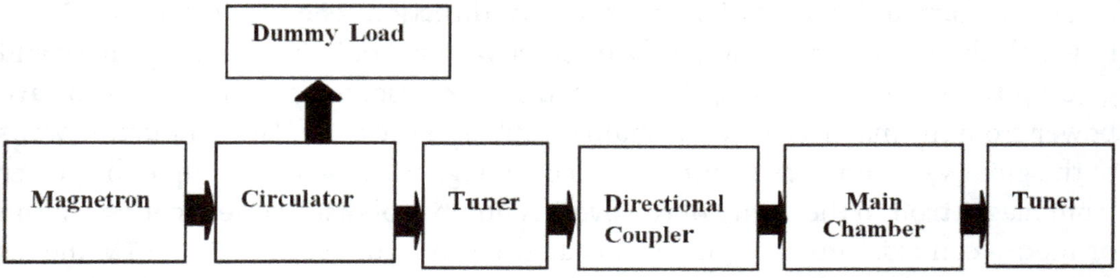

Figure 1. Block diagram of Microwave Processing Unit

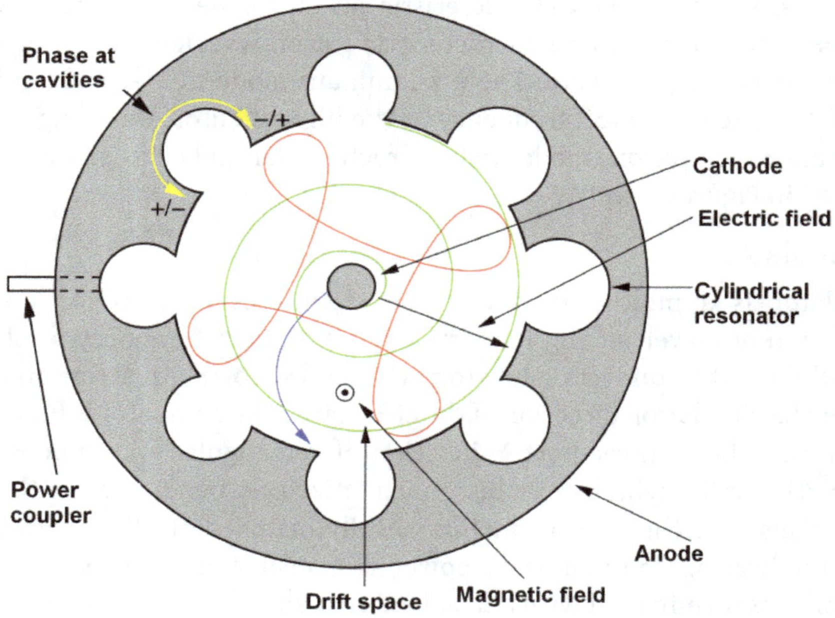

Figure 2. Construction and working of Magnetron

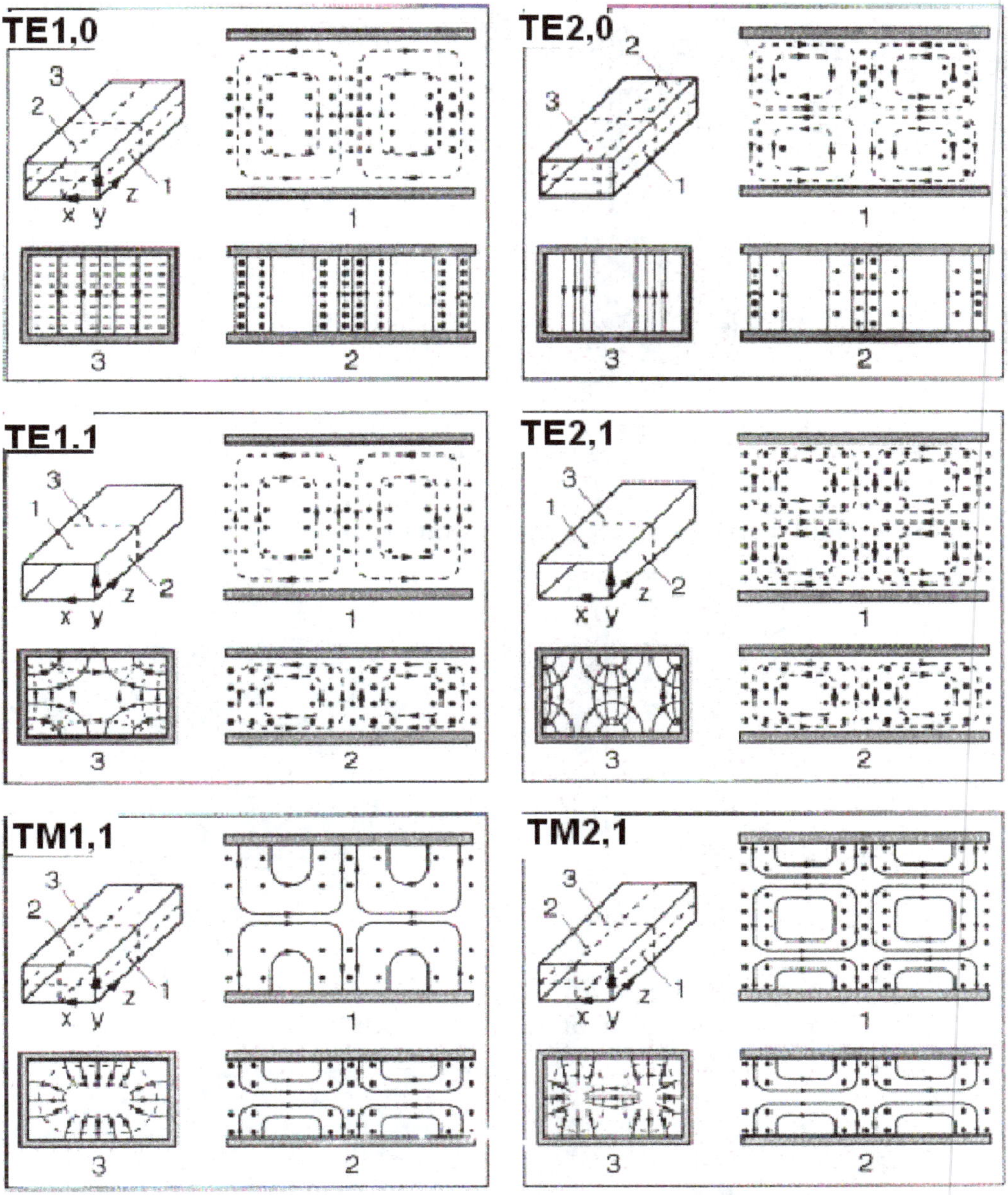

Figure 3. Field distribution for different modes in a rectangular wave guide

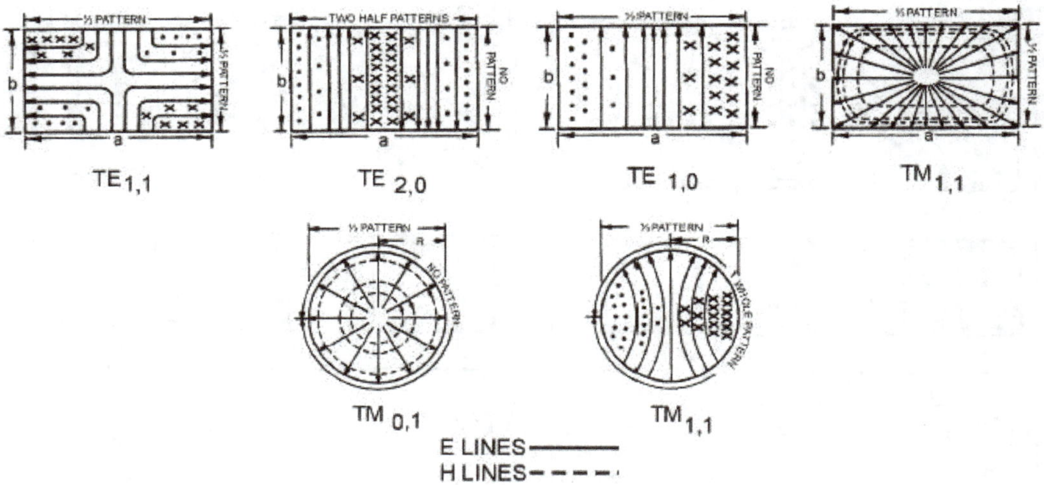

Figure 4. Field distribution for different modes in a Circular wave guide

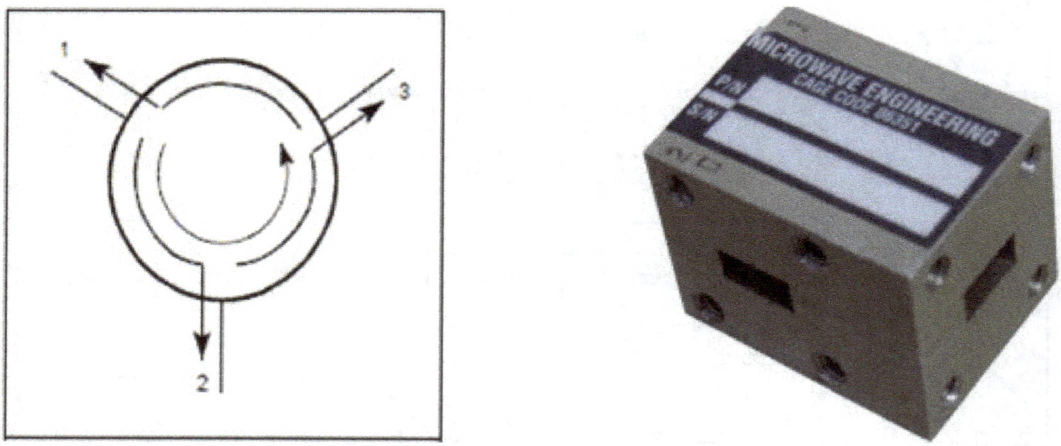

Figure 5. Schematic diagram of Circulator with its commercial design

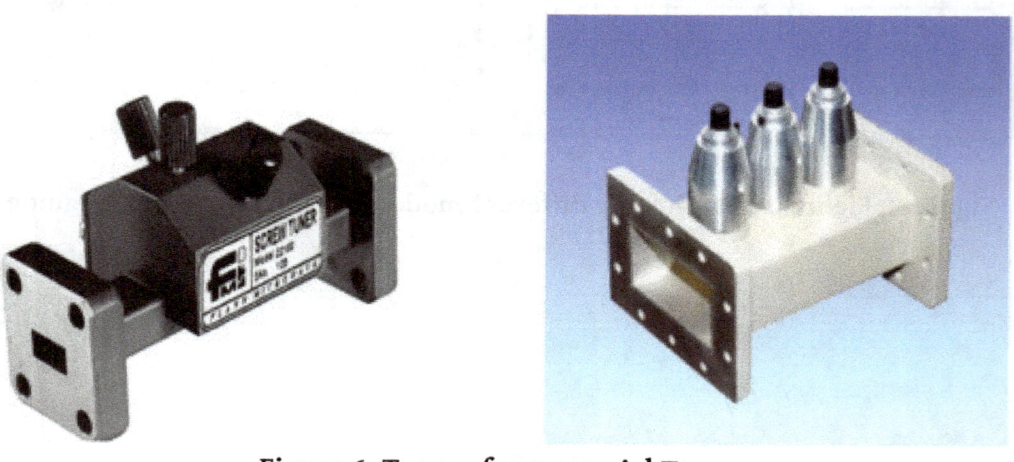

Figure 6. Types of commercial Tuner

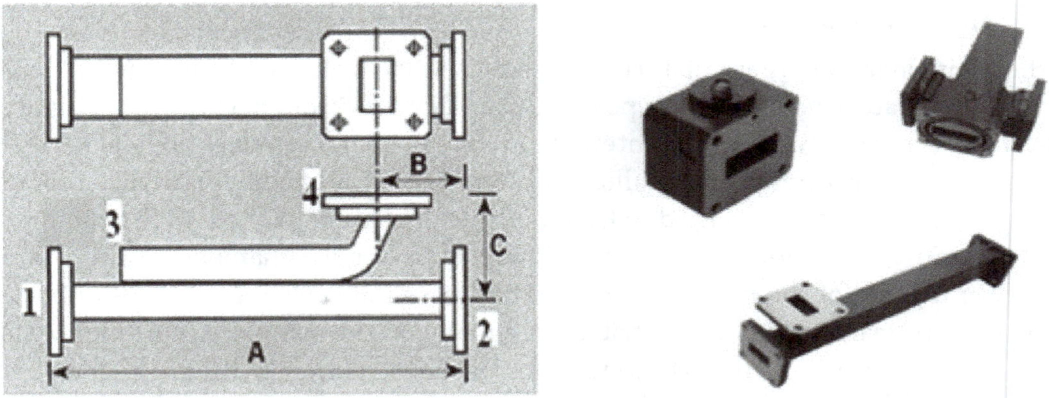

Figure 7. Schematic diagram of Directional coupler with commercial models

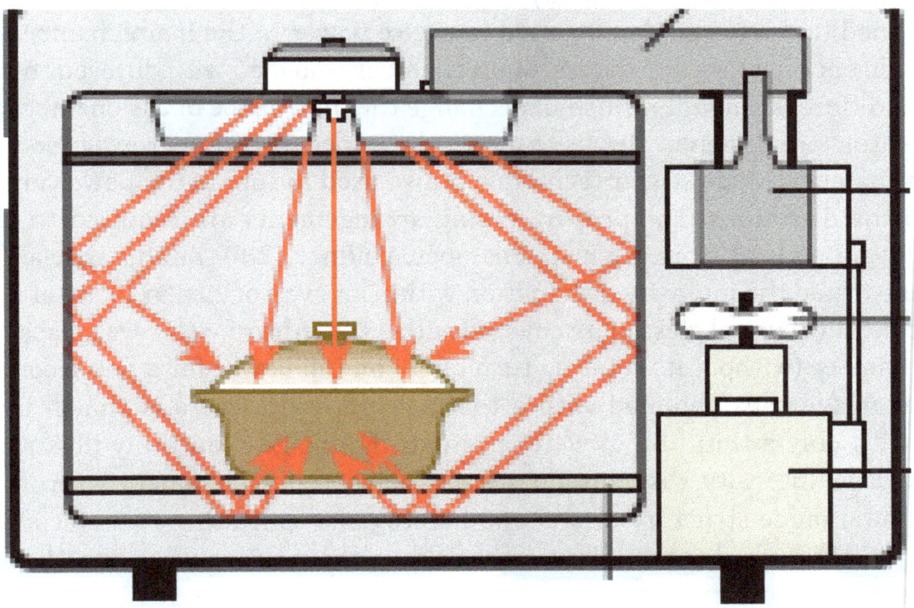

Figure 8. Main Microwave chamber

2.4. Directional coupler

A directional coupler is a four-port waveguide junction wherein power incident at port 1 couples to port 2, with a portion of the incident power also coupling to port 4, but no power couples to port 3. Additional power directed towards the back entering port 2 couples to port 1, with a portion of its coupling to port 3 and no coupling to port 4 (figure 7). The line 1-2 is referred to as the primary line, while the line 3-4 is designated as the linked line. All ports of the directional coupler exhibit optimal matching. The directional coupler measures incident and reflected power from a mismatched load.

2.5. The Main chamber

The Main chamber where material to be microwave processed is placed, is like a cavity resonator and microwave power is coupled to it through wave guides (figure 8). Once the microwave power enters the chamber, standing waves are formed due to multiple reflections from metallic walls. The size of the chamber is of the order of wavelength of microwave used so that multiple modes are excited. In the common domestic ovens the power distribution is uneven inside the chamber resulting into uneven heating of the material. The rotating turn table, therefore, is used inside the oven to make the power distribution equally on the sample. According to the need of microwave–materials interaction, the construction of the main chamber can be modified. In some applications multi chamber ovens are used and energy is coupled through a slot, an array of resonant slots, a radiating horn etc. In some cases modified wave guides are used to excite power in the main chamber. These are called applicators or exciters. Main chamber is loaded with different materials and in different amounts which may change the impedance of the chamber. These applicators are designed always to match the impedance of microwave source and main chamber. Mode stirrer technique is also used to spread the power uniformly in all the directions. The proper cooling arrangements are required to exhaust the unwanted heat from the main chamber. Vollmar (2004) in his special feature demonstrated the intensity distribution within an oven of $29x29x19cm^3$ at a height of 8 cm. A horizontal glass plate covered with a thin film of water was placed inside the chamber (without its rotating turn table) on full power for a few seconds. The colour image thus obtained with a thermal infrared camera is shown in figure 9 a. With only a thin film of water present, the image beautifully illustrates the microwave intensity distribution in a nearly empty chamber. A pronounced horizontal mode structure also shows heterogeneous heating of material. Cheng et al. (2001) studied the distribution of fields in the microwave cavity. The electric (E) field was found maximum in the centre, where the magnetic field is minimum. On the other hand maximum magnetic field is found near the wall with minimum electric field (Figure 9b).

3. MICROWAVE HEATING

3.1. Microwave versus conventional heating

Traditional heating often employs a furnace or oil bath to elevate the temperature of reactor walls through convection or conduction. The center of the sample requires significantly more time to reach the desired temperature. Conversely, microwaves infiltrate the substance, generating heat by direct interaction between the microwave and the substrate. Additionally, microwave heating offers advantages over conventional approaches, including volumetric heating, accelerated reaction rates, enhanced chemical yields, reduced energy consumption, and varied reaction selectivity. Combustion synthesis is a process employed to achieve compositional consistency in powders. Figure 10 demonstrates the advantages of microwave heating compared to traditional methods in the manufacture of LaCrO3 powder as uniform fine particles (Park et al. 1998). The mixed solution of the primary components was evenly partitioned and maintained for combustion separately on a hot plate and in a microwave oven. The resultant combustion products were subjected to analogous treatment, and their morphology was analyzed using scanning electron microscopy (SEM). The product resulting from hot plate combustion comprises rigid agglomerates generated by the two-dimensional connectivity of spherical particles. Conversely, the product of microwave-induced combustion has a smaller size and diminished agglomeration. Kappe and Dallinger (2006) tested the temperature of a 5 ml amount of ethanol and noted that the temperature increases uniformly throughout the volume when heated with microwave radiation, whereas the sample in an oil-heated tube in contact with the vessel wall heats up first (figure 11). Gedye et al. (1988) conducted various organic processes in sealed Teflon vessels using a microwave oven and under conventional reflux settings, observing an approximate rate enhancement ranging from 5 to 1240 times. Derivatives that typically necessitated 2-3 hours of preparation are efficiently synthesized in 2-3 minutes using a microwave oven.

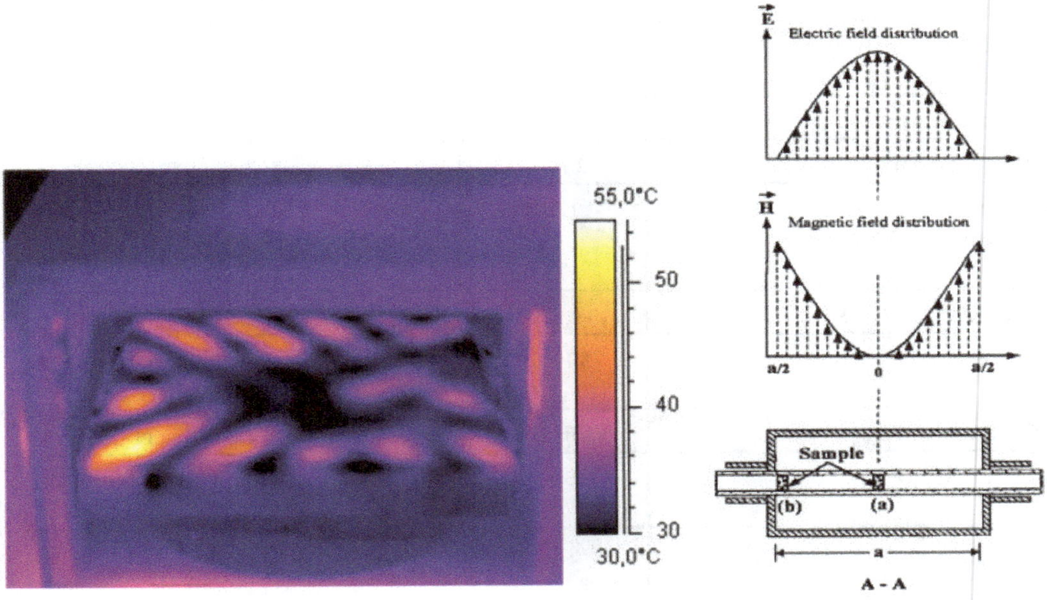

(a) (b)

Figure 9. a) Infra red thermal imaging inside microwave oven (Vollmer 2004)(b) The schematic of the microwave field distribution (E & M)in the microwave cavity (Cheng et al.2001)

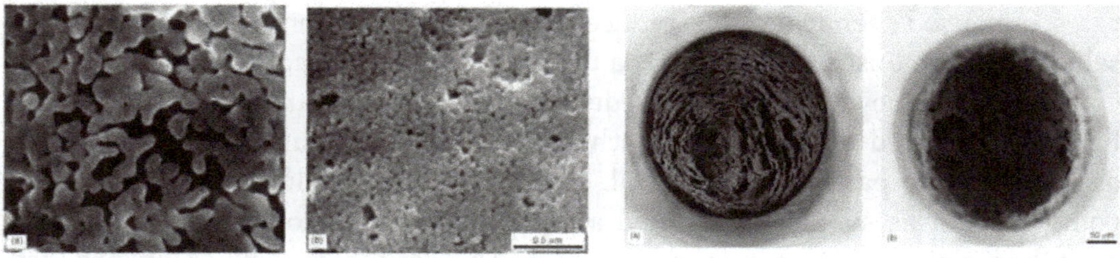

(a) (b) (a) (b)

Figure 10. Morphology with SEM micrograph of the combustion process (a) hot plate (b) microwave radiation (Park et al. 1998)

A few reactions which were carried out using microwave heating and compared with conventional heating indicating time and energy efficiency of the technique is compiled in Table 1.

Table 1. Comparison of the reaction times using microwave versus conventional heating.

Compound Synthesized	Reaction Time - Microwave	Reaction Time -Conventional	References
Esterification (benzoic acid with methanol)	5 min.	8h	Gedye et al (1988)
4-nitrobenzyl ester	2 min	1.5 h	Gedye et al (1988)
$CuBi_2O_4$	5 min	18h	Jones & Akridge (1995)
Bi_2Pd(Intermetallic)	4 min.	12 h	Lekse et al. (2007)
Ag_3In(intermetallic)	2 min.	48 h	Lee & So. (2000),
Layered Al and Zn double hydroxide with Na-dodecyl sulfate	1-2 h	2-3 days	Hussein et al. (2000)
Bronzes (Na_xWO_3)	13-15 min.	-	Guo et al. (2005)
Ti N	30 min	-	Vaidhyanathan & Rao (1997)
Cubanite $CuFe_2S_3$	3min	3 days	Chandra et al. 2010
$La_{2-x}Sr_xMn_2O_4$	30 s	-	Mingos & Baghurst. 1991
High Tc superconductors YBCO	12 h	72 h	Binner & Al-Dawery (1998)
Zeolite synthesis	170 ^{0}C in 30s	170^{0}C in 60min	Jansen 2004

Compound Synthesized	Reaction Time - Microwave	Reaction Time -Conventional	References
MgB_2	11min	-	Dong et al. 2007
$NaAlH_4$	2h	8h	Krishnan et al. 2009
$La_{0.2}Sr_{0.8}Mn_{0.8}Fe_{0.2}O_{3+\Delta}$	3h	3 days	Chandra (unpublished)

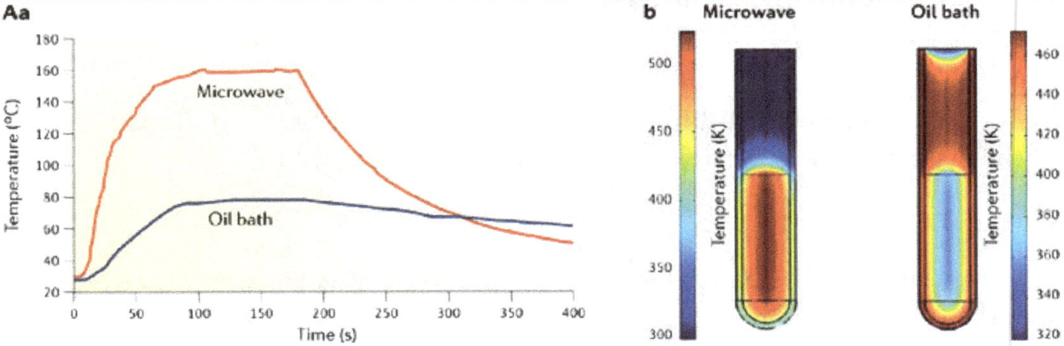

Figure 11. Temperature profiles for ethanol under microwave radiation and open vessel oil bath condition and temperature gradient 1 min. after heating (Kappe and Dallinger 2006).

3.2. Microwave interaction with materials

Two critical considerations in selecting the frequency of microwave radiation for heating materials are (i) power absorption within the substance and (ii) penetration depth. In electromagnetism, materials are classified into two categories: (i) conductors and (ii) insulators or dielectrics. The differentiation between them is not particularly pronounced. A single substance can function as a conductor at one segment of the electromagnetic spectrum and as a dielectric at another. Maxwell's theory posits that the ratio ($\sigma/\omega\varepsilon$) serves as a distinguishing factor, where σ represents electrical conductivity, ω denotes angular frequency, and ε signifies permittivity or dielectric constant. For excellent conductors, this ratio significantly exceeds unity, but in dielectrics, it is considerably less than unity. In dielectrics, electrostatic fields can endure for extended periods, exhibiting significant resistance when subjected to static voltage, but in good conductors (metals), the static current encounters less resistance. At elevated frequencies, the behavior is contingent upon the frequency of the electromagnetic waves and the associated conductivity, permittivity, and permeability of the material. The transmission of electromagnetic radiation through a material medium is linked to the numerical values of permittivity or dielectric constant at that frequency.The absorption of electromagnetic energy depends upon complex permittivity ε of that material and can be expressed as

$$\varepsilon = \varepsilon' + j\varepsilon'' \quad (1)$$

where ε ' = the real component of permittivity

ε" = the imaginary component of permittivity

The real part or relative permittivity represents the degree to which an electric field may built up inside a material when exposed to the electric field while the imaginary part or dielectric loss is a measure of amount of the field transformed into heat.

The Loss angle δ, the phase difference between the electric field and the polarization of the material is related to the complex dielectric constant as

$$\tan \delta = \varepsilon'' / \varepsilon' \quad (2)$$

Thus the tan δ, the dissipation factor determines the ability of material to transform absorbed energy into heat.

In terms of microwave interaction, the materials can be classified into three categories:

1. Microwave reflectors e.g. metals
2. Microwave transmitters - transparent to microwave radiations e.g. fused quartz, ceramics, zircon etc.; tan $\delta < 0.1$
3. Microwave absorbers taking up energy from the microwave field and heating the materials rapidly; tan $\delta > 0.1$

The electromagnetic energy absorption in dielectric materials primarily is due to the existence of permanent dipole moment of the molecules which tend to orient and reorient under the influence of electric field of microwave. The reorientation loss mechanism originates from the inability of the polarization to follow extremely rapid reversals of the electric field. In the low frequency (up to 100 MHz) electric field, the dipoles easily follow the changes in the field and their orientation changes in phase with the field. At higher frequencies the inertia of molecules and their interactions with neighbours make changing orientation more difficult and the dipoles lag behind the field. As a result, the conduction current density has a component in phase with the field and therefore power is dissipated in the dielectric material. At very high frequencies (1- 10 THz), the molecules can no longer respond to the electric field. At GHz frequency (ideal working range) the phase lag of the dipoles behind the electric field absorbs power from the field and therefore pronounced as dielectric loss due to dipole relaxation.

Another important parameter for microwave heating,penetration depth D_{ph} is defined as the depth into the material where the power is reduced to ~ 1/3 of the original intensity (Jansen 2004) The absorption coefficient α of dielectric material is related to imaginary parts of dielectric constant ε" or the refractive index. The depth of penetration D_{ph} of electromagnetic waves in matter is related to α as $D_{ph} = 1/\alpha$. Thus a material with higher dissipation factor will have lower penetration depth. The wavelength of the radiation also has influence on penetration depth. For microwave prone materials the absorption coefficient at 2.45 GHz is moderate and depth of penetration is of the order of 10 cm to 1m which results in absorption of microwave everywhere in material. The intensity of heat evolution in a sample depends on electrophysical properties of the materials, frequency, intensity of the applied field, penetration depth of the electromagnetic waves into the substance under treatment and geometric size of the sample. The dielectric parameter of the matter and the penetration depth depend strongly on temperature and therefore varies during heating.

Water is an excellent example of polar molecule. The principle of microwave heating of polar molecules (for the case of water) is illustrated very nicely in figure 5 (Tsuji et al. 2005). These molecules absorb microwave energy rapidly and the rate of absorption varies with the dielectric constant of the material. The electric dipoles present in the dielectric materials respond to the applied electric field of microwave. Resistive heating occurs when the dipolar orientation is unable to respond to the applied microwave field resulting into a phase lag.

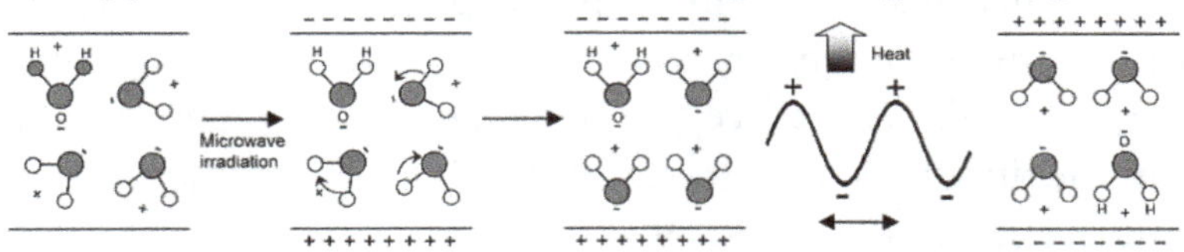

Figure 12. Heating mechanism of water due to microwave field. (Tsuji et al. 2005)

In general, the dipolar species in any medium possess a characteristic relaxation time τ and frequency dependent dielectric constant. If the dipolar relaxation is characterized by a single relaxation time, then dependency with frequency is given by Debye equations (Rao et al. 1999):

$$\varepsilon' = \varepsilon_{in} + (\varepsilon_s - \varepsilon_{in}) / (1+\omega^2\tau^2) \tag{3}$$

$$\varepsilon'' = (\varepsilon_s - \varepsilon_{in})\, \omega\tau / (1+\omega^2\tau^2) \tag{4}$$

Where ε_s and ε_{in} are the zero frequency and the infinite frequency dielectric constants respectively. ε" varies with frequency giving rise to a characteristic frequency at $\omega = 1/\tau$. For water, ε" being significant at 2.45 GHz, a rapid dissipation occurs as heating of water.

Microwave interaction with materials can be predicted if dielectric properties (relative dielectric constant ε' and relative dielectric loss factor ε") are known. The loss tangent $\tan\delta = \varepsilon''/\varepsilon'$ is an indicator of the ability of the material to convert absorbed energy into heat. A good absorber has $\tan\delta \geq 0.1$ while those with $\tan\delta \leq 0.01$ are transparent to microwave. Materials with high loss factor at the frequency of the incident radiation will heat at a faster rate from core to surface. Fused quartz, zircon, several glasses, ceramics and Teflon are good transmitter while in SiO_2,dielectric constant and losses do not show much dispersion hence no heating occurs. Tap water is microwave active as compared to distilled water. Alumina is microwave transparent at room temperature at 2.45 GHz but heats up efficiently at 1000º C becoming highly susceptible at 1500º C. SiC, on the other hand usually heat well at moderate frequency (Kubel 2005). Microwave susceptors are fabricated by depositing very thin layer of aluminium on polyster (PET) sheets. The thin layer of aluminium absorbs part of microwave energy creating currents in the metal. The thickness of the layer limits the currents hence preventing arcing however the currents are high enough to heat the susceptor to a high temperature. Such arrangements are popular in food industry. A list of materials useful during microwave heating process along with ε', ε", tanδ and D_{ph} is tabulated in Table 2.

3.3. Microwave interaction in metals

Skin depth and penetration depth are important parameters in microwave heating of metals. When microwaves are incident perpendicularly on the surface of the materials, its intensity decreases progressively due to dissipation inside the volume of the materials. Therefore the term 'penetration depth D_{ph} is defined as the distance in the direction of penetration at which the incident power is reduced to half of its initial value (Rao et al. 1999):

$$D_{ph}=3\lambda_0 / [\, 8.686\pi \tan\delta \, (\varepsilon')^{1/2} \,] \tag{5}$$

Where λ_0 is the wavelength of the microwave radiation.

In metals microwave propagation is usually described in terms of 'skin depth'. The skin depth is defined as the depth at which the magnitude of electric field drops to 1/e of the value at the surface and is given as (Newham et al. 1991):

$$\delta = 1/\, (\nu\pi\mu\sigma)^{1/2} \tag{6}$$

where ν is the microwave frequency (ω =2πν), μ is permeability of the free space and σ is the electrical conductivity.

The behaviors of bulk metal pieces and metal powder under microwave radiation are different. Skin depth in metal is very low and varies as $(\sigma)^{-1/2}$ showing less penetration of microwaves. In large metals and metal films, electric field gradients occur in the microwave cavity giving rise to electric discharge. In metal powder, due to eddy currents and plasma effects, very rapid heating takes place without any discharge.

Table 2. Various factors of the materials useful for microwave heating

Material	ε'	ε"	tan δ	D_{ph}(cm)	Reference
Silicon Carbide	10.5	11.0	1.048	0.28	Clark & Sutton (1996)
Ethylene Glycol	41	41	1.350	-	Kappe & Dallinger (2006)
Active C	7	2	0.286	-	Vos et al. 2003
Ethanol	24.3	6.08	0.250	-	Kappe & Dallinger (2006)
Methanol	32.7	20.9	0.240	-	"
Wood Wood (40% water)	1.591 5.1	0.033 1.12	0.0207 0.2196	3.57 4.0	Rattanadecho(2006) Starck et al (2005)
Soda Lime Glass	6.0	1.20	0.200	1.5	
Water (Distilled 25ºC) " (95ºC) Water (Deionized) Water (0.5% salt) Ice Water (Tap Water)	78.3 - 78.2 75.8 3.2 67.5	12.3 - 10.3 15.6 0.0029 6.0075	0.157 - 0.0009 0.089	1.5 5.7 1.68 1.09 1.80	Shulman (2002) ' Schiffmann (1986) "Starck et al.(2005) Hasna (2009)
SiC	10.4	0.9	0.0865	7	Starck et al (2005)
SiO_2	3.066	0.215	0.0701	-	Vos et al. 2003
Soot	~10	~3	0.03	-	Vos et al.2003
Lithium Disilicate Glass	7.5	0.17	0.023	1.30	Clark & Folz
Silicon Nitride	0.68	0.015	0.022	8.13	Starck et al (2005
Zirconia	-	-	0.015	100	Hasna (2009)
PVC	2.9	0.016	0.0055	200	Starck et al (2005)
Natural Rubber-With S (3.0pph) With S (3pph) & C black	2.27 3.366	0.004 0.0142	0.002 4.219	173400 274.0	Makul et al.(2010) "
Boron Nitride	4.35	0.0131	0.003	0.5	Starck et al (2005)
pyrex	4.0	0.005	0.0013	1.8	"
Aluminum Nitride	9.0	0.008	0.00089	1.18	"
Polyethylene	2.3	0.001	0.00040	2.57	"
PTFE	2.1	0.0006	0.0002857	4.7	"
Teflon	2.1	0.0003	0.00014	2.73 9200	" Shulman (2002)
Alumina γ – Al_2O_3	8.9 3.006	0.009 0.1720	0.00010	1.46	Starck et al (2005) Vos et al. 2003
Fused Quartz	3.8	0.0001	0.00003	1.9	-

Material	ε'	ε"	tan δ	D_{ph}(cm)	Reference
Cr_2O_3	10.3	~ .0025	0.0002427		Dube et al. 2007
Graphite	-	-	-	0.0038	Vatensev & Tretyakov 2007
Copper				0.00026	Vatensev & Tretyakov 2007
Epoxy resin				7300 4100	Vatensev & Tretyakov 2007 Shulman (2002)
Porcelain				56	Shulman (2002)
Quartz				1600	"

Metals are prohibited inside microwave oven due to discharges it produces inside damaging the magnetron supply. However microwaves are being used efficiently in melting of metals. Domestic microwave appliances are based on magnetron, cooled by a stream of air from a fan. The microwaves are reflected by the metal walls till all the energy is absorbed and converted into heat. To be used for metal casting, the domestic oven needs slight modifications (David Reid). The rotating glass plate must be removed, the vents admitting air into the chamber must be taped over and the air from the magnetron cooling redirected to the exterior. The insulation of the sample chamber, a critical factor, has two roles: one it contains the microwave energy within the shell so that the temperature rises to the melting point of the metal; second it protects the walls of the oven. Carbon, magnetite, ferrites are good susceptors of microwave radiation. Early experiments with Carbon (considered a good susceptor of microwaves) for melting of silver were very discouraging. The un-insulated crucible barely attained red hot. Though insulation helped but it was obvious to find a more efficient absorber. 8 mm of SiC paste mixed with clay, applied to inside of ceramic crucible also did not show any results even after 10 minutes of heating. Another crucible with clay-ferrite paste capped with carbon lined shell showed some improvement but silver did not melt. On lining the cap with magnetite-clay mixture and after 15 minutes of firing, silver started melting. A double susceptor- a carbon (graphite) loaded primary coat with magnetite was used successfully to melt small amount of cast iron. The caveat of metal melting is oxidation; hence all melting must be performed in an inert atmosphere or under vacuum.

Agrawal et al. (2006) attempted joining of metal/metal systems by inserting metal powder between metal/ ceramic pieces and irradiating with microwaves. The powder was heated preferentially and thus the steel joints were formed. Attempt of Microwave sintering of metal glasses without crystallization and with addition of Sn for sintering have been reported (Xie et al.,2009).

3.4. Role of magnetic field in microwave heating

So far the microwave interaction with nonmagnetic material is discussed in which electromagnetic energy loss is associated with complex permittivity of the material. The total microwave power dissipation in a material is given by

$$P = 2\pi \left[\varepsilon_0 \varepsilon' \tan\delta E^2 + \mu_0 \mu' \tan\varphi H^2 \right] \quad (7)$$

Where ε_0 and μ_0 are dielectric permittivity and magnetic permeability of free space, ε' and μ' are the real part of the dielectric permittivity and magnetic permeability of the sample, tan δ and tan φ are the dielectric and magnetic loss factor values and E and H are the electric and magnetic field values inside the sample. Although E and H appear in the equation, mostly it is assumed that entire reaction is due to the losses by the electric field vector. Cherradi et al.(1994) showed substantial contribution of magnetic field towards the heating of alumina (at high temperature), semiconductors and metals. Different materials seem to have different heating behaviours in the E and H microwave fields (Cheng et al. 2001). The high conductivity samples like metallic powders could be efficiently heated up in the magnetic field while the low conductivity samples (ceramics, alumina) got much higher heating in the pure electric field.

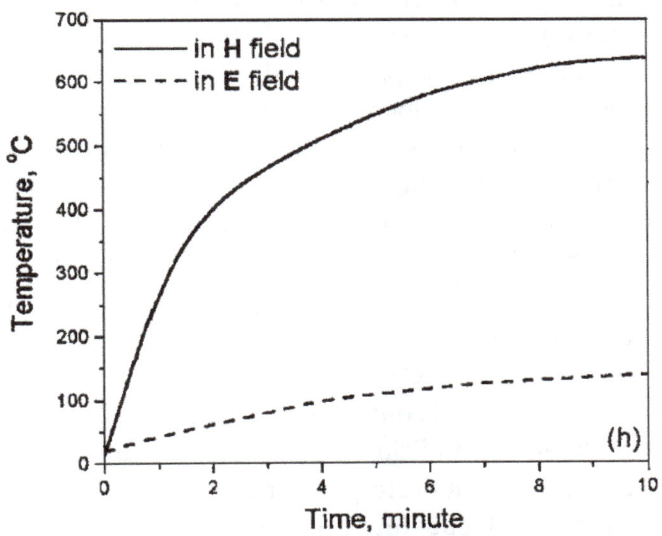

Figure 13. Comparison of the heating rate of powdered tungsten carbide- cobalt (WC-Co) sample (microwave power 200W) (Cheng et al. 2001)

For the same material (Cu), the compact sample absorbed lot of microwave energy in the magnetic field while the solid sample did not respond in the same condition. Roy et al.(2002) demonstrated that the common crystalline phase can be made non crystalline and hard magnets can be softened in the solid state by heating them in magnetic component of microwave radiation. The decrement in the resistance for polycrystalline LCMO films observed under influence of microwave radiation was explained by tunnelling of carriers through inter-grain boundaries (Lučun et al. 2007). The induced resistance depended on the sample position in the wave guide thereby near the narrow wall of the wave guide, magnetic component of the radiation had significant influence on the magneto-resistive properties of the film. While studying the dependence of electron paramagnetic resonance (EPR) of some manganites on microwave power, it was observed that the sample became unusually hot at high microwave power (Singh 2007) due to the interaction of the spin of e_g electron of Mn^{3+} with the magnetic component of microwave field.

4. MICROWAVE SYNTHESIS: AN EXAMPLE

The nature of the minerals as they exist at atmospheric pressure represents only a part of their real nature. The pressure and temperature conditions impart strong influence on the structure and characteristics of the mineral. As one travels from Earth's mantle to the core, both temperature and pressure conditions changes drastically. The range of pressure and temperature prevailing at the surface of the Earth can be compared to those on other planets of the solar system. For example Sulfide minerals occur in the Earth's as well as in the Mars's mantle. Determining the thermodynamical properties of sulfides by synthesizing minerals under various laboratory conditions and comparing their behavior with natural one would be very important to understand the paragenesis of their formation in the planetary system. One of the examples described here is orthorhombic cubanite $CuFe_2S_3$ which occurs in mineral deposits along with chalcopyrite ($CuFeS_2$) and pyrrhotite ($Fe_{1-x}S$) under certain conditions. Orthorhombic cubanite is very sensitive to temperature and pressure. At ~ 200º C or ~ 4 GPa, the orthorhombic structure transform to cubic isocubanite. The discovery of cubanite in CI chondrites and stardust in comet wild 2 clarified the mechanism of large scale mixing of low-temperature assemblages in early solar system. The phase relationship in Cu-Fe-S determines the co-existing phases as a function of temperature and point to the unique combination of temperature and composition (Yund and Kullerud 1966). Owing to the unique physicochemical conditions needed, orthorhombic cubanite is difficult to synthesize under normal laboratory environment. In this direction, synthesis of this rarely found mineral, cubanite ($CuFe_2S_3$) was successfully attempted using conventional as well as microwave heating. The rapid microwave heating seems to be advantageous over conventional technique showing a characteristic peak of cubanite at d= 3.22 Å (Pareek et al. 2008, Chandra et al. 2010,2011). It is reported that synthesis of orthorhombic cubanite depends not only on the preparatory conditions (controlled heating and cooling rate) but also on compositional variation. Figure 14 shows the dependency of synthesis process as well as proportional of sulfur. X-ray diffraction pattern of resistive heated sample slowly heated and cooled to room temperature (a) with stoichiometry Cu:Fe:S::1:2:3 has been analyzed as isocubanite (cubic cubanite), pyrrhotite and chalcopyrite while rapidly heated microwave synthesized sample (b) with Cu:Fe:S::1:2:1.5 showed orthorhombic cubanite indicated by the arrows in the figure 14 b. It is to be noted in figure 14b that the proportion of sulfur is an important component for the formation of orthorhombic cubanite. The sample was prepared with the same conditions but with increased proportion of Sulfur (Cu:Fe:S::1:2:3). The conspicuously absence of prominent pyrrhotite peak at ~ 55º and formation of cubic isocubanite is observed.

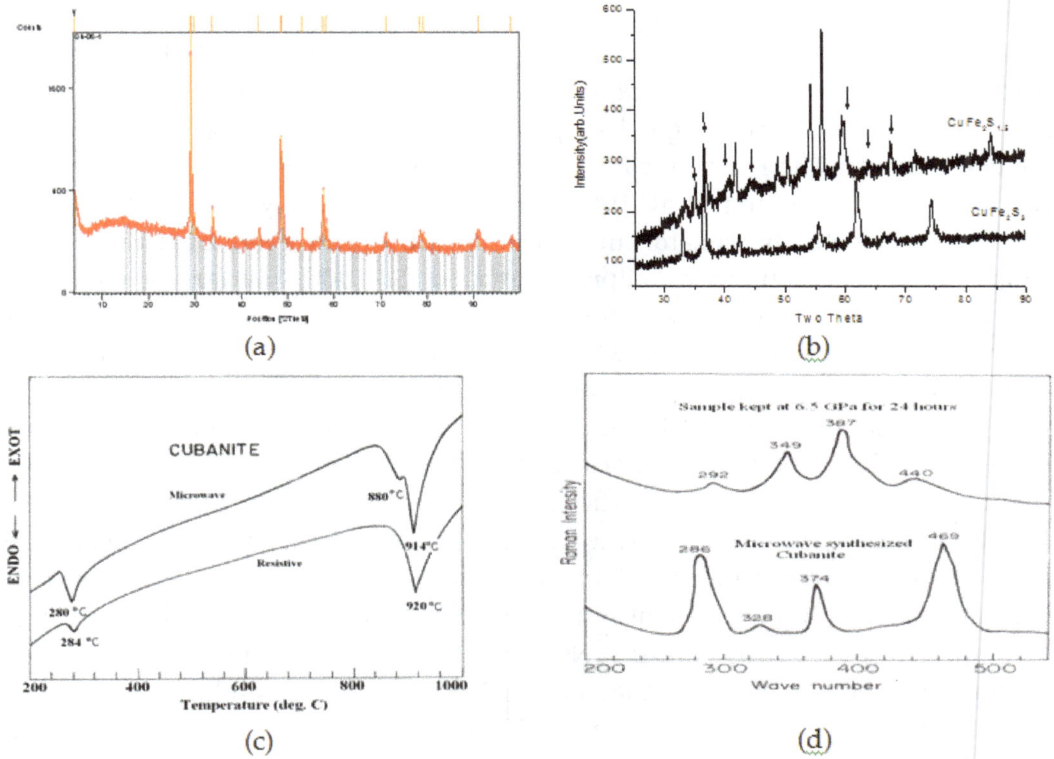

Figure 14. X-ray diffraction patterns of (a) resistive heated (b) microwave synthesized cubanite.The arrows mark the orthorhombic cubanite. (c) DTA and (d) Micro-Raman spectroscopic studies of - resistive heated and microwave synthesized cubanite

The intensity at endothermic peak at 280º C in the differential thermal analysis (DTA) patterns of synthetic cubanite samples prepared by resistive and microwave heating is a quantitative indicator of the presence of orthorhombic cubanite (Figure 14c). The peak at 880º C represents pyrrhotite phase. The absence of an endothermic peak in the region 550º C- 650º C confirms the absence of chalcopyrite in the sample. The Micro-Raman spectroscopic studies on the as prepared synthetic cubanite show strong Raman modes at 286,374 and 469cm^{-1} and a weak peak at 328cm^{-1}

corresponding to the desired cubanite (Fig 14d). The measurements repeated again after pressurizing the sample at 6.5 GPa for 24 h indicates isocubanite. Though x-ray diffraction pattern on conventionally synthesized sample does not indicate orthorhombic cubanite, Mössbauer spectroscopic studies confirm the presence of the phase through its sensitive parameters- magnetic hyperfine interaction and isomer shift. This is first ever study of successfully synthesized cubanite (Chandra et al. 2010).

The microwave dielectric heating effect originates from the natural ability of certain substances to efficiently absorb and to transform the electromagnetic energy into heat. If sufficient heat can be generated at a local level, then chemical reaction may be initiated at a rapid pace. To initialize the process, sometimes, it is necessary to introduce good susceptors of microwave radiation along with the constituents. The reaction can accelerate even if one of the constituents is a good susceptor. Baghurst et al. 1988 carried out pioneering work of the use of microwave dielectric heating for the synthesis of advanced ceramic materials- high temperature superconducting cuprates and $La_{2-x}Sr_xCuO_4$. The strong absorbing property of CuO (a good susceptor)leads to a rapid and effective pathway for the reaction achieving 700º C in 30 sec using domestic oven. However synthesis of ternary CMR perovskites [(La,Sr)MnO_3] are difficult to synthesis because La_2O_3, MnO_2 etc attain only 107º and 321º C respectively after 1800 s (Gibbons et al. 2000) while $SrCO_3$ does not absorb microwaves to any significant degree (Mingos & Baghurst 1991).The addition of efficient absorber (fine grained graphite) in the sample promotes the desired effect by significantly increasing the temperature of the reaction mixture. $Ca(NO_3)_2$ and $Mn(NO_3)_2$ are better receptors of microwave as compared to MnO_2. In the synthesis of mixed valent perovskites, these nitrates decompose to yield highly oxidising NO_2 facilitating the formation of Mn^{3+}/ Mn^{4+} (Gibbons et al. 2000). In the synthesis of $MnCo_2O_4$, the aqueous solution of nitrates of Mn and Co (non susceptible to microwave radiation) are mixed with carbon powder (good absorber of microwave)which not only acts as a surface enhancing agent but also decreases the decomposition temperature of the nitrates by~100º C (Nissinen et al 2003)

Bulk MgB_2 with highest T_c value is not easy to prepare because Mg is extremely volatile and susceptible to oxidation at elevated temperature. Usually Mg B_2 bulk sample are prepared in a sealed quartz / tantalum tubes or by high pressure sintering. However using Microwave heating, the evaporation and oxidation of Mg can be significantly reduced. Pellets of the constituent mixture in an alumina boat filled with SiC powder (used as MW susceptor) placed in a modified domestic oven operated at 800 W for 11 min resulted into good crystals of MgB_2 (Dong et al 2004)

Bosi et al (1992) predicted the possibility of improved microstructure on superconductors while processing the thin film via microwave heating. If the film thickness is less than the skin depth, "volumetric" heating similar to dielectric might be possible through microwave heating. Also in some thin film applications, where film needs to be deposited on the substrate (such as Si,a potential contaminant of superconductors) microwaves might be useful in the final post-processing stages because selective heating would heat the film keeping the substrate cool, suppressing the diffusion of silicon through the film. Conventional processing of bulk ceramic high T_c YBCO superconductors results in low and non-uniform oxygen contents throughout the sintered and annealed bodies, limiting the value of T_c. In contrast microwave heating offers the ability to achieve perfectly uniform and full oxygen content (x=7) and almost full densities (98%) throughout the bulk ceramic samples. The critical transition temperature was also improved, grain size was uniform and finer than in the conventionally processed sample. In addition the total processing time was approximately a factor of six faster. (Binner & Al-Dawery 1998)

Mercury sulphide is a useful material and is used widely in many fields such as ultrasonic transducers, image sensors, photoelectric devices, conventional preparation of the item is not popular because of the toxicity of mercury. Microwave heating found to be a fast (5-10 min), convenient, energy efficient and environment friendly route to the synthesis of nano-crystalline Hg S particles with controllable sizes. The size of the particles is found to be dependent on various solvent used e.g. Absolute ethanol, tetra-hydro-furan, distilled water, dimethyl formamide (DMF), and 20% DMF solutions respectively (Wang et al 2001)

The unprecedented flexibility of ink-jet printing makes it very suited for rapid prototyping applications. The ink-jet printing requires conducting tracks by using inks based on organic silver or copper precursors. The precursor is reduced to corresponding metal via a post-printing annealing step. Conventional radiation-convection-conduction heating is the most commonly used method wherein sintering temperatures are ~ 200º C. Therefore many potentially interesting substrate materials such as polymers cannot be used. The required long sintering time (~60 min.) also implied the non feasibility of the method for fast industrial application. Replacement of microwave heating fulfils the requirement for a fast, simple, cost effective and selective heating of only the printed components (Perelaer 2006)

In the manufacturing of diamonds, the processing through carbon is either by high temperature / high pressure (HTHP) or by high temperature/ low pressure deposition process such as Chemical Vapour Deposition (CVD). Crystalline diamonds predominantly composed of {100} and {111} faces was grown on a non-diamond substrate from a gaseous mixture of hydrogen and methane under microwave glow discharge condition (Kamo et al. 1983) Diamond wafers with a thickness ranging to several microns is a promising material for Infra-red optics as well as for x-ray and particle detector. To be economically viable, the chemical vapour deposition of diamond should be performed at a large area. Compared with hot filament CVD and other techniques, microwave plasma can provide a large and good quality diamonds using hydrocarbon radical and atomic hydrogen (Weidong 2006).

Useofmicrowaveheatingindryingofwoodisbecomingpopular.Rattanadecho(2006) studied the influence of irradiation times, working frequencies and sample size dependency of microwave heating patterns within the wood under various conditions. At 2.45 GHz, the power distribution as well as temperature distribution within the sample displays a wavy nature due to the thickness of the sample being close to the penetration depth. Most of the heating seems to occur at the centre of the test sample due to maximum electric field distribution.

5. HAZARDS OF MICROWAVES

Microwave radiation is a well-established source of significant health issues, including lack of appetite, irritation, pain, weariness, and headaches, collectively referred to as microwave syndrome. Dr. Neil Cherry, a pioneering researcher in the domain of radiofrequency radiation dangers, conclusively demonstrated through a comprehensive study that radiofrequency radiation induces sleep disturbances, reduces melatonin levels, and contributes to cancer in several bodily regions. The impact of microwave radiation on human biochemistry and physiology is contingent upon the frequency, intensity, and length of exposure. The safe limit for microwave radiations as per WHO and ICNIRP guidelines is ~2.5 mW/cm^2 or less. A microwave leakage meter should be employed to assess the level of microwave radiation surrounding the microwave heating system. If the radiation exceeds the prescribed limit, measures must be implemented to shield the system or appropriate absorbing materials should be utilized to reduce the radiation to acceptable levels. No aspect of life is entirely secure. The ultraviolet (UV) component of solar radiation is recognized to present significant dangers. It is advisable to consistently don microwave-safe gloves, an apron, and goggles, which are already available in the market, when operating near microwave heating devices.

6. CONCISE SUMMARY

In contemporary times, where scientific discovery and technology are intertwined, any novel synthesis approach that enhances the efficiency of new material synthesis or improves the thermodynamic characteristics of materials by sintering would be highly advantageous. Microwave heating is, in this regard, more rapid, environmentally sustainable, and has the potential to facilitate the synthesis of nanoparticles. This Green Chemistry methodology not only diminishes synthesis duration but also mitigates side reactions, hence enhancing yield and repeatability. An understanding of fundamental physical principles and practical elements of the microwave heating process in the chemical industry can significantly enhance the speed and efficiency of yield.

7. REFERENCES

1. Agrawal D. Cheng J. Peelamedu R. Fang Y. Roy R. 2006 Materia Japan. Bulletin of Japan Institute of Metals. 45 574 576 .

2. Baghurst D. R. Chippindale A. M. Mingos D. M. P. 1988 1988). Microwave synthesis for superconducting ceramics, Nature 332, 311.

3. Binner J. G. P. Al-Dawery I. A. H. 1998 Bulk YBCO high-T_c superconductors with uniform and full oxygen content via microwave process, Superconductor Science and Technology,11 449 457 .

4. Bilecka I. Niederberger M. 2010 Microwave chemistry for inorganic nanomaterial synthesis, Nanoscale 2 1358 1374 .

5. Bosi S. Beard G. Moon A. Belcher W. 1992 Microwave preparation of the $YBa_2Cu_3O_{7-\delta}$ superconductor. Journal of Microwave power and Electromagnetic Energy, 27 2 75 80 .

6. Chandra U. Parthasarathy G. Sharma P. 2010 Synthetic cubanite $CuFe_2S_3$:Pressure induced transformation to isocubanite. Canadian Mineralogist 48 1137 1147

7. Chandra U. Pooja sharma. Parthasarathy G. 2011 High-pressure electrical resistivity, Mössbauer, thermal analysis and micro-Raman spectrscopic investigations on microwave synthesized orthorhombic cubanite ($CuFe_2S_3$), Chemical Geology (in Press).

8. Cheng J. Rustum Roy. Dinesh Agrawal. 2001 Experimental proof of major role of magnetic field losses in microwave heating of metal and metallic composites, Journal of Materials Science Letters, 20 1561 1563 .

9. Cherradi G. Desgardin J. Provost Raveau B. 1994 In: Electroceramics IV, Vol. II (edited by Wasner R.,Hoffman S.,Bonnenberg and Hoffman C., RWTN), Aschen, 1219

10. Clark D. E. Sutton W. H. 1996 Microwave processing of materials, Annual Review of Material Science, 26 299 331 .

11. Dong C. Guo J. Fu G. C. Yang L. H. Chen H. 2007 Rapid preparation of MgB_2 superconductor using hybrid mirowave synthesis, Superconductor Science and Technology,17, L55 -L57.

12. Dube D. C. Agrawal D. Agrawal S. Roy R. 2007 High temperature dielctric study of Cr_2O_3 in microwave region, Applied Physics Letters, 90 124105 1 to 3

13. Gedye R. Smith F. Westaway K. Ali H. Baldisera L. Laberge L. Rousell J. 1986 The use of microwave ovens for rapid organic synthesis, Tetrahedron Letters, 27 279 282 .

14. Gedye R. N. Smith F. E. Westaway K. C. 1988 The rapid synthesis of organic compounds in microwave ovens, Canadian Journal of Chemistry 66 17 26 .

15. Giguere R. J. Bray T. I. Duncan S. M. Majetich G. 1986 Application of commercial microwave ovens to organic synthesis. Tetrahedron Letters, 27 4945 4948

16. Gibbons K. E. Jones M. O. Blundell S. J. Mihut A. L. I. Gameson I. Edwards P. P. Miyazaki Y. Hyatt N. C. Porch A. 2000 Rapid synthesis of colossal magnetoresistance manganites by microwave dielectric heating. Chemical Communications 159 160

17. Guo J. Dong C. Yang, Fu G. 2005 A green route for microwave synthesis of sodium tungsten bronzes Na_xWO_3 ($0<x<1$). Journal of Solid State Chemistry 178 58 63

18. Hasna A. M. 2009 Composite Dielectric heating and drying :The computational process, Proceedings of the World Congress on Engineering (WCE 2009). Vol I, 679 686 , 978-9-88170-125-1

19. Hyatt N. C. Porch A. 2000 Rapid synthesis of colossal magnetoresistance manganites by microwave dielectric heating, Chemical Communications, 159 160 .

20. Hussein M. Z. B. Zainal Z. Ming C. Y. 2000 Microwave-assisted synthesis of Zn-Al layered double hydroxide-sodium dodecyl sulphate nano composite, Journal of Materials Science letters, 19 10 879 883 .

21. Jansen K. 2004 Verified synthesis of zeolite materials, In: (2nd Revised edition), Microwave Technology in Zeolite Synthesis. http://www.ionline.org/synthesis/VS_2ndEd/MicrowaveTech.htm

22. Jones S. D. Akridge J. R. 1995 In: Handbook of solid state batteries and capacitors. Munshi, M.Z.A., Ed. World Scientific, Singapore, 209.

23. Kamo M. Sato Y. Matsumoto S. Setaka N. 1983 Diamond synthesis from gas phase in microwave plasma, Journal of Crystal Growth, 62(3), 642-644.

24. Kappe C. O. Dallinger D. 2006 The impact of microwave synthesis on drug discovery, Nature Reviews Drug Discovery, 5 51 63

25. Krishnan R. Agrawal D. Dobbins T. 2009 Microwave irradiation effects on reversible hydrogen desorption in sodium aluminium hydrides ($NaAlH_4$), Journal of Alloys and Compound, 470 250 255 .

26. Kubel,E., 2005 Advancements in Microwave heating technology, IndustrialHeating.com (January 43-53 43 53 .

27. Lee C. C. So W. W. 2000 High temperature silver- indium joints manufactured at low Temperature, Thin solid films, 366 196 201 .

28. Lekse J. W. Stagger T. J. Aitken J. A. 2007 Microwave metallurgy: synthesis of intermetallic compounds via microwave irradiation, Chemistry of Materials, 19 3601 3603

29. Lučun A. Butkute A. Maneikis O. Kiprijanovic O. Anisimovas F. Gradauskas J. Suziedelis A. Vengalis B. Kancleris Z.. Asmontas S. 2007 Magnetoresistance of polycrystalline $La_{0.7}Ca_{0.3}MnO_3$ films in a microwave magnetic field, Acta Physica Polonica 111 147 152

30. Makul N. Rattanadecho P. 2010 Microwave Pre-curing of natural rubber compounds using a rectangular wave guide, International Communication in Heat and Mass Transfer, 37 914 923 .

31. Metaxas A. C. Meredith R. J. 1983 Industrial Microwave Heating, Peter peregrinus Ltd., London, UK (for IEE).

32. Mingos D. M. P. Baghurst D. R. 1991 Application of microwave dielelctric heating effects synthetic problems in chemistry, Chemical Society Reviews, 20 1 1 47 .

33. Newham R. E. Jang S. J. Xu M. Jones F. 1991 Fundamental interaction mechanisms between microwave and matter, Ceramic transactions, 21, 51.

34. Nissinen T. Kiros Y. Gasik M. Leskela M. 2003 Microwave synthesis of $MnCo_2O_4$ nanoparticles on carbon, Chemistry of Materials, 15, 4974.

35. Pareek S. Rais A. Tripathi A. Chandra U. 2008 Mössbauer study of microwave synthesized (Cu,Fe) sulfide composite and correlation with natural mineral cubanite, Hyperfine Interaction, 186 113 120 .

36. Park H. K. Han Y. S. Kim D. K. Kim C. H. 1998 Synthesis of $LaCrO_3$ powders by microwave induced combustion of metal nitrate-urea mixture solution, Journal of Materials Science Letters, 17 785 787 .

37. Perelaer J. Berend-Jan de Gans. Schubert U. S., 2006 Ink-Jet printing and microwave sintering of conducting silver tracks, Advanced Materials, 18 2101 2104 .

38. Rattanadecho P. 2006 The simulation of microwave heating of wood using a rectangular waveguide: influence of frequency and sample size, Chemical Engineering Science, 61 4798 4811 .

39. Rao K. J. Vaidhyanathan B. Ganguli M. Ramakrishnan P. A. 1999 Synthesis of inorganic solds using microwaves, Chemistry of Materials, 11 882 895 .

40. Reid David, Melting metals in a domestic microwave. home.c.2i.net/metaphor/mvpage.html.

41. Roy R. Peelamedu R. Grimes C. Cheng J. Agrawal D. 2002 Major phase transformations and magnetic property changes caused by electromagnetic fields at microwave frequencies, Journal of Materials Research, 17 12 3008 3011 .

42. Schiffmann R. F. 1986 Food Product development for microwave processing, Food Technology, 40 6 94 98 .

43. Shulman H. S. 2002 Microwave heating ceramic. www.ceralink.com

44. Singh R. J. 2007 Heating of manganites by magnetic component of microwave., Indian Journal of Pure and Applied Physics, 45 454 458 .

45. Starck A. V. Muhlbauer A. Kramer C. Hand book. of thermoprocessing. Technologies . Fundamental processes. component safety. 2005 3-80272-933-1Verlag GmbH.

46. Tsuji M. Hashimoto M. Nishizawa Y. Kubokawa M. Tsuji T. 2005 Microwave-assisted synthesis of metallic nanostructures in solution, Chemical European Journal, 11 440 452 .

47. Vaidhyanathan B. Rao K. J. 1997 Synthesis of Ti,Ga, and V Nitrides: microwave-assisted carbothermal reduction and nitridation, Chemistry of Materials, 9 1196 1200 .

48. Vanetsev A. S. Tretyakov Yu. D. 2007 Microwave-assisted synthesis of individual and multicomponent oxides, Russian Chemical Reviews, 76 5 397 413 .

49. Vollmar M. 2004 Physics of microwave oven, Physics Education, 39 1 74 81 .

50. Vos B. Mosman J. Zhang Y. Poels E. Bliek A. 2003 Impregnated carbon as a susceptor material for low loss oxides in dielectric heating, Journal of Materials Science, 38 173 182 .

51. Wang H. Zhang J. Zhu J. 2001 A microwave assisted heating route method for the rapid synthesis of sphalerite- type mercury sulfide nanocrystals with different sizes, Journal of Crystal Growth, 233 829 836 .

52. Weidong M. Chuanxin W. Shenggao W. 2006 Microwave chemical deposited thick diamond film synthesis using CH_4/ H_2/ H_2O gas mixture, Plasma Science and Technology, 8 3 329 334 .

53. Xie G. Li S. Louzguin D. V. Cao Z. Yoshikawa N. Sato M. Inoue A. 2009 Fabrication of Ni-Nb-Sn metallic glassy-alloy powder and its microwave induced sintering behaviour, Journal of Microwave Power and Electromagnetic Energy, 43 17 22.

54. Yoshikawa N. 2010 Fundamentals and applications of microwave heating of metals, Journal of Microwave Power and Electromagnetic Energy, 44 1 4 11 .

55. Yund R. A. Kullerude G. 1966 Thermal stability of assemblages in Cu-Fe-S system, Journal of Petrology, 7 454 488 .

PART 2
MEDICAL SCIENCES

2 Chapter

NONINVASIVE THERMAL ASSESSMENT DURING MICROWAVE HEATING WITH A TINY RADIOMETER

1. INTRODUCTION: AN OVERVIEW

The utilization of microwaves in medicine has recently garnered renewed interest within the scientific community due to emerging techniques in breast cancer detection, such as microwave tomography and ultra-wide-band radar imaging, as well as high-power ablation, microwave angioplasty, and lipoplasty (Rosen et al., 2002). Nonetheless, the literature indicates that more established procedures utilizing RF/microwaves include hyperthermia and medical radiometry. In recent decades, several research groups have investigated numerous clinical uses of radio-thermometry for detecting thermal anomalies in subcutaneous and deeper regions of the human body.The possibility of using microwave radiometry for non-invasive thermometry was originally suggested back in the early 1970's (Edrich & Hardee, 1974; Enander & Larson, 1974) and the sensing principle was denoted *microwave thermography*. Prospected medical applications with highest potential include detection of breast cancer (often in conjunction with infrared thermometry) (Mouty et al., 2000), non-invasive temperature control of superficial (Ohba et al., 1995) and interstitial (Camart et al., 2000) hyperthermia, and control of brain temperatures of newborn infants during mild hypothermia (Hand et al., 2001; Maruyama et al., 2000). Additional applications, which have been investigated, comprise detection of inflammatory arthritis (MacDonald et al., 1994), extravasation rate of drugs (J. Schaeffer & Carr, 1986), changes of blood flow (Gabrielyan et al., 1992) or amount of lung water (Iskander et al., 1984) as well as post-mortem or cerebral temperature monitoring (Al-Alousi et al., 1994).

Microwave radiometry in clinical medicine aims at deriving information on internal body temperature patterns by measurement of natural thermal black-body radiation from tissue in the lower part of the microwave region (<5 GHz). Knowledge on such thermal patterns can give valuable information in clinical disease detection and diagnosis as well as providing quantitative temperature feedback in monitoring of thermal therapeutic processes.

The application of microwave radiometry raises several issues in determination of subcutaneous temperature heterogeneities. Most important is how to optimize design of the receiver hardware by identification of mutual compatible microwave devices with sufficiently low noise figures. The extremely low power levels (-174 dBm/Hz at 37o C body temperature) of electromagnetic thermal noise limit the applicability of microwave radiometers as the usable thermal signal competes both with external electromagnetic interference (EMI) and the internal noise produced by the hardware. When operated in, or close to, certain communication bands, proper electromagnetic shielding is imperative to stabilize the signal. As for internal noise interference, certain design schemes can be utilized to minimize the overall system noise.

Advances in controllable heating equipment, combined with improvements in non-invasive as well as invasive tissue thermometry techniques (Rhoon & Wust, 2005), have increased the feasibility of using hyperthermia (41-45 o C for 60 min) as a complementary agent in the treatment of cancer. The effectiveness of hyperthermia treatment is related to the induction of elevated temperatures integrated over time and is often quantified by cumulative equivalent minutes at 43 o C (CEM43o C). The optimal hyperthermia dose for combinations with radiation or chemotherapy is widely accepted to be 60 min treatment at 43 o C throughout 90% of measured tumor target temperatures, which represents a CEM43o CT90 thermal dose of 60 min (Sapareto & Dewey, 1984). Temperature homogeneity within the target tissue is an advantage to improve the efficacy of hyperthermia, since a narrow temperature window should bring the response-related minimum tumor temperature closer to the maximum tolerable temperature (45 o C) which avoids undue patient discomfort and complications. Maintaining the targeted temperature distribution within the therapeutic window is a major challenge in hyperthermia in order to attain effective radio- and/or chemosensitization of the treated volume. The therapy goal of obtaining cytotoxic temperatures (>42 o C) for periods of 1 h represents the classical dosimetry view and is supported in studies by Dewhirst et al. (1984) and Kapp and Cox (1993). In recent years, however, the biologic rationale for hyperthermia has been subject to discussion (Corry & Dewhirst, 2005). Physiological and cellular effects of mild temperature hyperthermia (39-42 o C for 1-2 h) are currently under investigation (Calderwood et al., 2005; Dewhirst et al., 2005). The importance of mild temperature hyperthermia related to positive thermo-radiotherapy treatment response is documented in published clinical trials (Jones et al., 2005; Thrall et al., 2005).

In practice, general characterization of tissue temperature distributions with bulk properties like volume average temperature or maximum temperature and location of the maximum can provide very useful realtime feedback information and clinical documentation to an operator. This information is fed back into the heating system for adjustment of antenna power in order to maintain a therapeutic temperature range and avoid side effects from excessive temperatures. Modeling with numerical computations can alternatively or complementary provide additional insight for dose-planning, although unknown patient-specific data (e.g. blood perfusion and power absorption characteristics) limit the reliability of such simulations. Microwave applicators are commonly used for heating of cancerous tissue. Hyperthermia applied to recurrent breast cancer resulted in a complete response rate increase from 31% to 65% in tumors of size smaller than 3 cm (Van der Zee et al. 1999). Multi-institutional randomized trials on the same disease all showed similar improvement from adding hyperthermia to radiotherapy (Jones et al., 2005; Vernon et al., 1996). Hyperthermia applicators used in previous efforts have in general been rather bulky and restricted to heating of smaller regions and flat anatomy (Lee, 1995; Stauffer, 2005). However, low cost and expandable printed circuit board (PCB) array construction facilitates development of antennas without the shortcomings of early superficial heating devices. Now lightweight and low profile multi-element array designs have been introduced such as the current sheet applicator (CSA) (Gopal et al., 1992), the contact flexible microstrip applicator (CFMA-12) (Lee et al., 2004), the microstrip spiral applicator (Lee, 1995), the annular aperture and horseshoe applicators (Carlier et al., 2002), and the Dual Concentric Conductor applicator (DCC) (Rossetto & Stauffer, 2001).

In the field of bio-electromagnetics, thermal considerations are crucial to evaluate dosimetry levels for bio-experiments. A multitude of factors determine the degree of temperature homogeneity in tissue during hyperthermia. This includes heterogeneity in tissue composition which in turn implies local variations in blood flow rate together with thermal and electric conductivity. As a result power absorption, effective heating depth, and efficiency of coupling electromagnetic power into the volume under treatment will vary spatially.

While the vast majority of thermal dosimetry for hyperthermia has been performed using invasive temperature probes to sample a small number of points, there are a number of non-invasive approaches under investigation which can quantify more complete 2-D and 3-D temperature distributions. These techniques include infrared thermography (Tennant & Anderson, 1990), computerized axial tomography (Rutt et al., 1986), ultrasound time-of-flight tomography techniques (Seip & Ebbini, 1995), electrical impedance tomography (Moskowitz et al., 1995), microwave tomography (Chang et al., 1998), magnetic resonance imaging (Samulski et al., 1994), and microwave radiometry (Ohba et al., 1995).

Whereas infrared thermography is able to map thermal emissions from the body surface only, radiometry in the lower microwave region has the potential to detect thermal radiation emitted by subcutaneous tissue up to a depth of several centimeters. Multispectral radiometry has been under investigation as a measurement technique to provide information on temperature depth distributions in tissue. As the radiometric signal strength from a volume element at a particular depth of the medium is correlated with frequency, a multi-frequency scan of a broader band can (at least in principle) be used to map depth temperature gradients. An important limitation of the radiometric observation principle is however the extremely weak signal level of the thermal noise emitted by the lossy material. Consequently, requirements of long integration time (~5 s) and wide integration bandwidth (~500 MHz) result in a maximum of 5-6 radiometric bands per antenna within the usable frequency scan range. Even with long data acquisition and signal post-processing times combined with increasingly complex system hardware for multifrequency radiometry, only a few data samples of depth related information are available. For a single antenna system, these sparse data sets can nevertheless produce useful depth-temperature information. Combined with a priori information on antenna radiation patterns and tissue dielectric and thermal properties, viable estimates of average temperature in selected tissue volumes can be obtained (Bardati et al., 1991; Camart et al., 2000; Jacobsen & Stauffer, 2003). However, for a full scan of multiple frequencies in a multiple antenna system, the temperature scanning time increases rapidly and can lead to unacceptable cooling during the radiometric listening period while microwave heating power is switched off. Thus from clinical, practical, economical, and technical considerations, the number of radiometric bands should be kept to a minimum (preferably single-band) for temperature monitoring of clinical hyperthermia procedures. The chapter is structured as outlined below. Section 2 delineates the fundamental radiometric theory, encompassing the detection system, signal model, and inversion algorithm, as well as the experimental configuration. Section 3 presents experimental phantom findings, encompassing heating and radiometric antenna configurations, recorded antenna return loss, and radiometric temperature assessments of a dynamic heating procedure. Ultimately, Section 4 assesses the findings, while Section 5 formulates conclusions based on the research.

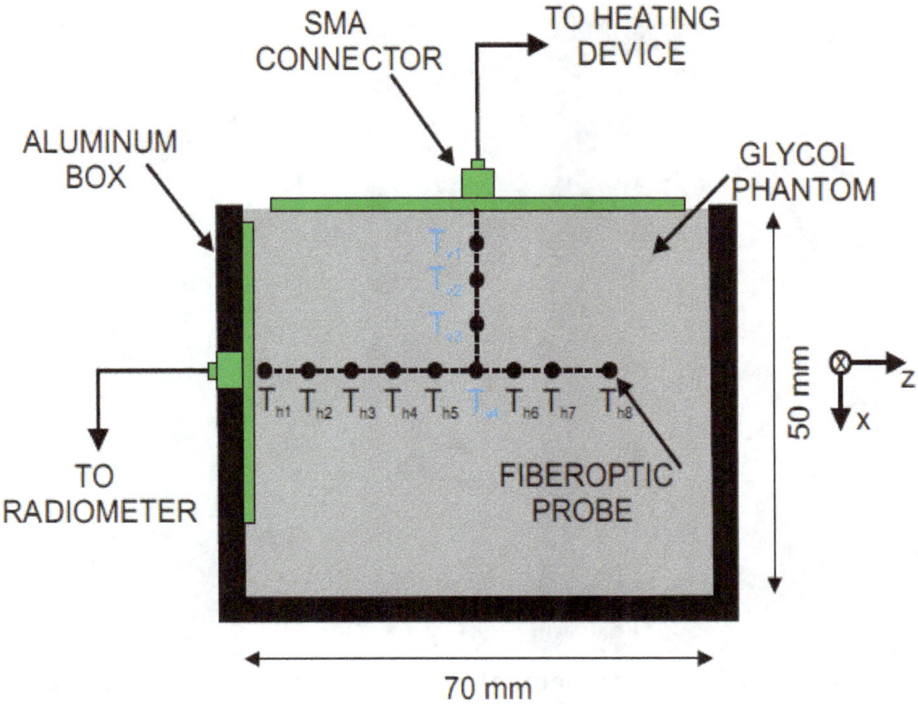

Fig. 1. Pitch-catch mode setup for 915 MHz heating and 3.5 GHz radiometric temperature reading.

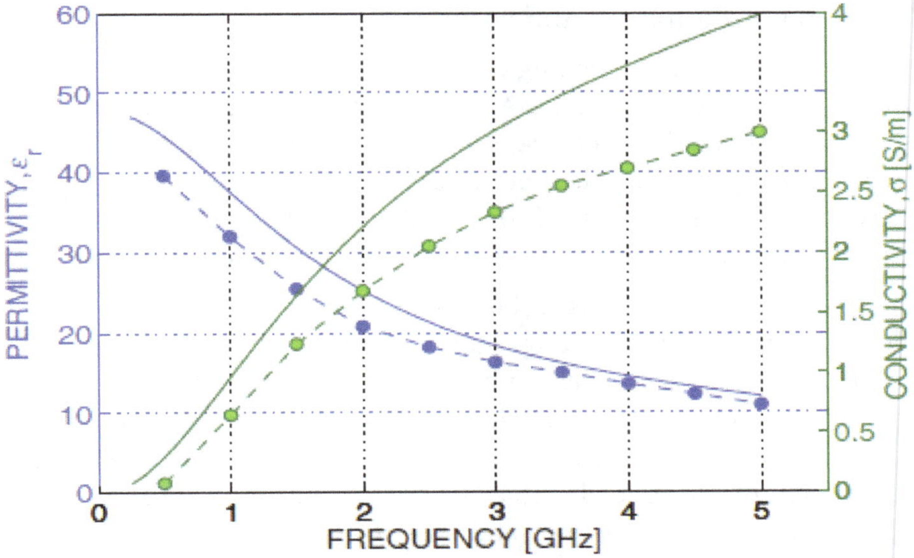

Fig. 2. Dielectric properties of glycol based phantom. Solid line: 90% glycol-10% water mixture. (●): solidified phantom of same mixture.

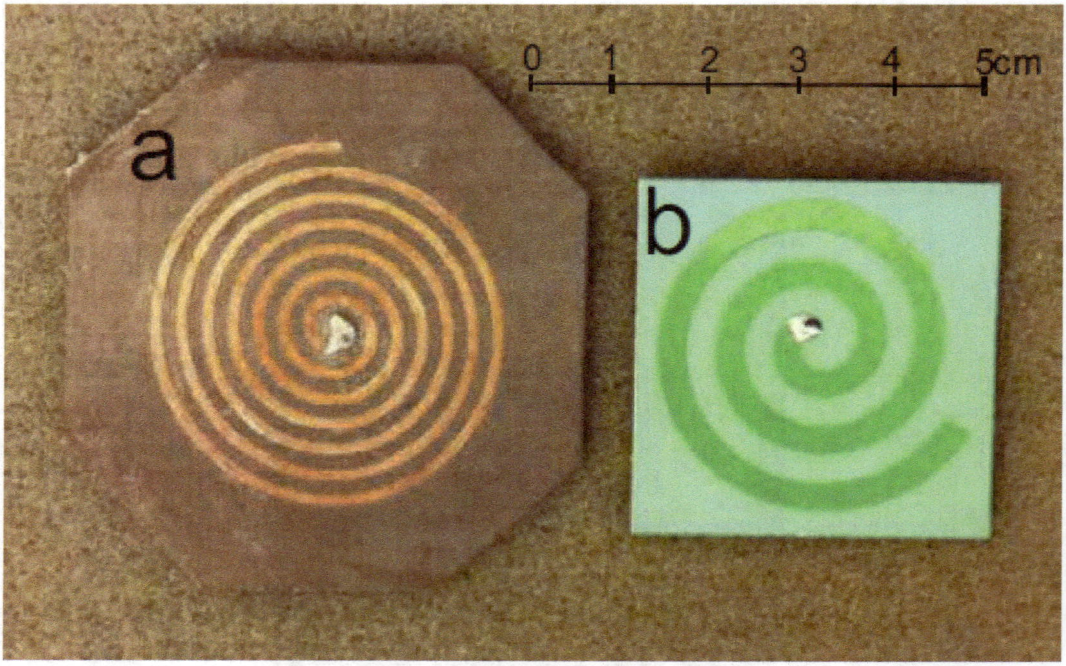

Fig. 3. Antennas used in experiment. a) 915 MHz heating spiral antenna, and b) 3.5 GHz radiometric spiral antenna.

2. METHODS

2.1 Antenna configuration and setup

Fig. 1 depicts the combined heating and temperature measurement configuration considered in the present study. A glycol-water phantom load was solidified with gelatine in a metal box of size 50×70×50 mm^3. We notice from Figure 2, that solidification using gelatine (low permittivity material) lowers both the overall permittivity and conductivity of the applied fluidic phantom. The phantom dielectric properties is similar to that of fat infiltrated mammary tissue.

The five sidewalls of the supporting box, of highly electrically conducting material, provide electromagnetic shielding against competing electromagnetic sources (e.g. cellular phones). However, the aluminum box was not thermally insulated from the surroundings and thus participates in the cooling process as the ambient temperature will influence thermal fluxes within the phantom. Furthermore, the phantom structure was topped with a 50 50 mm^2 spiral heating antenna operated at the medical 915 MHz frequency and driven by a 9 W power source. Arranged in a pitch-catched mode, a smaller 35 35 mm^2 spiral antenna was mounted inside one adjacent sidewall to monitor temperature gradients within the heated volume. The small spiral antenna was connected to a 3.5 GHz single band radiometer with an integration bandwidth of 500 MHz. A photograph of the respective antennas is shown in Figure 3 and measured return loss is depicted in Figure 4. Notice that the small spiral is well matched at the radiometric frequency of 3.5 GHz ($S11$ better than -10 dB) whereas the heating antenna has a return loss of -14 dB at 915 MHz. For correlation with radiometric readings, two perpendicular 1-D temperature profiles (see Figure 1) were established by means of twelve **fiberoptic** probes rack mounted in a LumiTherm X5R system 1. The custom made Dicke radiometer used in the experiments has been described in the literature before (Klemetsen et al., 2011). Using miniature surface mount devices (amplifiers, switches and circulators), the front-end printed circuit board is only 40×50 mm^2 in size. With a detector intergration time of 2 secs and integration bandwidth of 600 MHz, the brightness temperature resolution and theoretical lower accuracy are as low as $\sigma T_B = 0.029°$ C.

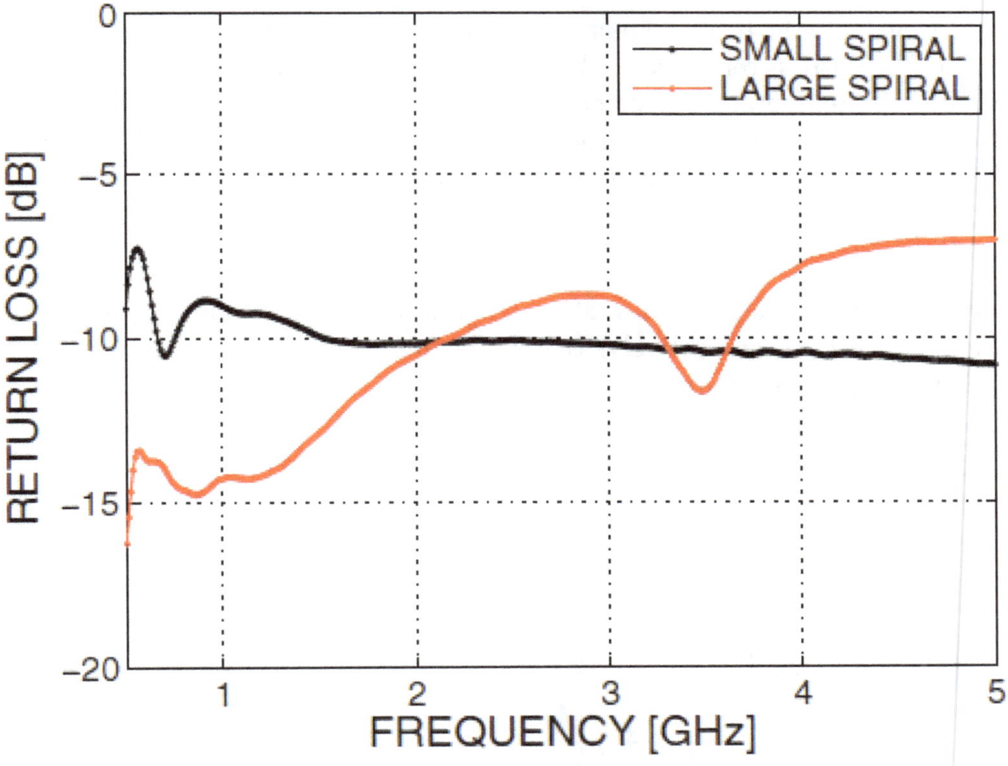

Fig. 4. Measured antenna return loss vs frequency.

2.2 Brightness temperature integral evaluation

Various approaches have been suggested to convert measured radiometric brightness temperatures to an actual temperature distribution within the volume under observation. These methods include parametric (Hand et al., 2001; Ohba et al.,

1995) and non-parametric (Bardati et al., 1993; Jacobsen & Stauffer, 2003) modeling. Below we present a 3-D parametric model which leads to a proportionality between the observed and sought parameter.

Consider the general definition of the brightness temperature T_B:

$$T_B = \int_{\Xi} T(\underline{r}) W(f_c, \underline{r}) dV \tag{1}$$

where W is the weighting function (WF) dependent on the spatial variable r and center frequency f_c within the integration bandwidth. Ξ is the volume under investigation and dV is an infinitesimal volume element.

The weighting function is normalized according to:

$$\int_{\Xi} W(f_c, \underline{r}) dV \equiv 1 \tag{2}$$

Next, we evaluate the term in Equation (1), which describes the contribution to the lossy load brightness temperature by introducing a parametric model. We assume that both the temperature distribution and weighting function are separable in space with respect to a Cartesian coordinate system; i.e. $T = T_x(x) \cdot T_y(y) \cdot T_z(z)$ and $W = W_x(x) \cdot Wy(y) \cdot Wz(z)$. Hence, the integral in Equation (1) expands to:

$$T_B = \int_{L_x} T_x(x) W_x(f_c, x)\, dx \int_{L_y} T_y(y) W_y(f_c, y)\, dy \int_{L_z} T_z(z) W_z(f_c, z)\, dz \tag{3}$$

where $L_{x,y,z}$ are the spatial integration lengths within the tissue volume Ξ in x-, y-, and z-direction, respectively. Using the following model functions for the respective factors in Equation (3):

$$T_x = \exp\left(-x/\gamma_{T_x}\right) \tag{4}$$

$$T_y = \exp\left(-y^2/\sigma_{T_y}^2\right) \tag{5}$$

$$T_z = \frac{T_{\max}}{1 + \alpha(z - z_m) + \beta(z - z_m)^2} \tag{6}$$

$$W_x = \frac{1}{\sqrt{\pi}\sigma_{E_x}} \exp\left(-x^2/\sigma_{E_x}^2\right) \tag{7}$$

$$W_y = \frac{1}{\sqrt{\pi}\sigma_{E_y}} \exp\left(-y^2/\sigma_{E_y}^2\right) \tag{8}$$

$$W_z = \frac{1}{d} \exp\left(-z/d\right) \tag{9}$$

where $\sigma_{Ex,y}$ are the lateral spatial widths of the weighting function in the x, y-direction, d is the (z-direction) spiral antenna 1/e-sensing depth, γT_x and σT_y are the lateral widths of the temperature distribution in the x and y-direction, T_{max}

is the maximum temperature in tissue, and $\{z_m, \alpha, \beta\}$ is a parameter set modeling curvature and asymmetry of the 1-D depth temperature profile. Notice that the required normalization $\int_\Xi W(\underline{r})dV$ is satisfied through the prefixed normalization factors. The 3-D model given in Equations (4)-(9) is a direct generalization of the 1-D model used by Jacobsen and Stauffer (2003) and similar to the 3-D model in Jacobsen and Stauffer (2007).

By carrying out the integration along all directions, and assuming that $L_x = L_y = (-\infty, \infty)$ and $L_z = (0, \infty)$, it is readily shown:

$$T_B = \frac{1}{d\sqrt{1+\sigma_{E_y}^2/\sigma_{T_y}^2}} \frac{1}{\sqrt{\pi\sigma_{E_x}^2}} \int_{-\infty}^{\infty} \exp(-x/\gamma_{T_x}) \exp(-x^2/\sigma_{E_x}^2)\, dx \cdot \tag{10}$$

$$\int_0^{\infty} \frac{T_{\max}}{1+\alpha(z-z_m)+\beta(z-z_m)^2} \exp(-z/d)\, dz \tag{11}$$

which might be semi-analytically expressed:

$$\begin{aligned} T_B = {} & \frac{T_{\max}}{d} \frac{1}{\sqrt{1+\sigma_{E_y}^2/\sigma_{T_y}^2}} \frac{1}{\sqrt{\alpha^2-4\beta}} \cdot \exp\left(-\frac{\sigma_{E_x}^2}{4\gamma_{T_x}^2}\right) \cdot \\ & \left[\exp\left(\frac{1}{2\beta d}(\alpha - 2\beta z_m - \sqrt{\alpha^2-4\beta})\right) E_1\left(\frac{1}{2\beta d}(\alpha - 2\beta z_m - \sqrt{\alpha^2-4\beta})\right) - \right. \\ & \left. \exp\left(\frac{\alpha}{2\beta d}(\alpha - 2\beta z_m + \sqrt{\alpha^2-4\beta})\right) E_1\left(\frac{1}{2\beta d}(\alpha - 2\beta z_m + \sqrt{\alpha^2-4\beta})\right)\right] \end{aligned} \tag{12}$$

$$= \nu(\alpha, \beta, z_m, \gamma_{T_x}, \sigma_{T_y}, \sigma_{E_y}, \sigma_{E_x}, d) T_{\max} \tag{13}$$

Table 1. Parameter values of radiometric antenna.

Characteristic	Parameter Nominal value
WF sensing depth	d 6.0 mm
WF lateral width, x-direction WF lateral width, y-direction	σE_x 7.0 mm σE_y 7.0 mm

where $E_1(z) = \int_1^\infty \exp(-zt)t^{-1}dt$ is the exponential integral of first order (Gradshteyn, 1980) which can be tabulated. Here, $\nu \leq 1$ is a proportionality factor that relates the observed and sought parameter.

ν can be derived given that the thermal, dielectric, and geometric parameters are a priori known for the setup. From the numerically generated brightness temperature *TB*, and using Equation (13), the maximum temperature within the volume can be calculated.

Applying the commercial software package CST Microwave Studio[2], full wave simulation of the radiometric antennas' radiation properties was used to establish the model parameters in Equations (7-9). Numerical values of these runs for the specific phantom are depicted in Table 1.

3. RESULTS

Figure 5 displays the probe-measured temperatures within the phantom as a function of time. We notice in Figure 5(a) the marked dynamic heating bursts displayed by T_{v1} for times less than 2000 secs. The probe closest to the heating antenna reaches a maximum temperature of 40°C at 1900 secs. Deeper down into the phantom, lower temperatures are observed as intuitively expected. Maximum temperatures at depth are also slightly time-delayed compared to more shallow temperatures due to some minor effects of thermal conduction.

As for the horizontal probes, it is seen in Figures 5 (b) and (c) that the maximum temperature along the z-axis occurs at a depth of 20mm and 30mm (27.2°C). Since probe T_{v4}, located in between T_{h5} and T_{h6}, shows a maximum of 27.4°C, the z-profile horizontal maximum is located (as anticipated) at the boresight symmetry line of the heating antenna.

To get an overview of the dynamic heating process taking place within the phantom, perpendicular to and in the observation direction of the radiometric antenna, temperature profiles along the z- and x-axis are plotted in Figure 6 at selected points in time. The shape of the profiles in both directions is modeled accurately by Equations (4) and (6). Notice the smooth shape of the model function in the z-direction with a maximum plateau underneath the center of the heating antenna. In addition, the x-direction temperature gradient is significant (up to 1°C/mm), especially subsequent to the heating sequences taking place for times less than 1800 secs. During the cooling process, the decay curves flatten and eventually become spatially constant.

Table 2 quantifies the parameter values obtained for the snapshots in Fig 6. Observe the relatively small values of γT_x in the heating phase. During the cooling process (time≥2150 secs), large values of γT_x model the x-direction temperature profile, meaning that it is virtually flat (homogeneous temperature) lateral to the radiometric observation axis.

Based on Table 1 and Equations (4)-(9), the brightness temperature T_B can be estimated through Equation (12) for the dynamic run discussed above.

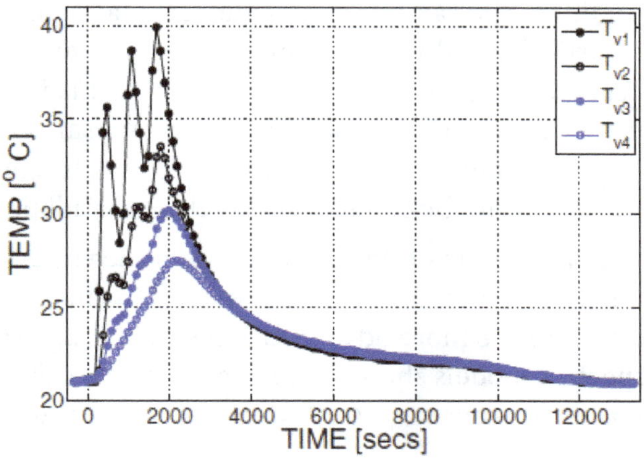

(a) Vertical probes in Fig. 1

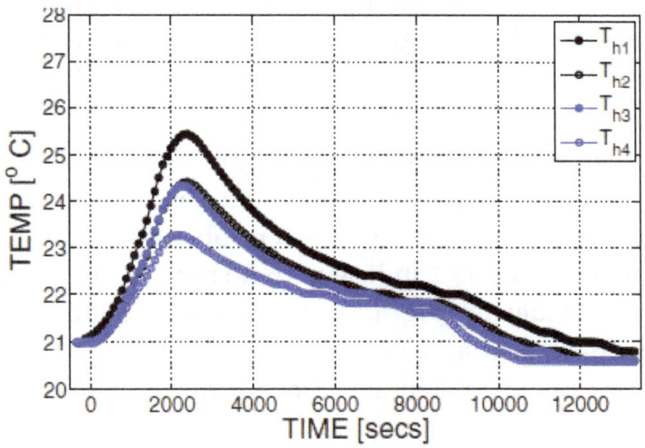

(b) Horizontal probes (first 4) in Fig. 1

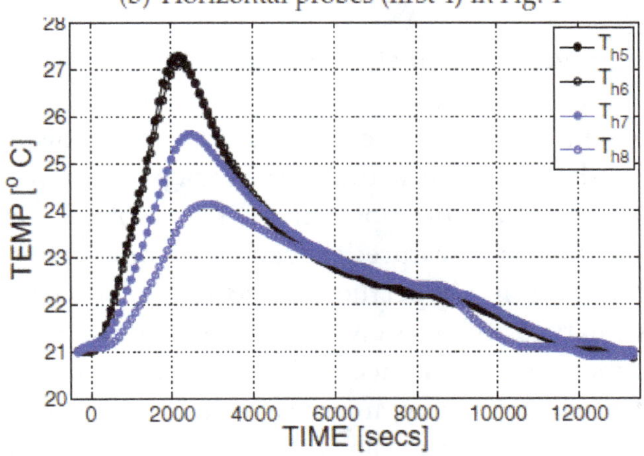

(c) Horizontal probes (last 4) in Fig. 1

Fig. 5. Fiberoptic probe temperature measurements vs time.

To evaluate the validity of the radiometric model, comparisons are made with the measured temperatures derived from the miniature radiometer. Figure 7 (upper panel) shows this comparison for two different models. This includes a simplified 1-D parametric model where variations along the transversal axis (x-direction) are neglected. In this case it is implicitly assumed that $\gamma T_x = \infty$. In the other 2-D model, also temperature variations in the transverse direction are adopted into the model.

From the deviation plot in Figure 7 (lower panel) we notice improved performance (smaller deviation between measurement and model) in the heating phase (times less than 2000 secs) when the more advanced 2-D model is applied. A quantitative analysis of the respective models shows an improvement in the standard deviation of the instantaneous error from σ_{T1-D} =0.28o C in the 1-D case to σ_{T2-D} =0.21o C in the 2-D case.

During the cooling phase, $\gamma T_x \gg 1$ and the difference in performance between the 1-D and 2-D model should be negligible. This is verified in Figure 7 (lower panel) as only marginal deviations in performance are observed for times greater than 3000 secs.

One pertinent question is to what extent the maximum temperature within the heated volume can be predicted given a measured brightness temperature; information that can be valuable to a system operator. The numerical span of the model parameters in Table 2 indicates that there exists no unique parameter combination that provides an overall proportionality factor ν in Equation (13). However, as seen in Figure 8, a constant ν still provides correlated estimates between radiometrically estimated and probe measured maximum temperatures within the heated volume.

4. DISCUSSION

Hyperthermia has proven to be an effective adjunct to standard cancer treatment methods. Breast cancer is among the most common cancers in women, and the local management of this disease continues to be a growing concern. Given the uncertainties related to the actual electromagnetic power deposition and the consequent temperature distribution in a variable, heterogeneous tissue configuration, a form of temperature feedback control is imperative for the safe and reliable operation of any hyperthermic system. Microwave radiometry, due to its non-invasive nature and entirely passive sensing concept, has the capacity to provide thermometry of subcutaneous tissues at depths of several centimeters. In contrast to infrared radiometry, microwave radiometry lacks high-resolution thermographic mapping of tissue due to lateral spatial resolution constraints imposed by antenna dimensions and separation, typically measured in centimeters.

The limitation, along with the requirement for adequate shielding to prevent interference from competing electromagnetic sources, imposes constraints on the implementation of this observational approach.

Table 2. Numerical parameter values of temperature profiles in Figure 6.

Time [s]	$\alpha[mm^{-1}]$	$\beta[mm^{-2}]$	z_m [mm]	γT_x [mm]
500	0.0192	0.00519	23.1	5.1
750	-0.0355	0.00538	19.7	11.2
1170	-0.0422	0.00411	18.3	9.8
1650	0.0632	0.00215	38.1	11.1
2150	-0.0571	0.00118	0.65	27.1
3100	-0.0355	0.00183	17.6	154.0
3642	-0.0141	0.00155	24.3	∞
6135	-0.0154	0.00092	27.3	∞
9700	-0.0290	0.00468	26.1	∞

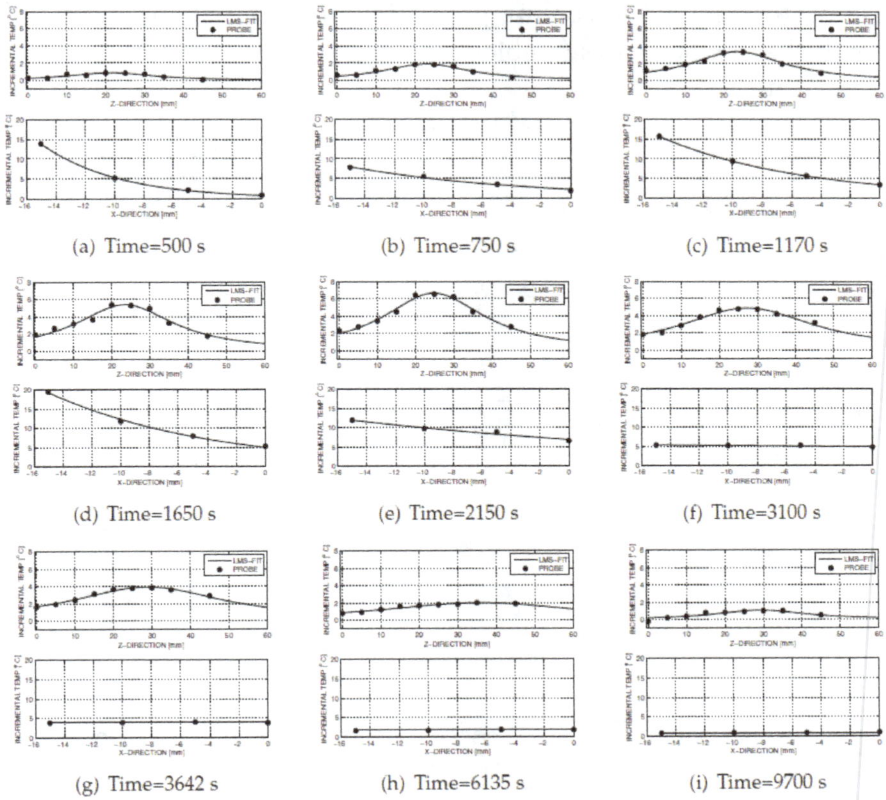

Fig. 6. Fiberoptic temperature profiles in phantom along two orthogonal axes for different times. Solid line: Fitted model in Equations (4) and (6).

Since medical microwave radiometry only offers sparse temperature data sets, either a priori experimental knowledge (Mizushina et al., 1993) or complementary thermal modeling (Bardati et al., 1993) have been proposed to help solve the underdetermined inverse problem of characterizing temperature distributions at depth in living tissue.

In the present study we evaluate the robustness of a proposed temperature estimator scheme based on single-band radiometric scanning. A bistatic antenna pitch-catch setup on a glycol based phantom was used to test the performance of a single-band miniature microwave radiometer. One large spiral antenna was used to heat the volume under investigation using the medical 915 MHz frequency. A similar, but smaller, spiral antenna was also put flush to the phantom to observe thermal radiation. Since no blood perfusion is present in the phantom, only 9 W was required to heat the volume up to therapeutic hyperthermia temperatures(>40°C). The heating was conducted in several burst to mimic the clinical scenario of interspersed heating and radiometric temperature reading. In order to investigate long term stability of the radiometric instrumentation, temperature readings were also performed in the cooling phase, in which the phantom interchanged energy with the surroundings (ambient temperature of 21o C) back to the initial pre-heating state.

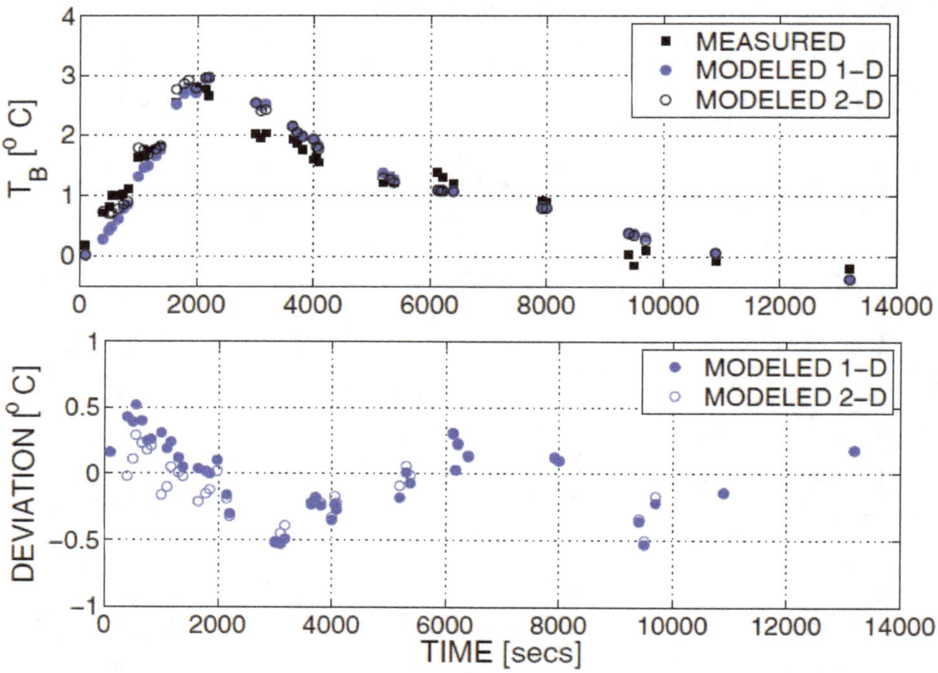

Fig. 7. Differential radiometric brightness temperature versus time. Measured and 1-D & 2-D models (upper panel), deviation between models and measurements (lower panel).

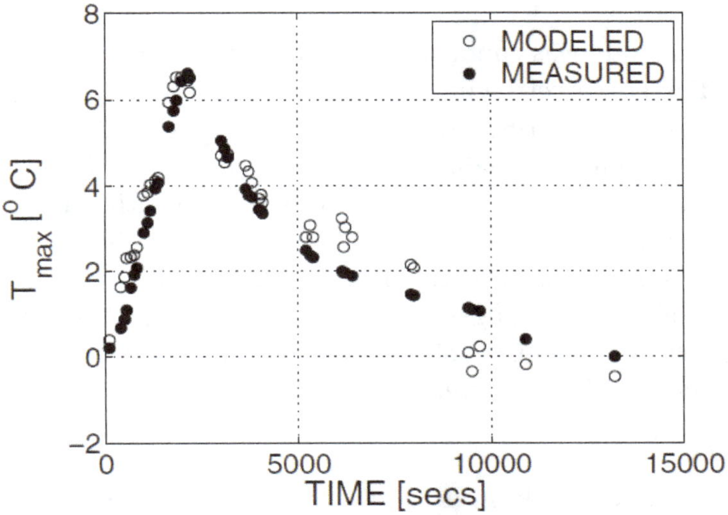

Fig. 8. Measured and radiometrically modeled (ν=0.65 in Equation (13)) maximum temperature within heated phantom vs time.

Overall, the concordance between the modeled and measured brightness temperatures was satisfactory, with deviations typically less than ±0.3°C of the maximum T_B temperature elevation of 3.0°C. Since the proportionality factor $\nu < 1$ in Equation (13), the maximum temperature elevation in the phantom is somewhat higher than the maximum observed brightness temperature. According to Figure 8, the maximum increase along the radiometric observation axis is about 6.5°C occuring at z=25 mm (see Figure 6) 2150 secs into the experiment.

The pitch-catch experimental setup produces temperature gradients in two directions. Hence, the radiometric weighting function is generally 2-D in cases where temperature equilibrium is not obtained along the transversal x-axis. Thus, one pertinent question is whether the proposed 2-D model performs better than the simplified 1-D model in such cases. This is indeed the case as can be observed in the heating phase of the experiment (Figure 7) displaying an improvement of typically 0.2°C in accuracy. However, when the transversal temperature gradient can be neglected, the 1-D and 2-D models perform equally well.

The capacity to forecast the maximum temperature within the heated volume based on a single-band radiometric measurement was examined. A significant correlation was noted between the measured and estimated maximum temperature during the heating phase. During the initiation of the cooling process, significant deviations were noted due to the more extreme values of the model parameters (refer to Figure 8). Fortunately, the cooling phase is less representative of a clinical hyperthermia scenario, when more stable circumstances with temperatures over 40°C are anticipated.

5. CONCISE SUMMARY

An experimental study was conducted to assess the accuracy of mono-frequency microwave radiometry in retrieving maximum bulk temperatures of a heated glycol-based phantom, utilizing a tiny radiometer operating at 3.5 GHz. A rapid retrieval method, utilizing a singular brightness temperature, was employed to forecast the dynamic brightness temperature throughout both the heating and cooling phases of the experiment. In instances of extreme temperature inhomogeneity within the phantom, more sophisticated temperature distribution models were required to forecast the observed changes. As the temperature distribution approached homogeneity near the conclusion of the experiment, one-dimensional temperature profiles were adequate to accurately represent the results.

6. REFERENCES

Al-Alousi, L. M., Anderson, R. A. & Land, D. V. (1994). A non invasive method for postmortem temperature measurement using a microwave probe, *Forensic Science Int.* 64: 35–46.

Bardati, F., Brown, V. J. & Bernardo, G. D. (1991). Multi-Frequency Microwave Radiometry for Retrieval of Temperature Distributions in the Human Neck, *The Journal of Photographic Science* 39: 157–160.

Bardati, F., Brown, V. J. & Tognolatti, P. (1993). Temperature Reconstruction in a Dielectric Cylinder by Multi-Frequency Microwave Radiometry, *Journal Electromagnetic Waves and Applications* 7(1): 1549–1571.

Calderwood, S. K., Theriault, J. R. & Gong, J. (2005). How is the immune response affected by hyperthermia and heat shock proteins?, *International Journal Hyperthermia* 21: 713–716.

Camart, J. C., Despretz, D., Prevost, B., Sozanski, J. P., Chive´, M., & Pribetich, J. (2000). New 434 MHz interstitial hyperthermia system monitored by microwave radiometry: theoretical and experimental results, *Int Journal of Hyperthermia* 16: 95–111.

Carlier, J., Camart, J. C., Dubois, L. & Priebetich, J. (2002). Modeling of planar applicators for microwave thermotherapy, *IEEE Transactions Microwave Theory Techniques* 50(12): 3036–3042.

Chang, J. T., Paulsen, K., Meaney, P. & Fanning, M. (1998). Non-invasive thermal assessment of tissue phantoms using an active near field microwave imaging technique, *International Journal Hyperthermia* 14(6): 513–534.

Corry, P. & Dewhirst, M. (2005). Special Issue: Thermal Medicine, Heat Shock Proteins and Cancer, *International Jornal of Hyperthermia* 8: 675–790.

der Zee, J. V., Holt, B. V. D., Rietveld, P. J. M. & et al. (1999). Reirradiation combined with hyperthermia in recurrent breast cancer results in worthwhile local palliation, *British Journal Cancer* 99: 483–490.

Dewhirst, M., Sapareto, S. & Connor, W. G. (1984). Importance of minimum tumor temperature in determining early and long term responses of spontaneous canine and feline tumors to heat and radiation, *Cancer Research* 44: 43–50.

Dewhirst, M. W., Vujaskovic, Z., Jones, E. & Thrall, D. (2005). *International Journal of Hyperthermia* 8: 779–790.

Edrich, J. & Hardee, P. C. (1974). Thermography at millimeter wavelength, *Proceedings of IEEE* 62: 1391–1392.

Enander, B. & Larson, G. (1974). Microwave radiometric measurements of the temperature inside a body, *Electronic Letters* 10: 317–318.

Gabrielyan, E. S., Khachatryan, L. A., Nalbaudyan, S. G., & Gregorian, F. A. (1992). Microwave method for determining cerebral blood flow", *Methods (New York: Plenum)* pp. 713–715.

Gopal, M. K., Hand, J. W., Lumiori, M. L. D., Alkhairi, S., Paulsen, K. D. & Cetas, T. C. (1992). Current sheet applicator arrays for superficial hyperthermia of chestwall lesions, *International Journal of Hyperthermia* 8(2): 227–240.

Gradshteyn (1980). *Table of Integrals, Series, and Products*, Academic Press, San Diego.

Hand, J. W., Leeuwen, G. M. J. V., Mizushina, S., de Kamer, J. B. V., Maruyama, K., Sugiura, T., Azzopardi, D. V., & Edwards, A. D. (2001). Monitoring of Deep Brain Temperature in Infants Using Multi-Frequency Microwave Radiometry and Thermal Modeling, *Physics in Medicine and Biology* 46: 1885–1903.

Iskander, M. F., Durney, C. H., Grange, T., & Smith, C. S. (1984). Radiometric Technique for Measuring Changes in Lung Water, *IEEE Trans. Microwave Theory Techiques* MTT-32: 554–556.

J. Schaeffer, A. El Mahdi, A. H. & Carr, K. (1986). Detection of extravasation of antineoplastic drugs by microwave radiometry, *Cancer Letters* 31: 285–291.

Jacobsen, S. & Stauffer, P. (2003). Nonparametric 1-D Temperature Restorartion in Lossy Media Using Tikhonov Regularization on Sparse Radiometry Data, *IEEE Transactions Biomedical Engineering* 50(2): 178–188.

Jacobsen, S. & Stauffer, P. R. (2007). Can we settle with single-band radiometric temperature monitoring during hyperthermia teatment of chestwall recurrence of breast cancer using a dual-mode transceiving applicator, *Physics in Medicine and Biology* 52(4): 911–928.

Jones, E. L., Oleson, J. R., Prosnitz, L., Samulski, T., Vujaskovic1, Z., Yu, D., Sanders, L. & Dewhirst, M. (2005). Randomized trial of hyperthermia and radiation for superficial tumors, *Journal of Clinical Oncology* 23: 3079–3085.

Kapp, D. S. & Cox, R. S. (1993). Cumulative minutes of isoeffective hyperthermia with T90=43o C is the treatment parameter most predictive for outcome in patients with metastatic adenocarcinoma of the breast and single tumor nodules per treatment field, *International Journal of Radiation Oncology Biology Physics* 27: 315.

Klemetsen, O., Birkelund, Y., Jacobsen, S. K., Maccarini, P. & Stauffer, P. R. (2011). Design of Medical Radiometer Front-End for Improved Performance, *Progress In Electromagnetic Research B* 27: 289–306.

Lee, E. R. (1995). Electromagnetic superficial heating technology, *in* P. Fessenden & C. C. Vernon (eds), *Thermoradiotherapy and Thermochemotherapy*, Springer-Verlag, pp. 193–217.

Lee, W. M., Gelvich, E. A., der Baan, P. V., Mazokhin, V. N. & Rhoon, G. C. V. (2004). Assessment of the performance characteristics of a prototype 12-element capacitive contact flexible microstrip applicator (CFMA-12) for superficial hyperthermia, *International Journal of Hyperthermia* 20(6): 607–624.

MacDonald, A. G., Land, D. V. & Sturrock, R. D. (1994). Microwave thermography as a non-invasive assessment of disease activity in inflammatory arthritis, *Clinical Rheumatology* 13: 589–592.

Maruyama, K., Mizushina, S., Sugiura, T., van Leeuwen, G. M. J., Hand, J. W., Marrocco, G., Bardati, F., Edwards, A., Azzopradi, D. & Land, D. (2000). Feasibility of Noninvasive Measurement of Deep Brain Temperature in Newborn Infants by Multifrequency Microwave Radiometry, *IEEE Trans Microwave Theory Techniques* 48: 2141–2147.

Mizushina, S., Shimizu, T., Suzuki, K., Kinomura, M., Ohba, H. & Sugiura, T. (1993). Retrieval of Temperature-Depth Profiles in Biological Objects from Multi-Frequency Microwave Radiometric Data, *Journal Electromagnetic Waves and Applications* 7(11): 1515–1547.

Moskowitz, M. J., Ryan, T. P., Paulsen, K. D. & Mitchell, S. E. (1995). Clinical implementation of electrical-impedance tomography with hyperthermia, *International Journal Hyperthermia* 11(2): 141–149.

Mouty, S., Bocquet, B., Ringot, R., Rocourt, N. & Devos, P. (2000). Microwave radiometric imaging (MWI) for the characterization of breast tumors, *The European Journal of Applied Physics* AP 10: 73–78.

Ohba, H., Kinomura, M., Ito, M., Sugiura, T. & Mizushina, S. (1995). Multi-frequency microwave radiometry for noninvasive thermometry using a new temperature profile model function, *Trans. IEICE Electron.* E-78: 1071–1081.

Rhoon, G. C. V. & Wust, P. (2005). Special Issue: Non Invasive thermometry for thermotherapy, *International Journal of Hyperthermia* 21(6): 489–600.

Rosen, A., Stuchly, M. A. & Vorst, A. V. (2002). Applications of RF/Microwaves in Medicine, *IEEE Trans. Microwave Theory Techniques* 50: 963–974.

Rossetto, F. & Stauffer, P. R. (2001). Theoretical characterization of dual concentric conductor microwave applicators for hyperthermia at 433 MHz, *International Journal Hyperthermia* 17(3): 258–270.

Rutt, B. K., Fike, J. R. & Stauffer, P. R. (1986). *In: 34th Meeting of Radiation Research Society; 1986; Las Vegas* p. 35.

Samulski, T. V., Clegg, S. T., Das, S., Macfall, J. & Prescott, D. M. (1994). Application of a new techology in clinical hyperthermia, *International Journal Hyperthermia* 10(3): 389–394.

Sapareto, S. A. & Dewey, W. C. (1984). Thermal dose determination in cancer therapy, *International Journal of radiation Oncology, Biology & Physics* 10: 787–800.

Seip, R. & Ebbini, E. S. (1995). Noninvasive Estimation of Tissue Temperature Response to Heating Fields Using Diagnostic Ultrasound, *IEEE Transactions on Biomedical Engineering* 42(8): 828–839.

Stauffer, P. (2005). Evolving technology for thermal therapy of cancer, *Internation Journal of Hyperthermia* 21(8): 731–744.

Tennant, A. & Anderson, A. P. (1990). A robot-controlled microwave antenna system for uniform hyperthermia treatment of superficial tumours with arbitrary shape, *International Journal of Hyeprthermia* 6(1): 193–202.

Thrall, D. E., LaRue, S. M., Yu, D., Samulski, T., Sanders, L., Case, B., Rosner, G., Azuma, C., Poulsen, J., Pruitt, A. F., Stanley, W., Hauck, M. L., Williams, L., Hess, P. & Dewhirst, M. (2005). Thermal dose is related to duration of local control in canine sarcomas treated with thermoradiotherapy, *Clinical Cancer Research* 11: 5206–5214.

Vernon, C. C., Hand, J. W., Field, S. B., Machin, D., Whaley, J. B., van der Zee, J., van Putten, W. L. J., van Rhoon, G. C., van Dijk, J. D. P., Gonzalez-Gonzalez, D., Liu, F. F., Goodman, P. & Sherar, M. (1996). Radiotherapy with or without hyperthermia in the treatment of superficial localized breast cancer: Results from five randomized controlled trials, *International Journal Radiation Oncology Biology Physics* 35(4): 731–744.

3 Chapter

THE FUNCTION OF MICROWAVE HEATING IN ANTIGEN RETRIEVAL WITHIN CRYOSECTIONS OF FORMALIN-FIXED MAMMALIAN TISSUES

1. INTRODUCTION: AN OVERVIEW

Immunocytochemistry is a prevalent technique for examining the localization of proteins in mammalian tissues and cultured cells. Tissues and cells intended for immunocytochemical investigations are typically fixed and stored using 2%-10% p-formaldehyde (PFA, formaldehyde, formalin), subsequently embedded in paraffin or frozen, and then sectioned. PFA has been utilized as a fixative in histology studies, preferred above alcohols, since the late 19th century. It is lauded for its efficacy over a wide range of doses and situations without causing "overfixation." Furthermore, compared to alcohol-fixed tissues, PFA-fixed tissues exhibit reduced tissue deformation and enhanced staining with traditional dyes, including hematoxylin and aniline dyes. The effects of PFA fixation were demonstrated to be dependent on pH, temperature, and protein type (Fox et al., 1985).

While PFA fixation produces superior outcomes for tissues stained with traditional dyes, it frequently fails to deliver satisfactory results in immunocytochemical studies. Although PFA fixation preserves antigenicity, it denatures proteins and can compromise tissue and cellular integrity by damaging structural components and obscuring antibody binding sites (epitopes) (McNicol & Richmond, 1998). The intensity of this phenomena can impact the accessibility of epitopes, so complicating or rendering impossible the interaction of antibodies with their binding sites. Despite the longstanding recognition of the issue of epitope masking, the mechanisms by which PFA alters proteins in tissues remain ambiguous. Epitope masking typically arises from PFA-induced cross-linking of reactive sites on the same protein and/or neighboring proteins through bridges generated by methylene, amino, imino, aromatic, and other reactive groups (Fraenkel-Conrat et al., 1948, 1949). Nonetheless, this masking occurs at varying rates among different proteins; in tissues preserved for extended durations, distinct proteins exhibit divergent reactions, and the efficiency of consecutive labeling is contingent upon the specific protein. In the absence of any unmasking technique, the proteins vimentin and desmin were not stained with their respective monoclonal antibodies after one day of tissue fixation. The lymphocyte antigens LN1, LN2, and LN3 exhibited weak staining with polyclonal antibodies in cells fixed in formalin for three days, whereas thyroglobulin and carcinoembryonic antigen demonstrated robust staining with polyclonal antibodies in tissues even after 14 days of formalin fixation (Leong & Gilham, 1989). These phenomena can be somewhat elucidated by the antigen's affinity for an antibody, wherein monoclonal antibodies, due to their greater specificity compared to polyclonal antibodies, possess fewer reactive binding sites (McNicol & Richmond, 1998). Seshi et al. (2004) provided further elucidation on the epitope masking issue by employing engineered peptides to replicate the antibody binding sites on the target proteins. The authors categorized the peptides into three distinct groups based on their sensitivity to PFA: (a) PFA-sensitive peptides containing tyrosine and arginine, (b) peptides devoid of tyrosine and arginine, which were resistant to PFA, and (c) tyrosine-containing peptides, which exhibited sensitivity to PFA only in the presence of an adjacent arginine-containing peptide. Additional studies indicate that PFA treatment can alter peptides based on their amino acid sequences (Metz et al., 2004), and that calcium and some other divalent cations may influence epitope masking by forming stable complexes with proteins during fixation (Morgan et al., 1994). Recent studies indicate that both the reactive sites (amino groups, tryptophan, tyrosine, lysine, etc.) and their spatial arrangement within the three-dimensional protein structure may significantly influence the masking of epitopes after PFA fixation (Toews et al., 2008). Alterations in protein shape may render epitopes inaccessible to antibodies in immunocytochemical analyses (McNicol & Richmond, 1998).

To address the issue of masked epitopes resulting from PFA fixation, it is necessary to unmask or recover these epitopes by various antigen retrieval techniques prior to the application of specific antibodies. These treatments are typically employed with segments of paraffin-embedded tissues and involve heating at elevated temperatures, treatment with diverse alcohols, detergents, or high pressure, or various combinations of these methods. Antigen retrieval is often characterized as a "high temperature heating technique to restore the antigenicity of tissue sections obscured by formalin fixation" (Shi et al., 2001); however, in certain instances, concealed epitopes can be reinstated solely through detergent treatment, without the application of heat (Brown et al., 1996). Effective antigen retrieval procedures can markedly enhance staining in archival tissues preserved in formalin for extended periods (Hann et al., 2001).

Despite the longstanding application of antigen retrieval methods, the underlying mechanisms remain ambiguous. Heat, in conjunction with buffers of differing pH values, appears to be a fundamental element in most antigen retrieval techniques, purportedly disrupting protein cross-links established during fixation. Chelating drugs can, in certain instances, assist by disrupting the calcium-protein complexes formed by fixation (Gown, 2004; Leong et al., 2007). In paraffin-embedded tissue sections, a microwave oven was determined to be superior to a conventional oven as a heat source (Shi et al., 1991). Conversely, a study by Cuevas et al. (1994) involving 80 distinct antibodies demonstrated that microwave heating of paraffin sections markedly improved staining for 41 antibodies, thereby establishing the microwave as a crucial instrument in immunocytochemistry. Besides conventional and microwave ovens, alternative heat sources include pressure cookers (Norton et al., 1994) and hotplates for heating tissue sections (Hann et al., 2001) or entire tissues prior to sectioning (Ino, 2003). In addition to heating, certain antigen retrieval techniques involve the application of high vapor pressure, various detergents, buffers with differing pH and concentration, and various metal ions, which may be utilized individually or in various combinations, with or without heating (Shi et al., 2001).

Rigorous antigen retrieval protocols, encompassing microwave heating and alcohol treatment, are standard procedures for unmasking epitopes in paraffin-embedded, PFA-fixed tissue sections (Cattoretti et al., 1993; Cuevas et al., 1994; Hoetelmans et al., 2002; Leong et al., 2002), whereas the application of these protocols to cryosections of PFA-fixed tissues is less prevalent. The availability of epitopes in tissue cryosections is thought to be less problematic than in paraffin-embedded tissue sections. The application of sodium dodecyl sulfate (SDS) to tissue cryosections, without heating, has proven effective in augmenting immunostaining with certain antibodies (Brown et al., 1996; Sabolić et al., 1999), suggesting that unmasking techniques may facilitate the exposure of antibody binding epitopes in cryosections as well. We have recently demonstrated that heating cryosections of PFA-fixed rat and mouse organs in a microwave oven enhances immunostaining with specific antibodies for various organic anion transporters (Bahn et al., 2005; Breljak et al., 2010; Brzica et al., 2009b; Ljubojević et al., 2004 & 2007; Yokoyama et al., 2008), glucose (Balen et al., 2008; Sabolić et al., 2006), and sulfate (Brzica et al., 2009a).

To illustrate the significance of diverse unmasking protocols for immunocytochemical analyses in tissue cryosections, we utilized cryosections of rat kidney and liver tissues, which were fixed with 4% PFA in vivo. We applied various antigen retrieval protocols, both with and without microwave heating, and assessed the intensity of immunostaining for several representative proteins localized in the cell membrane (cell adhesion molecule 105 (CAM105), megalin (GP330), Na/K-ATPase, aquaporin 1 (AQP1)), cytoplasm (metallothionein), cell membrane and intracellular organelles (vacuolar H+-ATPase (V-ATPase)), and cytoskeleton (actin, tubulin). Utilizing V-ATPase as a model, we examined the impact of heating duration and varying microwave power levels on the intensity and dispersion of antibody staining. The immunostaining was conducted using indirect immunofluorescence, wherein specific epitopes on proteins were initially labeled with appropriate primary polyclonal or monoclonal antibodies, followed by labeling with fluorescent molecule-conjugated secondary antibodies and subsequent fluorescence microscopy.

2. MATERIAL AND METHODS

2.1. Antibodies and other material

Primary antibodies used in these studies were monoclonal or polyclonal affinity purified antibodies raised against the holoproteins or defined, protein-specific peptide sequences. Some of them were purchased commercially, and some were non-commercial, but their use in immunocytochemistry has been described in our previous publications (down indicated in brackets). The commercial antibodies against the following proteins were used: Na/K-ATPase (monoclonal anti-peptide antibody against the α1 subunit of the human protein, Santa Cruz Biotechnology, CA, USA), metallothionein (monoclonal antibody against the polymerized MT1 and MT2 horse holoproteins; Dako North America, CA, USA), actin (monoclonal anti-peptide antibody; Millipore, MA, USA), and α-tubulin (monoclonal antibody against the see urchin filament; Sigma, MO, USA). The non-commercial antibodies against the following proteins were used: megalin (polyclonal antibody against the rat holoprotein; characterized in Abbate et al., 1994 and Sabolić et al., 2002), CAM105 (polyclonal antibody against the rat holoprotein; characterized in Sabolić et al., 1992a), AQP1; polyclonal antibody against the rat holoprotein; characterized in Sabolić et al., 1992b), and V-ATPase (polyclonal anti-peptide antibody against the 31 kDa ("E") subunit; characterized in Herak-Kramberger et al., 2001). Cryosections were incubated in an optimal concentration of the antibody, as defined in preliminary experiments (data not shown), in a refrigerator over night (ON; 12-14 hours).

Secondary antibodies were the CY3-labeled goat anti-rabbit (GARCY3) or donkey anti-mouse IgG (DAMCY3), purchased from Jackson ImmunoResearch (West Grove, PA, USA), and FITC-labeled goat anti-rabbit (GARF) or goat anti-mouse (GAMF) IgG, purchased from Kirkergaard & Perry Laboratories (Gaithersburg, MD, USA). These antibodies were applied in the following concentrations: GARCY3 – 1.5 μg/mL, DAMCY3 – 1.25 μg/mL, GARF – 2.5 μg/mL, and GAMF – 5 μg/mL, at room temperature (RT) for 1 hour. All incubations with the antibodies were performed in the moist chamber.

The following noncommercial solutions have been prepared and used:

1. PBS (phosphate buffered saline (in mM): 137 NaCl, 2.7 KCl, 8 Na_2HPO_4, 2 K_2PO_4, pH 7.4) PFA (4% w/v *p*-formaldehyde in PBS)
2. HS-PBS (high salt PBS; PBS containing a triple (411 mM) concentration of NaCl)
3. BSA (bovine serum albumin; 1% w/v in PBS)
4. Citrate buffer (10 mM citrate buffer, pH 3, pH 6, and pH 8)
5. Sucrose solution (30% w/v sucrose in PBS)
6. SDS (detergent sodium dodecyl-sulphate; 1% w/v in PBS)
7. T-X-100 (Triton-X-100; 0.1%, 0.5%, and 2% w/v in PBS)
8. Alcohols xylol and propanol were used undiluted (100%), whereas ethanol was used in concentrations of 60%, 70%, and 96%, prepared v/v in distilled water.

The ingredients in these solutions, and other chemicals used in this study were of the highest purity, and were purchased from Sigma (St. Louis, MO, USA). The SuperFrost-plus microscope slides were purchased from Thermo Scientific (Menzel-Gläser, Braunschweig, Germany). The fluorescence fading retardant (Vectashield) was purchased from Vector Laboratories Inc., CA, USA.

2.2. Animals, tissue fixation, cryosections

Adult male Wistar strain rats were anaesthetized and their circulatory system was perfused *in vivo* using the Masterflex pump (Cole-Parmer, IL, USA) *via* the left ventricle of the heart, first with aerated (95% O_2/5% CO_2) and temperature equilibrated (37º C) PBS for 2-3 min to remove blood, and then with 150 mL of fixative (4% PFA) for 4-5 min. Kidneys and the largest liver lobe were removed, cut in ~1 mm thick slices, postfixed in the same fixative at 4º C for 24 hours, and then washed with four abundant volumes of PBS to remove PFA. The tissues were then rinsed with, and stored refrigerated in PBS containing 0.02% NaN_3 until use. For cryosectioning, tissues were soaked in 30% sucrose solution at 4º C overnight, embedded in OCT-medium (Tissue-Tek, Sakura, Japan), and frozen at –25º C. Four µm-thick cryosections were sliced in a Leica CM 1850 cryostat (Leica instruments GmbH, Nussloch, Germany), collected on SuperFrost-plus microscope slides, dried at room temperature for 2-3 hours, and stored refrigerated until use.

2.3. Protocols of antigen retrieval

In order to reveal optimal protocols for unmasking the antibody-binding epitopes, tissue cryosections were treated with different protocols and then subjected to immunostaining procedure with specific antibodies. The immunostaining in the absence of any antigen retrieval (Protocol A) was compared with that following treatment with SDS (Protocol B), following heating in a microwave oven in strongly acidic (pH 3), weakly acidic (pH 6), and alkaline (pH 8) citrate buffers (Protocols C, D, and E, respectively), and following treatment in alcohols and consecutive heating in a microwave oven in the same citrate buffers (Protocols F, G, and H, respectively). The incubations were performed either at room temperature (RT) or in a refrigerator (4º C). Heating of cryosections was performed in a conventional microwave oven with the variable power settings (max. 800 W). Therefore, in these studies the following protocols were applied:

1. Protocol A, untreated cryosections (steps): Rehydration in PBS (15 min/RT) - incubation in BSA (30 min/RT) - incubation in primary antibody (refrigerator/ON) - rinsing in HS PBS (2x5 min/RT) - rinsing in PBS (2x5 min/RT) - incubation in secondary antibody (60 min/RT) - rinsing in HS PBS (2x5 min/RT) - rinsing in PBS (2x5 min/RT) – covering with Vectashield and the cover glass, and sealing lateral openings with nail polish.
2. Protocol B, SDS-treatment (steps): Rehydration in PBS (15 min/RT) - incubation in SDS (5 min/RT) – rinsing in PBS (4x5 min/RT) - incubation in BSA (30 min/RT) - incubation in primary antibody (refrigerator/ON) – other steps are the same as listed in protocol A.
3. Protocol C, microwave heating (steps): Rehydration in PBS (15 min/RT) – immersion in citrate buffer, pH 3, and heating in a microwave oven (4x5 min/800 W, with intermittent addition of the evaporated buffer; cryosections should be fully immersed in the buffer all the time) - cooling down of cryosections in the buffer (20 min/RT) – rinsing in PBS (3x5 min/RT) – incubation in 0,5% T-X-100 (15 min/RT) – incubation in 2% T-X-100 (30 min/RT), - rinsing in PBS (2x5 min/RT) – incubation in BSA (30 min/RT) - incubation in primary antibody (refrigerator/ON) - incubation in 0.1% T-X-100 (10 min/RT) – rinsing in PBS (2x5 min/RT) - incubation in secondary antibody (1 hour/RT) – incubation in 0.1% T-X-100 (10 min/RT) – rinsing in PBS (2x5 min/RT) covering with Vectashield and the cover glass, and sealing with nail polish.
4. Protocol D, microwave heating (steps): The same as protocol C, except the citrate buffer had pH 6.
5. Protocol E, microwave heating (steps): The same as protocol C, except the citrate buffer had pH 8.
6. Protocol F, alcohol treatment and microwave heating (steps): Immersion (without rehydration) and incubation in 100% xylol (30 min/RT) - incubation in 100% propanol (5 min/RT) – incubation in 96% ethanol (5 min/RT) – incubation in 70% ethanol (5 min/RT) – incubation in 60% ethanol (5 min/RT) – rinsing in destilled H_2O (5 min/RT) - rehydration in PBS (15 min/RT) - immersion in citrate buffer, pH 3, and heating in a microwave oven (4x5 min/800 W, with intermittent addition of the evaporated buffer; cryosections should be fully immersed in the buffer all the time) – the consecutive steps are the same as listed in protocol C.
7. Protocol G, alcohol-treatment and microwave heating (steps): The same as the protocol F, except the citrate buffer had pH 6.

8. Protocol H, alcohol-treatment and microwave heating (steps): The same as the protocol F, except the citrate buffer had pH 8.

In a supplementary series of experiments, antigen retrieval-dependent immunostaining utilizing the anti-V-ATPase antibody in cryosections of PFA-fixed rat kidney cortex was employed to evaluate (a) the effects of heating duration at a constant power, and (b) the effects of power variation at a constant heating duration. To evaluate time-dependency, cryosections of the renal cortex were subjected to microwave heating at 800 W for intervals of 5 min, 2x5 min, 3x5 min, and 4x5 min, adhering to the settings specified in the optimum procedure. To illustrate the power-dependent effects, cryosections of the renal cortex were subjected to microwave heating in 4x5 minute cycles at 80 W, 160 W, 400 W, 560 W, and 800 W, adhering to the settings specified in the optimum protocol. In both instances, the control cryosections were prepared in a similar manner, but without heating.

The fluorescence was analyzed using an Opton III RS fluorescence microscope (Opton Feintechnik, Oberkochen, Germany) and captured with a Spot RT slider camera and software (Diagnostic Instruments, Sterling Heights, MI, USA). Every collection of photos, pertaining to the retrieval technique for a particular antibody/antigen pair, was captured using identical recording conditions. The pictures were loaded into Adobe Photoshop 6.0 for processing, compilation, and annotation. In certain instances, the fluorescence associated with CY3 or FITC was transformed into black and white mode utilizing the same program.

3. RESULTS

3.1. Antigen retrieval of epitopes specific for proteins located in cell membrane domains

In these experiments, cryosections of the PFA-fixed rat kidney or liver tissues underwent the above-listed unmasking protocols, followed by immunostaining of proteins known to be largely localized in the specific cell membrane domains of the nephron epithelium and hepatocytes: cell adhesion molecule CAM105, an ectoprotein expressed in the luminal (apical/brush-border (BBM)) membrane of the proximal tubules and the plasma membrane of peritubular capillaries (Sabolić et al., 1992a), megalin, a glycoprotein expressed in the BBM and subapical vesicles of the proximal tubules (Abbate et al., 1994; Kerjaschki & Farquhar, 1983; Sabolić et al., 2002), Na/K-ATPase, an integral membrane protein expressed in the contraluminal (basolateral (BLM)) cell membrane of the proximal and other tubules (Kashgarian et al., 1985; Sabolić et al., 1999) and in the hepatocyte sinusoidal membrane, and AQP1, a transmembrane protein expressed in both luminal and contraluminal membranes of the proximal tubules and thin descending limbs (Nielsen et al., 1993; Sabolić et al., 1992b).

3.1.1. CAM105

Previous studies showed localization of CAM105 in the BBM of S1 and S3, but not S2 segments of the rat kidney proximal tubules, and also in the plasma membrane of peritubular capillaries (Sabolić et al., 1992a). The present study was performed on cryosection of the kidney cortex, thus showing the protein expression in proximal tubule S1 segments and peritubular capillaries (Fig. 1). As shown in Fig. 1A, in the absence of any antigen retrieval conditions, a limited staining was observed in the BBM of S1 and in peritubular capillaries (arrowheads), but not in S2. In the SDS-treated cryosections (B), the staining distribution was similar to, but the intensity was weaker than that in (A). In cryosections treated with heating in a microwave oven (C-E), the staining intensity dependent on buffer pH, being weak at pH 3 (C), intermediate at pH 6 (D), and strong at pH 8 (E). Localization of the staining in the BBM of S1 and in the plasma membrane of peritubular capillaries, remained unchanged. Alcohol treatment plus microwave heating resulted in strong reduction of the staining intensity, to almost none at pH 3 (F) and pH 6 (G), while peritubular capillaries remained stained. However, at pH 8 (H), the staining in both location was present, but the intensity was weaker than in E. In summary, the CAM105-related staining of peritubular capillaries was present in all conditions, while the staining of BBM was dependent on the applied protocol. The best staining of CAM105 was achieved after using protocol E.

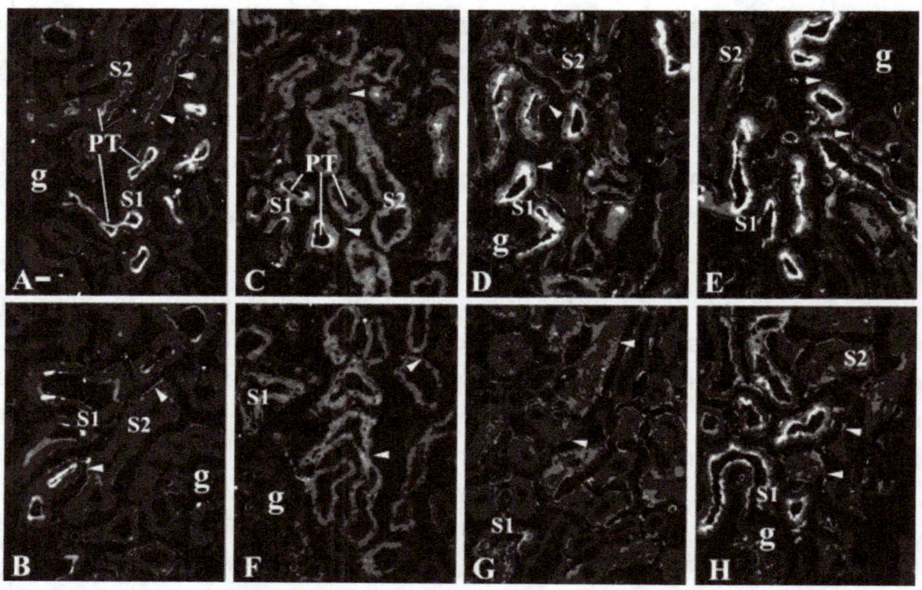

Figure 1. Distribution and intensity of immunostaining of CAM105 in cryosections of the PFA-fixed rat kidney cortex tissue that underwent various antigen retrieval protocols listed in the section 2.3. g, glomeruli; PT, proximal tubules (S1, initial segment; S2, convoluted segment); arrowheads, peritubular capillaries. Bar, 20 µm.

3.1.2. Megalin

Immunostaining of megalin in the kidney cortex in cryosections treated with different antigen retrieval protocols is demonstrated in Fig. 2, A-H. A relatively strong staining of the proximal tubule apical domain (BBM and subapical vesicles) was present already in untreated cryosection (A), while SDS-treatment resulted in an additional intensity (B). Protocols with the microwave heating (C-E) showed the pH-dependent staining intensity with a sequence: pH 3 > pH 6 > pH 8, where the staining pattern and intensity at pH 3 (C) were similar to those in control conditions (c.f. A). Protocols with the alcohol treatment followed by heating (F-H) caused overall weaker staining when compared to protocols C-E, while displaying the same pH-dependency. Therefore, megalin immunostaining was not improved with heating cryosections at any pH without or with alcohol pretreatment, and the protocol with SDS-treatment alone (B) appears to be the best antigen-revealing method for megalin epitopes in the proximal tubules.

3.1.3. Na/K-ATPase

Fig. 3 (A-H) shows effects of different unmasking protocols on immunostaining of Na/K-ATPase in cryosections of the rat kidney cortex. In control conditions, Na/K-ATPase was stained weakly in the BLM of proximal tubules and slightly stronger in distal tubules and cortical collecting ducts (A). Following the SDS-treatment, an enhanced staining was observed in the distal tubules and collecting ducts, but the staining in proximal tubules remained unchanged (B). Protocols with the microwave heating at pH 3 (C) and pH 6 (D) led to an enhancement in staining in all tubule profiles, particularly in the proximal and distal tubules (pH 3 < pH 6), whereas at pH 8 (E), the staining was weak in all tubules. In alcohol+heating-treated cryosections (F-H), the overall staining intensity was weaker than, but the staining pattern was similar to that in C-E. Therefore, protocol D represents the optimal antigen revealing approach for immunostaining of Na/K-ATPase along the rat nephron.

In order to check if the same protocol is valid for unmasking the epitopes of Na/K-ATPase in hepatocytes, cryosections of the PFA-fixed liver tissue underwent treatment with various protocols (Fig. 4, A-H). In control cryosections (A), and in cryosections treated with SDS (B), no significant staining of Na/K-ATPase was observed. In the microwave heating-treated cryosections (C-E), protocols C and E were only slightly unmasking (C < E), whereas the protocol D was better, exhibiting a fair staining of the hepatocyte sinusoidal membrane. Following alcohol treatment+microwave heating (F-H), protocols F and H were also weak, whereas protocol G was strongly unmasking, showing the strongest staining intensity in the hepatocyte sinusoidal membrane (arrows), whereas the bile canaliculi remained unstained (arrowheads). Therefore, unlike the situation in kidneys, where the best antigen revealing conditions were in protocol D, the best conditions for unmasking Na/K-ATPase in the liver cells were in protocol G. This experiment indicates that the unmasking conditions of a common antigen in tissue cryosection of different organs can be different.

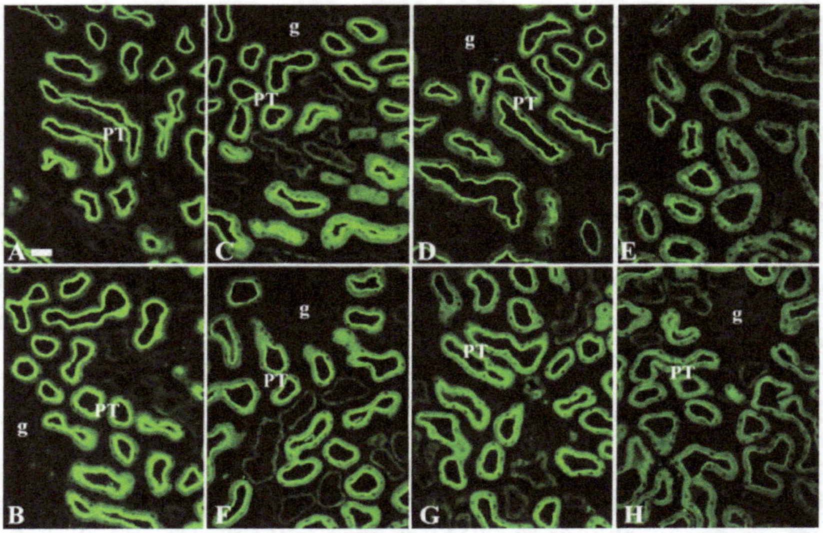

Figure 2. Distribution and intensity of immunostaining of megalin in cryosections of the PFA-fixed rat kidney cortex tissue that underwent various antigen retrieval protocols listed in the section 2.3. g, glomeruli; PT, proximal tubules. Bar, 20 µm.

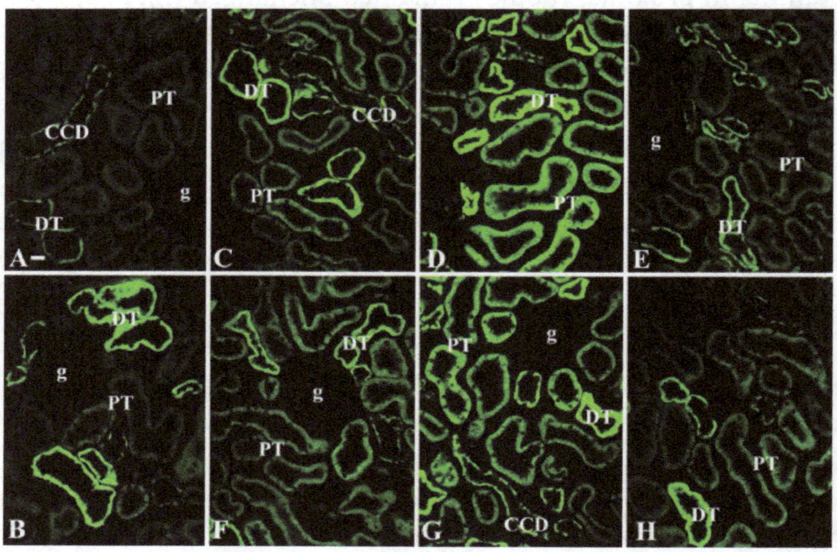

Figure 3. Distribution and intensity of immunostaining of Na/K-ATPase in cryosections of the PFA-fixed rat kidney cortex tissue that underwent various antigen retrieval protocols listed in the section 2.3. g, glomeruli; PT, proximal tubules; DT, distal tubules; CCD, cortical collecting ducts. Bar, 20 µm.

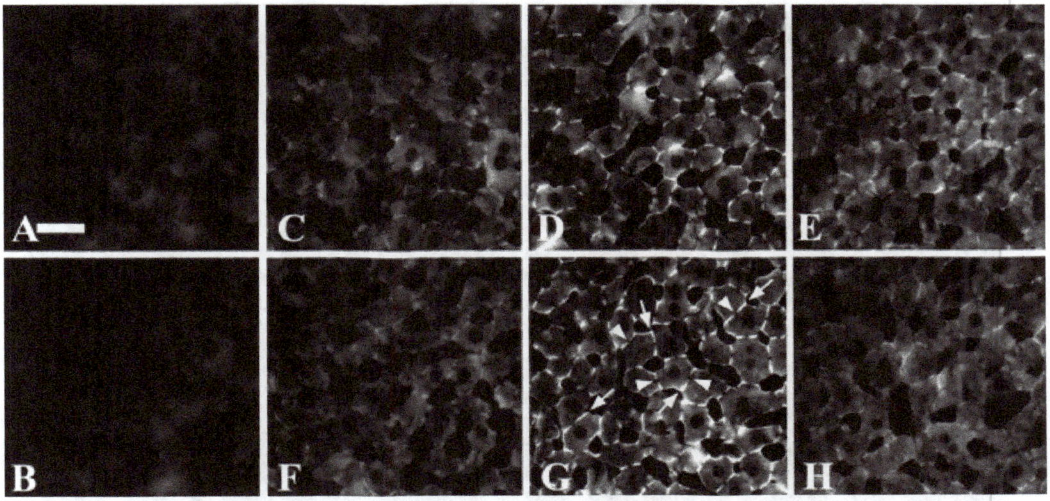

Figure 4. Distribution and intensity of immunostaining of Na/K-ATPase in cryosections of the PFA-fixed rat liver tissue that underwent various antigen retrieval protocols listed in the section 2.3. Arrows, the stained sinusoidal membrane; arrowheads, unstained bile canaliculi. Bar, 20 µm.

3.1.4. AQP1

AQP1 was previously localized to the luminal and contraluminal cell membrane domains of the proximal tubules and thin descending limbs (Nielsen et al., 1993; Sabolić et al., 1992b). Fig. 5. shows effects of various antigen retrieval protocols on immunostaining of AQP1 in cryosections of the rat kidney cortex, where only the proximal tubules are expected to be positive for this water channel. In cryosection treated with protocol A, a fair staining of AQP1 was observed in both BBM (arrowheads) and BLM (arrows) in the proximal tubules (A), whereas in the SDS-treated cryosections, the overall staining was much weaker (B). Following microwave heating (C-E), the staining was pH-dependent, being in both membrane domains strongest at pH 3 (C), weaker and present largely in the BBM at pH 6 (D), and negligible in both membrane domains at pH 8 (E). Treatment of cryosections with alcohols+microwave heating (F-H) resulted in distinct staining pattern; at pH 3, the overall staining was relatively weak, and present largely in the BBM (F), at pH 6, the staining was hardy visible in both membrane domains (G), whereas at pH 8, the staining in the BLM was stronger than in the BBM (H). Therefore, this experiment indicates that protocol C is the best for unmasking the AQP1 epitopes in both BBM and BLM of the proximal tubules, protocols D and F preferably unmask these epitopes in the BBM, whereas protocol H preferably unmasks the same epitopes in the BLM.

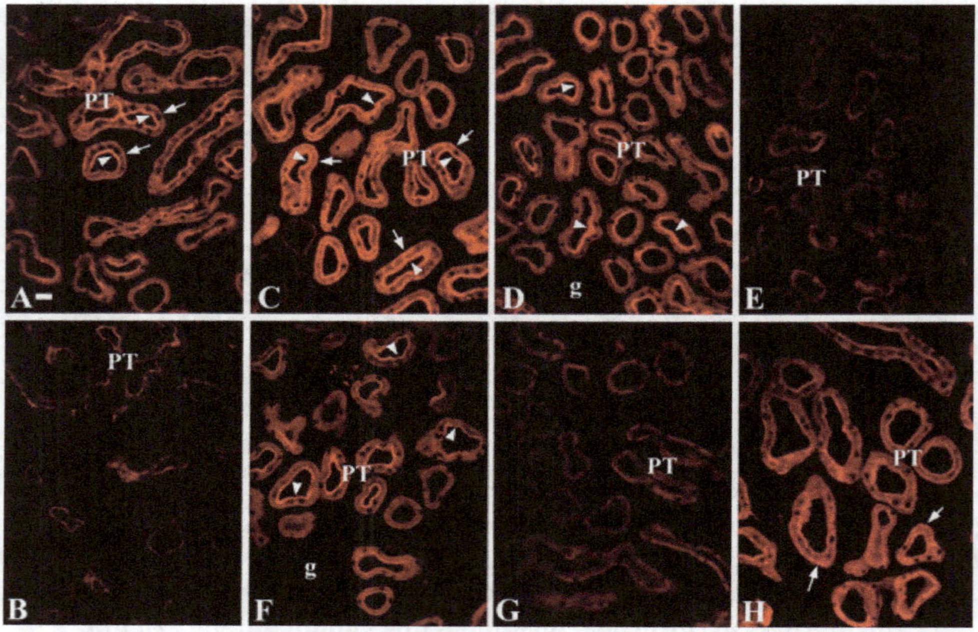

Figure 5. Distribution and intensity of immunostaining of AQP1 in cryosections of the PFA-fixed rat kidney cortex tissue that underwent various antigen retrieval protocols listed in the section 2.3. g, glomeruli; PT, proximal tubules; arrows, BLM; arrowheads, BBM. Bar, 20 μm.

3.1.5. V-ATPase

V-ATPase is localized in the BBM, subapical vesicles, and in various intracellular acidic organelles in the proximal tubule cells and in other cell types along the rat nephron (Brown et al., 1988). As shown in Fig. 6, a negligible staining of V-ATPase was recorded in control (A) and in the SDS-treated (B) cryosections of the kidney cortex. However, the microwave heating at pH 3 caused a strong unmasking effect, resulting in bright staining of the proximal tubule apical domain and intercalated cells in the cortical collecting ducts (C).

In cryosections treated with protocols D and E, the overall staining was much weaker than in those treated with protocol C, but intercalated cells in the collecting ducts were generally stained better than the proximal tubules. Compared with the staining after protocol C, treatment of cryosections with alcohols combined with microwave heating (F-H) resulted in overall weaker staining in all tubule profiles and in all conditions, with the pattern pH 3 < pH 6 < pH 8. Therefore, protocol C appeared to be an optimal unmasking approach for immunocytochemical presentation of V-ATPase in cryosections of the PFA-fixed kidney tissue.

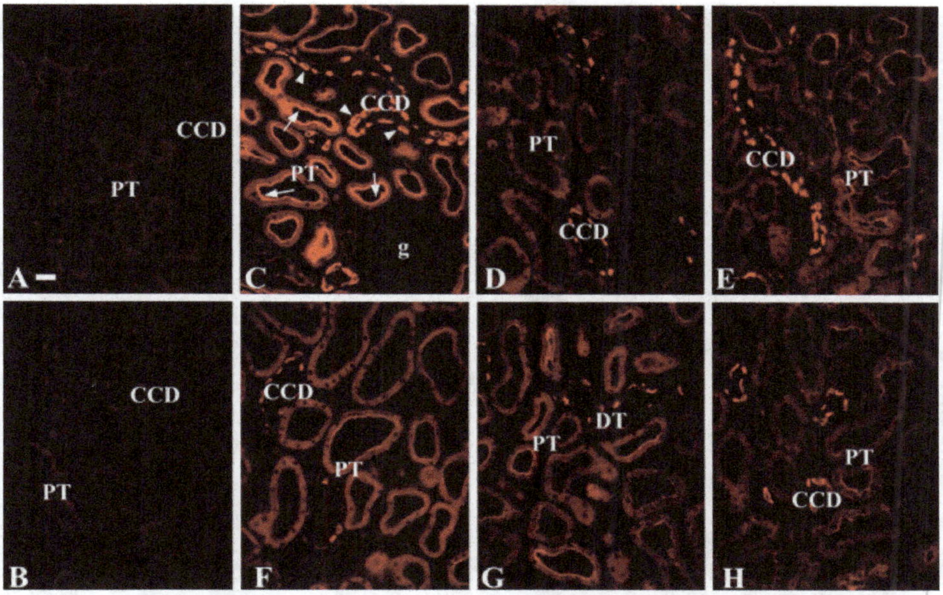

Figure 6. Distribution and intensity of immunostaining of V-ATPase in cryosections of the PFA-fixed rat kidney cortex that underwent various antigen retrieval protocols listed in the section 2.3. g, glomeruli; PT, proximal tubules; DT, distal tubules; CCD, cortical collecting ducts; arrows, apical cell domain (BBM and subapical organelles); arrowheads, intercalated cells in CCD. Bar, 20 μm.

3.2. Antigen retrieval of epitopes specific for cytoplasmic protein metallothionein

In the rat kidney, metallothionein is expressed in the cytoplasm of various cells localized along the nephron. In the kidney cortex, a variable expression of this protein has been detected in the proximal tubule cells (Sabolić et al., 2010, and unpublished data).

In cryosections processed with protocol A, only a weak staining was observed in the cell cytoplasm of some proximal tubules (A). The SDS-treatment increased the number of metallothionein-positive tubules, and slightly increased the staining intensity (B). Following heating in the microwave oven at pH 3, the staining was further increased, exhibiting the cells with a variable staining intensity from weak (arrowheads) to strong (arrows) (C). The intensity and the heterogeneity in the staining intensity was additionally enhanced in cryosections treated with the microwave heating at pH 6 (D), but significantly diminished after heating at pH 8 (E). In cryosections treated with protocols F-H, in all these conditions the overall staining, particularly that in H, was weaker than in D. In conclusion, protocol D appears to be optimal for immunolocalization of metallothionein in cryosections of the PFA-fixed rat renal cortex.

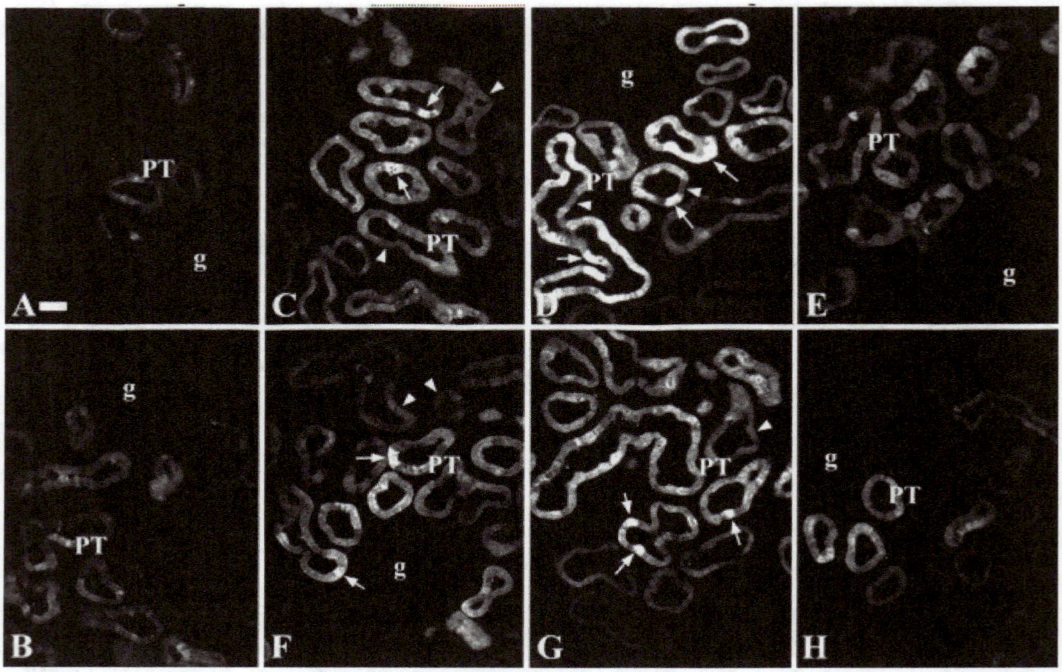

Figure 7. Distribution and intensity of immunostaining of metallothionein in cryosections of the PFA-fixed rat kidney cortex that underwent various antigen retrieval protocols listed in the section 2.3. g, glomeruli; PT, proximal tubules; arrows, strongly-stained cells; arrowheads, weakly-stained cells. Bar, 20 µm.

3.3. Antigen retrieval of epitopes specific for cytoskeletal proteins

3.3.1. Actin

Fig. 8, A-H shows effects of various protocols of antigen retrieval on actin-related immunostaining in cryosections of the PFA-fixed rat kidney cortex. Control (A) and the SDS-treated cryosections (B) showed no significant staining of actin in any cortical structure. Following heating in the microwave oven (C-E), the staining intensity gradually increased with increasing pH (pH 3 < pH 6 < pH 8) in the glomeruli, proximal tubule BBM, and in individual cells (probably intercalated cells) of the cortical collecting duct.

Cryosections treated with protocols that included treatment with alcohols and microwave heating (F-H) exhibited the same, pH-dependent staining pattern, but the staining intensity was higher in all conditions. Therefore, pretreatment with alcohols, followed by microwave heating in alkaline condition, as present in protocol H, enabled the strongest immunostaining of actin in cryosections of the PFA-fixed rat kidney cortex.

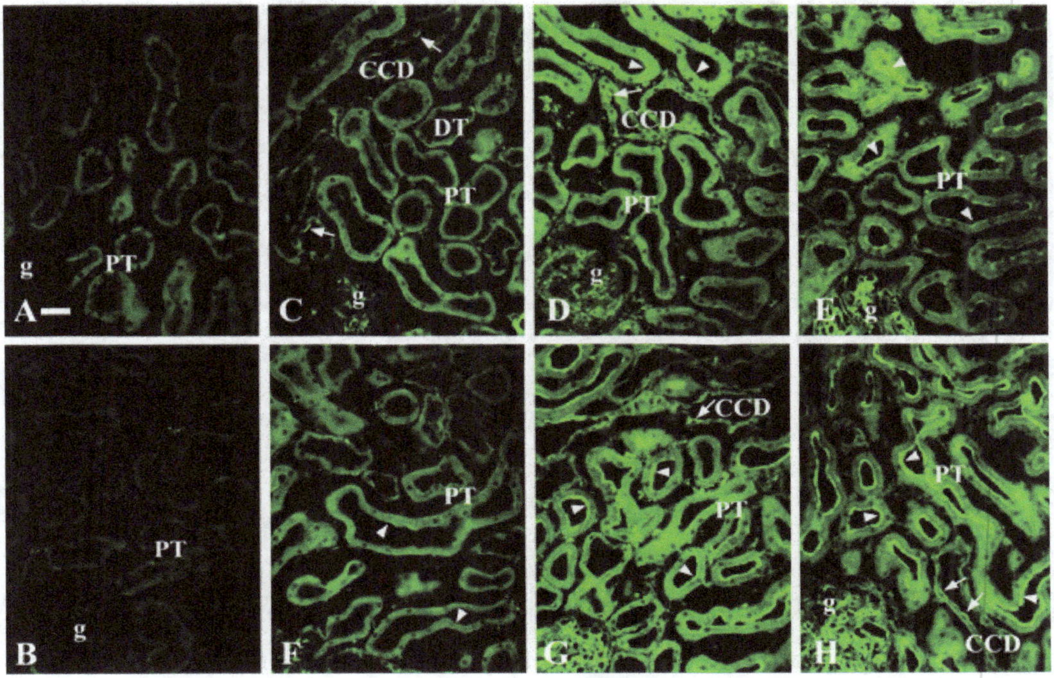

Figure 8. Distribution and intensity of immunostaining of actin in cryosections of the PFA-fixed rat kidney cortex that underwent various antigen retrieval protocols listed in the section 2.3. g, glomeruli; PT, proximal tubules; DT, distal tubules; CCD, cortical collecting duct; arrows, the actin-positive individual (probably intercalated) cells in CCD; arrowheads, the actin-positive BBM in proximal tubules. Bar, 20 μm.

3.3.2. α-Tubulin

Effects of different antigen retrieval protocols on immunostaining with the anti-α-tubulin antibody was performed in cryosections of the PFA-fixed rat kidney outer stripe (Fig. 9, A-H). As shown in our previous publication (Sabolić et al., 2002), in the proximal tubule cells α-tubulin forms the apicobasally-oriented bundles of microtubules, which are particularly rich in the cells of proximal tubule straight segments (S3) in the outer stripe zone. A weak staining of α-tubulin in the S3 segments was evident in cryosections that had been processed with protocol A. The staining was somewhat stronger in the SDS-treated cryosections (B). Heating in the microwave oven did not significantly affect the staining intensity at pH 3 (C), heating at pH 6 resulted in a bright staining of microtubule bundles (D), whereas after heating at pH 8 (E), the staining was weaker than in D.

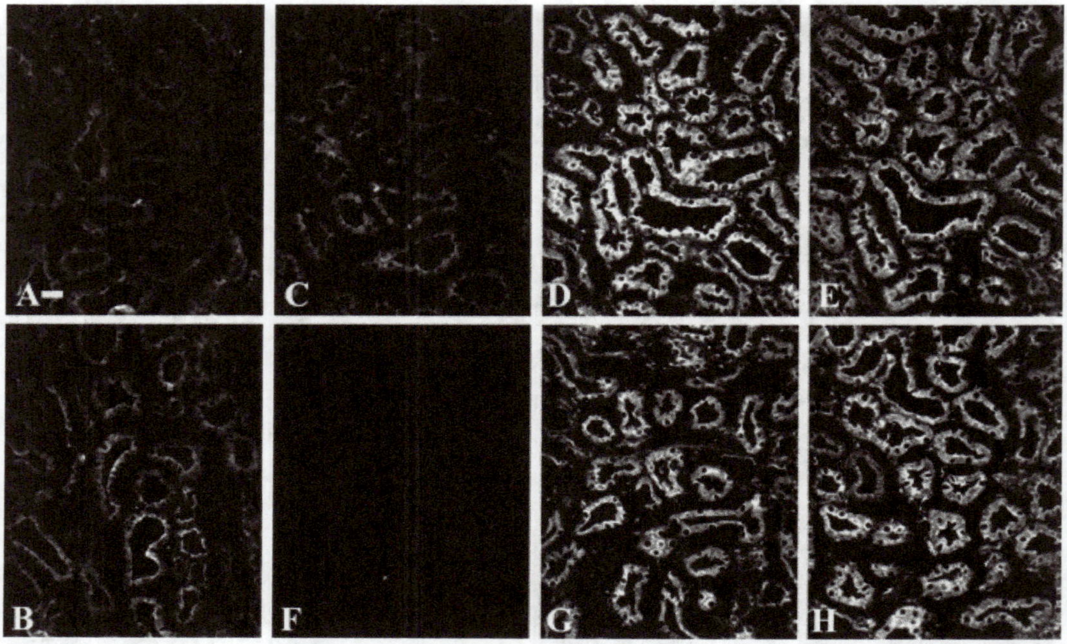

Figure 9. Immunostaining of α-tubulin in cryosections of the PFA-fixed rat kidney outer stripe following various antigen retrieval protocols listed in the section 2.3. The staining is localized in intracellular bundles of microtubules largely in the proximal tubule S3 segments. Bar, 20 μm.

Alcohol treatment combined with heating at pH 3 unmasked virtually nothing (F), whereas similar protocols at pH 6 (G) and pH 8 (H) resulted in a relatively strong and similar staining, yet weaker then that elicited with protocol D. Therefore, protocol D was found to be optimal for unmasking α-tubulin epitopes in cryosections of the PFA-fixed rat kidney tissue.

Various antigen retrieval protocols were also applied on cryosections of the PFA-fixed liver tissue (Fig. 10, A-H) in order to compare the staining pattern with that in the kidney (c.f., Fig. 9). A very weak staining was observed in control conditions (A), and the SDS-treatment resulted in only limited enhancement of staining (B). After heating cryosections in a microwave oven at pH 3, no significant staining was obtained (C). However, heating at pH 6 resulted in strong staining of microtubules in hepatocytes (D), whereas at pH 8 (E), the staining was weaker than at pH 6. Similar pattern, but at the lower level of staining intensity was observed in cryosections that had been treated with protocols F-H. Therefore, unlike the situation with Na/K-ATPase (c.f., Fig. 3 **and** Fig. 4), where the optimal unmasking protocols in cryosection of the kidney and liver tissues were different, the optimal protocols for unmasking α-tubulin in cryosections of both organs were similar.

3.4. Impact of constant heating power during a specified duration and of variable exposure to constant heating power

The preceding experiments showed that for most antigens to be localized by their antibodies, the optimal unmasking effect was achieved with one of the protocols that included heating of cryosections in a microwave oven at the power of 800 W for 20 min. Megalin was the only antigen that did not need heating for optimal presentation. These data thus indicate an indispensable role of microwave heating in unmasking epitopes for immunocytochemical presentation of a variety of antigens in cryosections of the PFA-fixed tissues. In the following two experiments we tested the importance of (a) microwave heating power at the fixed time, and (b) duration of the microwave heating at the fixed power, for unmasking efficiency using V-ATPase as the representative antigen and general conditions in protocol C. The conditions in this protocol were found to be optimal for unmasking epitopes of V-ATPase in cryosection of the PFA-fixed rat kidney cortex (c.f. Fig. 6).

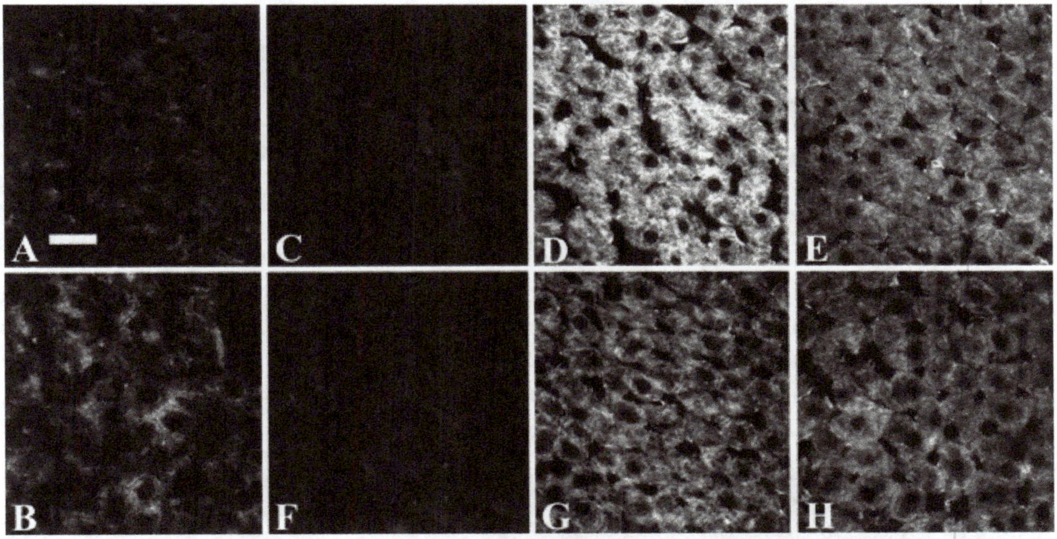

Figure 10. Immunostaining of α-tubulin in cryosections of the PFA-fixed rat liver following various antigen retrieval protocols listed in the section 2.3. Bar, 20 μm.

Fig. 11. shows the effects of microwave heating of cryosections in citrate buffer, pH 3, at different power settings (0-800 W) for 20 min on the V-ATPase immunostaining. Incubation of cryosections for 20 min without heating resulted in a very weak staining of the cells in collecting ducts (a). After heating at 80 W (b) and 160 W (c), the staining intensity of the collecting duct cells was slightly increased, while other tubule profiles remained unstained. At 400 W (d), the staining intensity of the individual collecting duct cells was stronger, and a weak staining of the proximal tubule BBM was also observed. The overall staining in both collecting duct cells and proximal tubule BBM further increased at 560 W (e), and the brightest staining was recorded after heating at 800 W (f).

Effects of heating duration (0-20 min) at the fixed power (800 W) are shown in Fig. 12. Heating was applied in cycles, 5 min each. An incubation of cryosections at room temperature (heating duration 0 min) resulted in weak staining of collecting ducts (a). Heating for 5 min (one cycle) only slightly increased the staining intensity of the cells in collecting ducts (b). After heating for 10 min (two cycles), a limited staining was observed in the proximal tubule BBM, while the staining intensity of the collecting duct cells further increased (c). Three cycles of heating (15 min) revealed a strong staining in both nephron segments (d), whereas four heating cycles (20 min) resulted in an additional, but slight increase in the staining intensity (e). From the data in these two experiments we conclude that both the heating power and duration of heating in a microwave oven are crucial for an optimal retrieval of hidden antibody-binding epitopes in cryosections of the PFA-fixed tissues.

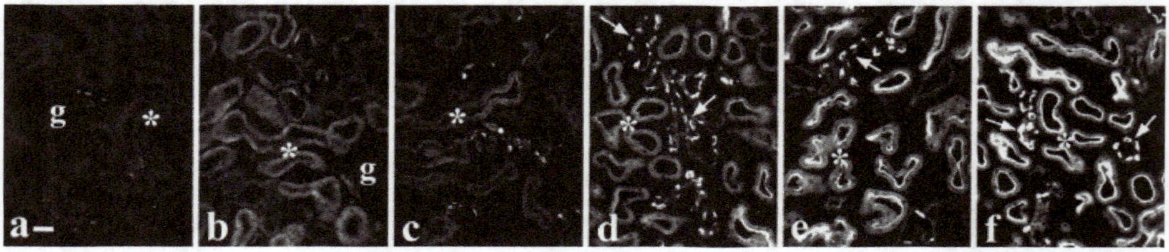

Figure 11. Effect of power setting of microwave oven on the staining pattern and intensity of V-ATPase in cryosections of the PFA-fixed rat kidney cortex. (a) Without heating, and heating at (b) 80 W, (c) 160 W, (d) 400 W, (e) 560 W, and (f) 800 W for 20 min, other conditions being in protocol C. g, glomeruli; asterisks, proximal tubules; arrows, cortical collecting ducts. Bar, 20 μm.

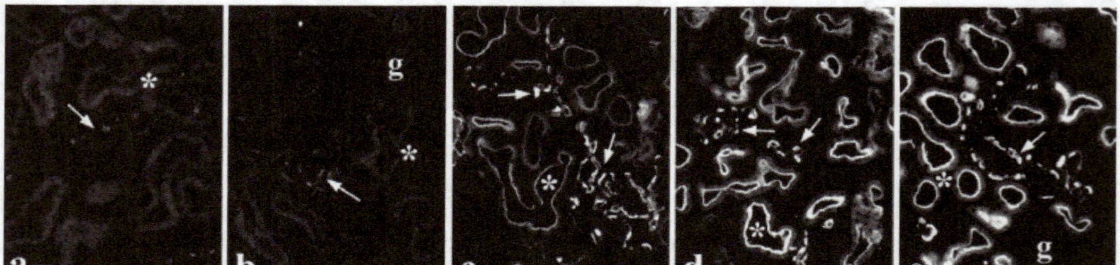

Figure 12. Effect of duration of microwave heating at the fixed power setting (800 W) on the staining pattern and intensity of V-ATPase in cryosections of the PFA-fixed rat kidney cortex. (a) Without heating, and heating for (b) 5 min, (c) 10 min, (d) 15 min, and (e) 20 min, other conditions being in protocol C. g, glomeruli; asterisks, proximal tubules; arrows, cortical collecting ducts. Bar, 20 μm.

4. DISCUSSION

An effective immunocytochemistry relies significantly on the expertise gained from experimental investigations aimed at identifying ideal circumstances for presenting certain antibody-binding epitopes. Our experience suggests that the predictability of these conditions for any antibody and tissue or cells is uncertain, necessitating the testing of multiple antigen retrieval procedures with each new antibody and tissue prior to commencing comprehensive immunocytochemical studies.The SDS-treatment of cryosections of the PFA-fixed tissues and cultured cells was previously demonstrated as a beneficial technique for unmasking epitopes of several membrane-bound proteins (Alper et al., 1997; Brown et al., 1996; Sabolić et al., 1999). However, using similar cryosections of the rat and mouse tissues, we have recently applied a few harsh antigen retrieval protocols that are commonly used to unmask hidden epitopes in the PFA-fixed, paraffin-embedded sections, and efficiently demonstrated the expression of various transporters in the respective organs (Bahn et al., 2005; Balen et al., 2008; Breljak et al., 2010; Brzica et al., 2009a& 2009b; Ljubojević et al., 2004& 2007; Yokoyama et al., 2008). In these studies, in preliminary experiments we have always tested a battery of antigen-revealing protocols with each antibody and tissue cryosections until one, a protocol that resulted in the highest immunostaining intensity, was defined and further used. In most cases, the SDS-treatement had only a limited potency in unmasking the tested antigens, and one of the revealing protocols with microwave heating without or with alcohol pretreatment usually proved to be more efficient in revealing cellular localization of the protein-specific epitopes and the intensity of their staining. In the present study, we describe our experience in using various antigen revealing protocols, and the importance of microwave heating in these protocols in order to compare their efficiency in unmasking epitopes on the proteins that reside in specific cellular locations. The major data from these experiments are summarized in Table 1.

As shown in Table 1, except for megalin in the proximal tubule BBM, whose staining was slightly enhanced only following treatment of cryosections with SDS (protocol B), all other antigens were better stained after being treated with one of the protocols that included microwave heating, without (protocols C-E) or with pretreatment with alcohols (protocols F-H). Each tested antigen needed one distinct unmasking protocol to be optimally exposed, with an unpredictable preference for the buffer pH and presence of alcohols.

Table 1. Relative efficiency of various antigen retrieval protocols B-H in unmasking hidden epitopes in cryosections of the PFA-fixed tissues, as observed in the present study.

			Microwave heating					
Antigen	Cellular localization	SDS treatment	- Alcohol treatment			+Alcohol treatment		
			pH 3	pH 6	pH 8	pH 3	pH 6	pH 8
CAM105	PT – BBM	-1	-1	+2	**+3**	-2	-2	+2
Megalin	PT - BBM & SAV	+1	0	-1	-2	-1	-1	-2
Na/K-ATPase								
Kidney	BLM	+1	+1	**+3**	+1	+1	+2	+1
Liver	SM	0	+1	+2	+1	0/+1	**+3**	0/+1
AQP1	BBM & BLM	-2	**+2**	0/-1	-3	0/-1	-3	0/-1
V-ATPase	BBM & SAV & IO	0	**+3**	+1	+1	+1	+1	+1
Metallothionein	PT – Cytoplasm	+1	+2	**+3**	+1	+2	+2	+1
Actin	Cytoskeleton	0/-1	+1	+1	+2	+1	+2	**+3**
Tubulin								
Kidney	Cytoskeleton	+1	0	**+3**	+2	-3	+2	+2
Liver	Cytoskeleton	+1	-1	**+3**	+2	-1	+2	+1/+2

The efficiency was based on visual estimation of the immunostaining intensity, relative to that obtained for the respective antigen processed with protocol A. PT, proximal tubule; BBM, brush-border membrane; SAV, subapical vesicles; BLM, basolateral membrane; SM, sinusoidal membrane; IO, intracellular organelles. (-) Relative decrease in staining intensity, (0) similar staining intensity, and (+) relative increase in staining intensity, when compared to the intensity in untreated (control) cryosections. Conditions that resulted in relatively strongest unmasking effect and staining intensity were indicated in bold.

The beneficial effect of the buffer pH combined with heating in unmasking hidden antigens has been previously well documented with paraffin sections (Evers & Uylings, 1994; Grossfeld et al., 1996; Shi et al., 1995; Shi et al., 1997). Furthermore, also unlike megalin, whose epitopes were temperature sensitive and were poorly exposed and stained following microwave heating, all other antigens required heating in a microwave oven either without or with alcohol pretreatment for best unmasking effects. Since the benefits of microwave heating for antigen retrieval in paraffin sections have been described (Shi et al., 1991), various time-dependent and power-related protocols have been used to unmask the hidden epitopes for different antibodies. Most of them played with duration of heating at the boiling point of the buffers at the maximum power settings (Evers & Uylings, 1994; Hoetelmans et al., 2002; Munakata & Hendricks, 1993; Shi et al., 1991). Our data obtained with an optimal protocol for unmasking the V-ATPase epitopes following time-dependent and power-dependent heating in a microwave oven indicate the importance of such studies for defining optimal antigen retrieval conditions also with cryosections.

Our data indicate that the optimal protocols for unmasking specific epitopes can be different in cryosections of different tissues. Thus, the optimal conditions for revealing the Na/K-ATPase epitopes in cryosections of the kidney cortex were found in protocol D (microwave heating in buffer pH 6, without alcohol pretreatment), whereas the same epitopes in cryosections of the liver tissue were best exposed in using protocol F (microwave heating in buffer pH 6, with alcohol pretreatment). Similar data were recently found in the studies of sulfate anion transporter sat-1 in cryosections of the rat kidney and liver with a monoclonal antibody; the SDS-treatment was best in unmasking the sat-1 epitopes in cryosections of the kidney cortex, whereas in cryosections of the liver tissue, protocol D (microwave heating in buffer pH 6) was needed for optimal immunostaining (Brzica et al., 2009a). Nevertheless, certain epitopes were optimally extracted from various tissues using identical techniques, as demonstrated by tubulin epitopes, which were most effectively exposed using protocol D in cryosections of both renal and hepatic tissues. All these observations corroborate the previously reported findings that, under in vivo conditions, the microenvironment of each tissue might influence antigen masking, alongside the reactive sites of the protein and/or their three-dimensional architectures (Toews et al., 2008).

5. CONCISE SUMMARY

Based on the data obtained in the preceding experiments, we conclude that: (a) the antigen retrieval protocols, that are usually used with sections of the formalin-fixed, paraffin-embedded tissues, can be efficiently used with cryosections of the PFA-fixed tissues, (b) in order to unmask the hidden epitopes in tissue cryosection, in most, but not all cases the retrieval protocols that include microwave heating give the best unmasking effect and immunostaining intensity, (c) the unmasking effect with a defined protocol is epitope-related and dependent on the buffer pH, d) the unmasking efficiency is dependent on heating power (more power – better unmasking effect) and heating time in a microwave oven (longer heating – better unmasking effect); however, the heating power of 800 W, applied for 20 min, is enough to obtain the maximal unmasking effect and immunostaining intensity with a variety of antibodies, and (d) cryosections of different organs can respond differently to the antigen retrieval conditions required for the same epitope. The subsequent observation underscores the imperative of employing diverse antigen detection procedures with cryosections of each tissue individually and for each antigen-antibody pair prior to routinely commencing immunocytochemical studies. Thus, these investigations can achieve ideal staining intensity and distribution while greatly reducing the quantity of primary antibodies required.

6. REFERENCES

Abbate M. Bachinsky D. R. Mc Cluskey R. T. Brown D. 1994 Expression of gp330 and gp330/α2-macroglobulin receptor-associated protein in renal tubular differentiation, Journal of the American Society of Nephrology 4 12 2003 2015.

Alper S. Stuart-Tilley A. K. Biemesderfer D. Shmukler B. E. Brown D. 1997 Immunolocalization of AE2 anion exchanger in rat kidney, American Journal of Physiology Renal Physiology 273 (4): F601 -F614.

Bahn A. Ljubojević M. Lorenz H. Schultz C. Ghebremedhin E. Ugele B. Sabolić I. Burckhardt G. Hagos Y. 2005 Murine renal organic anion transporters mOAT1 and mOAT3 facilitate the transport of neuroactive tryptophan metabolites, American Journal of Physiology Cell Physiology 289 (5): C1075 -C1084.

Balen D. Ljubojević M. Breljak D. Brzica H. Žlender V. Koepsell H. Sabolić I. 2008 Revised immunolocalization of the Na^+-D-glucose cotransporter SGLT1 in rat organs with an improved antibody, American Journal of Physiology Cell Physiology 295 (2): C475 -C489.

Breljak D. Ljubojević M. Balen D. Žlender V. Brzica H. Micek V. Kušan M. Anzai N. Sabolić I. 2010 Renal expression of organic anion transporter Oat5 in rats and mice exhibits the female-dominant sex differences, Histology & Histopathology 25 11 1385 1402 .

Brown D. Hirsch S. Gluck S. 1988 Localization of proton-pumping ATPase in rat kidney, Journal of Clinical Investigation 82 6 2114 2126 .

Brown D. Lydon J. Mc Laughlin M. Stuart-Tilley A. Tyszkowski R. Alper S. 1996 Antigen retrieval in cryostat tissue sections and cultured cells by treatment with sodium dodecyl sulfate (SDS), Histochemistry & Cell Biology 105 4 261 267 .

Brzica H. Breljak D. Krick W. Lovrić M. Burckhardt G. Burckhardt B. C. Sabolić I. 2009a The liver and kidney expression of sulfate anion transporter sat-1 in rats exhibits male-dominant gender differences, Pflügers Archiv- European Journal of Physiology 457 6 1381 1392 .

Brzica H. Breljak D. Ljubojević M. Balen D. Micek V. Anzai N. Sabolić I. 2009b Optimal methods of antigen retrieval for organic anion transporters in cryosections of the rat kidney, Archives of Industrial Hygiene and Toxicology 60 1 7 17 .

Cattoretti G. Piler S. Parravicini C. Becker H. G. Poggi S. Bifulco Ckey. G. D'Amato L. Sabattini E. Feudale E. Reynolds F. Gerdes J. Rilke F. 1993 Antigen unmasking on formalin-fixed, paraffin embeded tissue sections, Journal of Pathology 171 2 83 98 .

Cuevas E. C. Bateman A. C. Wilkins B. S. Johnson P. A. Williams J. H. Lee A. H. Jones D. B. Wright D. H. 1994 Microwave antigen retrieval in immuno-cytochemistry: a study of 80 antibodies, Journal of Clinical Pathology 47 5 448 452 .

Evers P. Uylings H. B. M. 1994 Microwave-stimulated antigen retrieval is pH and temperature dependent, Journal of Histochemistry & Cytochemistry 42 12 1555 1563.

Fox C. H. Johnson F. B. Whiting J. Roller P. P. 1985 Formaldehyde fixation, Journal of Histochemistry and Cytochemistry 33 8 845 853 .

Fraenkel-Conrat H. Olcott H. S. 1948 Reaction of proteins with formaldehyde VI. Cross-linking of amino groups with phenol, imidazole or indole groups. Journal of Biological Chemistry 174 3 827 843 .

Fraenkel-Conrat H. Brandon B. A. Olcott H. S. 1949 The reaction of formaldehyde with proteins VII. Demonstration of intermolecular cross-linking by means of osmotic pressure measurements, Journal of Biological Chemistry 177 1 477 486 .

Gown A. M. 2004 Unmasking the mysteries of antigen or epitope retrieval and formalin fixation, American Journal of Clinical Pathology 121 2 172 174 .

Grossfeld G. D. Shi S. R. Ginsberg D. A. Rich K. A. Skinner D. G. Taylor C. R. Cote R. J. 1996 Immunohistochemical detection of thrombospondin-1 in formalin-fixed, paraffin-embedded tissue, Journal of Histochemistry & Cytochemistry 44 7 761 766 .

Hann C. R. Springett M. J. Johnson D. H. 2001 Antigen retrieval of basement membrane proteins from archival eye tissues, Journal of Histochemistry & Cytochemistry 49 4 475 482 .

Herak-Kramberger C. M. Breton S. Brown D. Kraus O. Sabolić I. 2001 Distribution of the vacuolar H^+ATPase along the rat and human male reproductive tract, Biology of Reproduction 64 6 1699 1707 .

Hoetelmans R. W. M. van Slooten H. J. Keijzer R. van de Velde C. J. H. van Dierendonck J. H. 2002 Comparison of the effects of microwave heating and high pressure cooking for antigen retrieval of human and rat Bcl-2 protein in formaldehyde-fixed, paraffin embedded sections, Biotechnic & Histochemisty 77 7 137 144 .

Ino H. 2003 Antigen retrieval by heating en bloc for pre-fixed frozen material, Journal of Histochemistry & Cytochemistry 51 8 995 1003 .

Kashgarian M. Biemsderfer D. Caplan M. Forbush B. I. I. I. 1985 Monoclonal antibody to Na/K-ATPase: Immunocytochemical localization along nephron segments, Kidney International 28 6 899 913 .

Kerjaschki D. Farquhar M. G. 1983 Immunocytochemical localization of the heymann nephritis antigen (gp330) in glomerular epithelial cells of normal lewis rats, Journal of Experimental Medicine 157 2 667 686 .

Leong A. S. Gilhmam P. N. 1989 The effect of progressive formaldehyde fixation on the preservation of tissue antigens, Pathology 21 4 266 268 .

Leong A. S. Lee E. S. Yin H. Kear M. Haffajee Z. Pepperall D. 2002 Superheating antigen retrieval, Applied Immunohistochemistry & Molecular Morphology 10 3 263 268 .

Leong T. Y. Leong A. S. 2007 How does antigen retrieval work, Advances in Anatomic Pathology 14 2 129 131 .

Ljubojević M. Balen D. Breljak D. Kušan M. Anzai N. Bahn A. Burckhardt G. Sabolić I. 2007 Renal expression of organic anion transporter OAT2 in rats and mice is regulated by sex hormones, American Journal of Physiology Renal Physiology 292 (1): F361 -F372.

Ljubojević M. Herak-Kramberger C. M. Hagos Y. Bahn A. Endou H. Burckhardt G. Sabolić I. 2004 Rat renal cortical OAT1 and OAT3 exhibit gender differences determined by both androgen stimulation and estrogen inhibition, American Journal of Physiology Renal Physiology 287 (1): F124 -F138.

Mc Nicol A. M. Richmond J. A. 1998 Optimizing immunohistochemistry: antigen retrieval and signal amplification, Histopathology 32 2 97 103 .

Metz B. Kersten G. F. A. Hoogerhout P. Brugghe H. F. Timmermans H. A. M. de Jong A. Meiring H. ten Hove. J. Hennink W. E. Crommelin D. J. A. Jiskoot W. 2004 Identification of formaldehyde-induced modifications in proteins (reaction with model peptides), Journal of Biological Chemistry 279 8 6235 6243 .

Morgan J. M. Navabi H. Schmid K. W. Jasani B. 1994 Possible role of tissue bound calcium ions in citrate-mediated high-temperature antigen retrieval, Journal of Pathology 174 4 301 307 .

Munakata S. Hendricks J. B. 1993 Effect of fixation time and microwave oven heating time on retrieval of the Ki-67 antigen from paraffin embedded tissue, Journal of Histochemistry & Cytochemistry 41 8 1241 1246 .

Nielsen S. Smith B. L. Christensen E. I. Knepper M. A. Agre P. 1993 CHIP28 water channels are localized in constitutively water-permeable segments of the nephron, Journal of Cell Biology 120 2 371 383 .

Norton A. J. Jordan S. Yeomans P. 1994 Brief, high-temperature heat denaturation (pressure cooking): A simple and effective method of antigen retrieval for routinely processed tissues, Journal of Pathology 174 4 371 379 .

Sabolić I. Breljak D. Škarica M. Herak-Krambeger C. M. 2010 Role of metallothionein in cadmium traffic and toxicity in kidneys and other mammalian organs, Biometals 23 5 897 926 .

Sabolić I. Čulić O. Lin H. S. Brown D. 1992a Localization of ecto-ATPase in rat kidney and isolated renal cortical membrane vesicles, American Journal of Physiology, 262 (2): F217 -F228.

Sabolić I. Herak-Krambeger C. M. Breton S. Brown D. 1999 Na/K-ATPase in Intercalated cells along the rat nephron revealed by antigen retrieval, Journal of the American Society of Nephrology 10 5 913 922 .

Sabolić I. Ljubojević M. Herak-Kramberger C. M. Brown D. 2002 Cd-MT causes endocytosis of brush-border transporters in rat renal proximal tubules, American Journal of Physiology Renal Physiology 283 (6): F1389 -F1402.

Sabolić I. Valenti G. Verbavatz J. M. Van Hoek A. N. Verkman A. S. Ausiello D. A. Brown D. 1992b Localization of the CHIP28 water channel in rat kidney, American Journal of Physiology Cell Physiology 263 (6): C1225 -C1233.

Seshi R. S. Vani K. Messana E. Bogen S. A. 2004 A molecular mechanism of formalin fixation and antigen retrieval, American Journal of Clinical Pathology 121 2 190 199 .

Shi S. R. Key M. E. Kalra K. L. 1991 Antigen retrieval in formalin-fixed, paraffin-embedded tissues: an enhancement method for immunohistochemical staining based on microwave oven heating of tissue sections, Journal of Histochemistry & Cytochemistry 39 6 741 748 .

Shi S. R. Imam A. Young L. Cote R. J. Taylor C. R. 1995 Antigen retrieval immunohistochemistry under the influence of pH using monoclonal antibodies, Journal of Histochemistry & Cytochemistry 43 2 193 201 .

Shi S. R. Cote R. J. Taylor C. R. 1997 Antigen retrieval immunohistochemistry: past, present, and future, Journal of Histochemistry & Cytochemistry 45 3 327 343 .

Shi S. R. Cote R. J. Taylor C. R. 2001 Antigen retrieval techniques: current perspectives, Journal of Histochemistry & Cytochemistry 49 8 931 937 .

Toews J. Rogalski J. C. Clark T. J. Kast J. 2008 Mass spectrometric identification of formaldehyde-induced peptide modifications under in vivo protein cross-linking conditions, Analytica Chemica Acta 618 2 168 183 .

Yokoyama H. Anzai N. Ljubojević M. Ohtsu N. Sakata T. Miyazaki H. Nonoguchi H. Islam R. Onozato M. L. Tojo A. Tomita K. Kanai Y. Igarashi T. Sabolić I. Endou H. 2008 Functional and immunochemical characterization of a novel organic anion transporter Oat8 (Slc22a9) in rat renal collecting duct, Cellular Physiology & Biochemistry 21 4 269 278 .

4 Chapter
A WAY OF DISINFECTION IN DENTISTRY THAT USES MICROWAVES

1. INTRODUCTION: AN OVERVIEW

The management of cross-infection has garnered attention in dentistry over recent decades, owing to concerns regarding the transmission of infectious diseases, including AIDS, hepatitis, tuberculosis, pneumonia, and herpes, among dental patients, dental personnel, dental offices, and dental prosthesis laboratories (Codino & Marshall, 1976; Infection control in the dental office, Council on Dental Materials and Devices, 1978; Sande et al., 1975). This concern began in the 1970s, when the microorganism *Mycobacterium tuberculosis* was isolated in patients' molds (Ray & Fuller, 1963) and it was found that *Mycoplasma pneumoniae* was transmitted to laboratory technicians when they performed wear on contaminated dentures (Sande et al., 1975). Several items, such as instruments (Tarantino et al., 1997), dental burs (Rizzo, 1993), impressions (Abdelaziz et al., 2004), gypsum casts (Berg et al., 2005; Davis et al., 1989), dental appliances (Rohrer & Bulard, 1985), and others are often heavily contaminated with microorganisms from saliva and blood. Powell et al. (1990) also observed that 67% of all the materials received in the laboratories, among them crowns, molds wax records, and dentures, were contaminated with pathogenic microorganisms. The continual exposure of dental professionals and instruments to saliva and blood throughout nearly all procedures is a persistent risk and potential factor in the transmission of infection. The employment of an insufficient disinfection protocol in the management of dental supplies endangers unsuspecting personnel and leads to significant preventable cross-contamination.

A dental practitioner bears a legal and ethical obligation to avert infections in patients and staff, as well as a vested interest in safeguarding themselves against disease transmission by patients.. The Centers for Disease Control and Prevention of the United States (Kohn et al., 2004) have recommended sterilization and disinfection procedures to guarantee the safety of dental treatment in the dental health care settings. Typical methods of sterilization and disinfection include dry heat at 160 to 180ºC for two hours, wet steam under pressure at 121ºC for at least 15 minutes (autoclaving), immersion in chemical solutions, gamma radiation, and ethylene dioxide gas. However, these methods can be time-consuming and some of them require special and costly equipment, knowledgeable operative personnel, and constant surveillance. In order to overcome these limitations, microwave irradiation has been recommended as a practical physical sterilization method (Cottone et al., 1991) that is as effective as autoclaving (Tate et al., 1995). The lethal effects of a high frequency electric field on microorganisms was first described in 1925 (Kahler, 1929) and the destructive effects of strong radio-frequency fields on microorganisms were investigated in 1954 (Brown & Morrison, 1954). In fact, there are studies that have pointed out microwaves as a method for disinfecting food (Culkin & Fung, 1975), contact lenses (Hiti et al., 2001; Rohrer et al., 1986), laboratory microbiologic materials (Border & Rice-Spearman, 1999; Latimer, 1977), items of intimate clothing contaminated with *Candida albicans* (Friedrich & Phillips, 1988), hospital garbage (Hoffman & Hanley, 1994), coloring matter used in the cosmetic industry (Jasnow, 1975), and instruments used in medicine (Rosaspina et al., 1994). Hence, interest in this area has been maintained and in 1985 the technology was applied to the sterilization of dental appliances (Rohrer & Bulard, 1985).

In dentistry, microwave irradiation has been used for several purposes, including disinfection of toothbrushes (Nelson-Filho et al., 2011; Spolidorio et al., 2011), tongue scrapers (Spolidorio et al., 2011), instruments (Tarantino, 1997), contaminated gauze (Border & Rice-Spearman, 1999; Cardoso et al., 2007), dental burs (Fais et al., 2009; Rizzo, 1993), composite polishing instruments (Tate et al., 1995, 1996), molds made from elastomers (Abdelaziz et al., 2004), and gypsum casts (Berg et al., 2005; Davis et al., 1989). Furthermore, microwave irradiation has been widely accepted for polymerizing acrylic resin (Ilbay et al., 1994), drying gypsum products and investment materials (Hersek et al., 2002; Luebke & Chan, 1985; Luebke & Schneider, 1985; Tuncer et al., 1993), and as a post-polymerization treatment for reducing the residual monomer contents of polymerized acrylic resins and its cytotocixity (Jorge et al., 2009; Nunes de Mello et al., 2003). Besides these purposes, some relevant applications of microwave energy in the dentistry field are the disinfection of removable dentures (Dovigo et al., 2009; D.G. Ribeiro et al., 2009; Rohrer & Bulard, 1985; Sanitá et al., 2009; Silva et al., 2006) and the use of this disinfection method to treat patients with oral candidiasis (Banting & Hill, 2001; Neppelenbroek et al., 2008; Webb et al., 2005).

The effectiveness of microwave irradiation appears to depend on the medium in which the dentures are submerged, the duration of exposure, the microwave oven's power level, and the specific bacteria involved. Moreover, when choosing a disinfecting method, one must meticulously evaluate its impact on the physical and mechanical qualities of the irradiated materials. Therefore, the formulation of distinct protocols must be tailored to each specific scenario, aiming to ensure uniform sterilization while preserving dental components.

This chapter aims to: 1. elucidate the applications of microwave irradiation in dentistry; 2. evaluate the microbiological efficacy and suggested protocols for microwave irradiation; 3. delineate the mechanisms by which microwaves affect microorganisms; and 4. examine the impact of microwave irradiation on the characteristics of dental materials and devices.

2. APPLICATIONS IN DENTISTRY FIELD

The recognition of the potential for transmission of numerous infectious microorganisms during dental procedures has led to an increased concern for infection control in dental practice. Approaches to the clinical use of microwaves for preventing cross-infection have shown relevant results. Devices and instruments used in dental offices have been identified as a source of cross-contamination among patients and from patients to dental personnel. With this in mind, investigations were undertaken to explore the efficacy of microwave irradiation in disinfecting dental mirrors (Tarantino et al., 1997) and handpieces (Rohrer & Bulard, 1985). In addition, dental burs, which may become heavily contaminated with necrotic tissues, saliva, blood, and potential pathogens during use, can also be sterilized by microwave irradiation (Fais et al., 2009; Rizzo, 1993; Rohrer & Bulard, 1985). In order to prevent cross-infection, microwave energy can also be used to the disinfection of finishing and polishing instruments (Tate et al., 1995, 1996). As any another device used in dental offices, finishing and polishing instruments routinely come into contact with patient's saliva and blood and may also act as a source of cross-contamination. In accordance with the studies of Tate et al. (Tate et al., 1995, 1996), these dental devices can be effectively sterilized by microwave irradiation.

Another common dental procedure that may cause cross-infection, especially between patients and dental laboratory personnel, is the making of impressions. Previous studies showed that the majority of impressions arriving at a dental laboratory were contaminated with bacteria and other microorganisms (Egusa et al., 2008; Powel et al., 1990; Ray & Fuller, 1963; Sande et al., 1975; Sofou et al., 2002), irrespective of whether they had been exposed to a disinfectant procedure or merely rinsed with tap water (Sofou et al., 2002). The study of Egusa et al. (2008) showed an extensive contamination of alginate impressions with oral streptococci, staphylococci, *Candida* spp., methicilin resistant *Staphylococcus aureus* (MRSA), and *Pseudomonas aeruginosa*. These results indicate that a large number of microbes are retained on impression materials and are viably transferred onto the surface of stone casts. In fact, it has been shown that microorganisms can be transferred from impressions to gypsum models (Egusa et al., 2008; Leung & Schonfeld, 1983) and that the dental casts are potential sources of microbial transmission (Egusa et al., 2008; Leung & Schonfeld, 1983; Sofou et al., 2002). Also, even a cast from a properly disinfected impression may subsequently become contaminated by a technician or clinician. Considering all these information, microwave irradiation has been suggested to the disinfection of both impressions (Abdelaziz et al., 2004) and dental casts (Berg et al., 2005; Davis et al., 1989). The clinical relevance of impressions and dental casts microwave disinfection is that this procedure can be performed quickly and repeatedly, without the use of toxic, pungent, or allergenic chemicals. However, the disinfection of impression materials hinders possible cross-contamination only at the time the cast is poured. Because casts become contaminated after the intra-oral adjustments of dental appliances, they must be regarded as the major vehicle for cross-contamination and should be disinfected using microwave energy throughout all phases of the dental treatment.

Another application of microwave irradiation in preventing cross-infection is the disinfection of disposable materials. Although there is little scientific support for this purpose, some reports documented the disinfection of contaminated gauze and swabs (Border & Rice-Spearman, 1999; Cardoso et al., 2007) with the use of microwave irradiation. Cardoso et al. (2007) and Border & Rice-Spearman (1999) demonstrated that a short period of exposure of contaminated gauze pieces and swabs to microwave energy (30 seconds) inhibit the growth of pathogenic microorganisms. The authors suggested that microwave oven could be used instead of an autoclave in a variety of clinical and research settings because the procedure is rapid and the equipment cost is minimal.

Among the purposes of microwave energy in cross-infection prevention, one of the most important is the disinfection of removable dentures. In addition to its contamination by the oral microorganisms, it has been reported that dentures are contaminated at various stages during their fabrication (Verran et al., 1996; Wakefield, 1980; Williams et al., 1985) and are capable of transmitting microorganisms to other materials, dental equipments, laboratory, and technicians (Kahn et al., 1982). Microscopic studies have also demonstrated that a biofilm similar to that formed on natural teeth is present on dentures (Budtz-Jorgensen & Theilade, 1983). Large quantity of *Candida* spp. (Budtz-Jorgensen & Theilade, 1983) and some bacterial species associated with systemic diseases have been found in removable dentures, with predominance of gram-positive bacteria, as *Staphylococcus* spp., *Streptococcus* spp., and *Actinomyces* spp. (Chau et al., 1995; Glass et al., 2001a; Monsenego, 2000). Gram-negative species, such as *Neisseria perfava, P. aeruginosa, Klebsiella pneumoniae*, and *Enterobacter cloacae*, have also been identified (Henderson et al., 1987; Latimer, 1977). In fact, in vivo studies (D.G. Ribeiro et al., 2009; Rossi et al., 1996) found *C. albicans*, *S. aureus*, *Streptococcus mutans*, and MRSA on the surfaces of dentures from patients, with *C. albicans* having the highest prevalence in these biofilms. Therefore, the denture can function as a reservoir of microorganisms, enabling the transmission of diseases in the dental office and from it to the prosthetic laboratories. Manipulation of dentures in the different dental procedures may also disseminate the microorganisms throughout the environment in the form of aerosols (Clifford & Burnett, 1995). These pathogens may be inhaled by the dentist, assistants, and laboratory technicians, resulting in cross-infection between patients and dental personnel. In the context of denture microwave disinfection, the first studies were performed in order to demonstrate the effectiveness of microwave irradiation in inactivating microorganisms adhered to complete dentures (Thomas & Webb, 1995; Rohrer & Bulard, 1985; Webb et al., 1998). Using several protocols, positive results were obtained by in vitro and in vivo studies, proving that microwave energy can be an effective method in the disinfection of dentures (Dovigo et al., 2009; Glass et al., 2001b; D.G. Ribeiro et al., 2009; Sanitá et al., 2009; Silva et al., 2006). Thus, performing microwave disinfection of a removable denture before it is transferred to a dental laboratory, and immediately before it is returned to the patient, provides a measure of infection control for all parties.

In the course of time, the fit of dentures progressively declines as a result of time-dependent changes in the supporting tissue. In this context, hard chairside reline acrylic resins or soft denture liners are proposed for permanent or temporary improvement of denture fit. These auto-polymerizing acrylic resins allow the clinician to reline a removable prosthesis directly in the mouth in intimate contact with a large area of oral mucosa. Although such liners improve the denture fit and offer comfort to some edentulous and partially edentulous patients, these materials may present a source of other problems. One of their greatest disadvantages is the ongoing task of hygiene maintenance. Denture reline materials, especially the soft liners, have been found to be more prone to microbial adhesion than acrylic resin base materials because of their surface texture and higher porosity (Nikawa et al., 1992). The rougher surface of a prosthesis with a soft liner exhibits greater colonization by *C. albicans* when compared with a conventional acrylic resin denture (Wright et al., 1985). As a result, the oral mucosa is more susceptible to infections. A study evaluated the effectiveness of microwave disinfection on three hard chairside reline resins and observed a consistent sterilization after microwave exposure (Neppelenbroek et al., 2003). In other investigations (Dixon et al., 1999; Mima et al., 2008), microwave irradiation resulted in sterilization of a hard chairside reline resin and soft denture liners contaminated by four pathogenic microorganisms. It is also important to emphasize that disinfection by microwaves promotes inactivation of the pathogenic microorganisms present on both the surface and inside the pores of the acrylic materials (Chau et al., 1995; Dixon et al., 1999).

By being a reservoir of pathogens, the tissue surface of the acrylic resin denture enhances the infective potential of microorganisms and favors the appearance of oral infections (Budtz-Jorgensen, 1990). Oral candidiasis, represented by denture stomatitis in denture wearers, is one of the most common manifestations of disease associated with the use of removable dentures. Denture stomatitis is mainly caused by microorganisms of the *Candida* genus and normally affects the palate of approximately 65% of denture wearer patients (Chandra et al., 2001). This infection is characterized by the presence of multiple hyperemic points in the mucosa subjacent to the removable dentures of patients and, in more advanced cases, diffused erthymatous areas and papillary hyperplasia of the palate may also be observed (Newton, 1962). Considering that microbial adherence and colonization on dental prostheses favor the appearance of denture stomatitis, the cleansing and disinfection of dentures is fundamental to prevent this disease. Recent studies have suggested denture disinfection by microwave irradiation for the treatment of denture stomatitis. The microwave regimes established in laboratorial studies provided the baseline for subsequent clinical trials. Banting & Hill (2001) conducted the first study that evaluated the effectiveness of microwave energy for denture disinfection as a co-adjuvant in the treatment of denture stomatitis. The authors observed that the method was effective in the reduction of the clinical signs of infection. These findings are in agreement with those found by Webb et al. (2005) a few years later. A more recent study conducted by Neppelenbroek et al. (2008) also evaluated the effectiveness of complete denture disinfection by microwave energy in the treatment of patients with denture stomatitis. In agreement to Banting & Hill (2001) and Webb et al. (2005), it was observed that disinfection of the dentures by microwaves was effective for the treatment of denture stomatitis (Neppelenbroek et al., 2008). Investigations are still being carried out to evaluate the effectiveness of microwave irradiation in the treatment of denture stomatitis (Silva et al., 2008; Vergani et al., 2008). Further tests have been performed in order to evaluate the effectiveness of denture irradiation in the treatment of diabetic denture wearer patients with denture stomatitis, showing promising outcomes (Sanitá et al., 2010, 2011). Based on the above mentioned clinical studies, the use of microwaves has shown important results for the treatment of denture stomatitis. This disease is one of the most frequent opportunistic infections found in denture wearers, including the diabetic patients, and it may extend regionally and result in a systemic infection that is associated with high mortality rates (Colombo et al., 1999; Meunier-Carpentier et al., 1981). Hence, the prevention of colonization of the oropharynx is critically important in preventing systemic infections due to *Candida*.

A recent study has shown that pathogenic microorganisms can adhere in toothbrushes and tongue scrapers made from stainless steel- and polystyrene-based injection-moulded plastic (Spolidorio et al., 2011). Thus, toothbrushes and tongue scrapers become contaminated after use and, if not properly disinfected, may be a reservoir of microorganisms that maintain their viability for a significant amount of time, ranging 24 hours to 7 days (Nelson-Filho et al., 2006). Microbial survival promotes reintroduction of potential pathogens in the oral cavity or dissemination to other individuals when cleaning devices are stored together or shared (Ankola et al., 2009). Hence, these cleaning devices should be regularly disinfected. With this in mind, investigations have demonstrated the efficacy of microwave irradiation for disinfection of toothbrushes and tongue cleaners (Nelson-Filho et al., 2011; Spolidorio et al., 2011), suggesting that this may be a practical and low-cost alternative method of disinfection that can be easily used in the oral hygiene care practices.

This section of the chapter describes the several applications of microwave energy in dentistry field. The use of microwave irradiation for rapid disinfection of different dental materials and appliances may be an important tool in the context of the prevention of cross-infection in dental practice. In addition, there is sufficient scientific evidence that the use of this physical method of denture disinfection is effective in the treatment of denture stomatitis.

3. MICROBIOLOGICAL EFFICACY AND SUGGESTED REGIMENS

The microbiological effectiveness of microwave irradiation has been documented in the literature and the effectiveness seems to be directly related to the protocol adopted. When defining a microwave irradiation protocol, the parameters to be considered are: the time of exposure; the level of power of the microwave oven; the material to be irradiated; the vehicle in which the material is immersed, and the type of microorganisms that colonize the material. Considering these parameters and the several applications of microwave energy in dentistry field, different protocols of microwave irradiation are available and have been tested.

There are some studies that evaluate microwave irradiation for the disinfection of dental air turbine handpieces (Rohrer & Bulard, 1985), mirrors (Tarantino et al., 1997), burs (Fais et al., 2009; Rizzo, 1993; Rohrer & Bulard, 1985), and finishing and polishing instruments (Tate et al., 1995, 1996). In the study of Rohrer & Bulard (1985), handpieces contaminated with a mixture of four aerobic bacteria (*Staphylococcus epidermis, S. aureus, K. pneumonia, Bacillus subtilis, Clostridium histolyticum*) and *C. albicans* were consistently sterilized with an exposure of 10 minutes at 720W to microwave irradiation when the materials were attached to a three-dimensional rotating device. These authors also observed that handpieces contaminated by both polio type 1 and herpes simplex type 1 viruses were consistently sterilized after an exposure to microwave irradiation. Another in vitro investigation proved that dental mirrors contaminated with *S. aureus*, *B. subtilis*, and *Bacillus stearothermophilus* can be sterilized by microwave irradiation at 600W for 4 minutes, with the instruments immersed in an aldehyde solution (Tarantino et al., 1997). A microwave regime of 10 minutes at 720W is sufficient to sterilize carbide and diamond burs contaminated with a mixture of *S. aureus*, *K. pneuminiae*, and *B. subtilis* (Rohrer & Bulard, 1985). In another effective protocol, carbide burs are individually placed in a loosely capped test tube with 10 mL of distilled water, transferred to the right lateral position inside the microwave oven, and then exposed to 5 minutes at 600W (Fais et al., 2009). A similar protocol consisting of 6 minutes of irradiation at 750W (Tate et al., 1995, 1996) can be adopted when disinfecting finishing and polishing instruments (Enhance finishing cups, L.D. Caulk Co., Milford, DE, USA and Min-Identoflex fine cups, Centrix Inc., Shelton, CT, USA).

Likewise, 10 minutes of irradiation at high power can sterilize customized impressions made from both vinyl polysiloxane (Cinch Platinum, Parkell, Farmingdale, NY, USA) and polyether (Impregum F, 3M ESPE AG Dental Products, Seefeld, Germany) rubber impression materials (Abdelaziz et al., 2004). Microwave irradiation can also be used for gypsum casts disinfection (Berg et al., 2005; Davis et al., 1989). In the studies of Davis et al. (1989) and Berg et al. (2005), molds were contaminated with pathogenic microorganisms and the stone casts obtained were submitted to microwave irradiation. While *Serratia marcescens* cells on casts were inactivated by microwave irradiation at 900W for 1, 5 or 20 minutes (Davis et al., 1989), *S. aureus* and *P. aeruginosa* were killed after 5 minutes of irradiation (Berg et al., 2005). Microwave irradiation was ineffective in killing *B. subtilis* transferred to stone casts (Davis et al., 1989). Although a complete inactivation was not obtained, the gypsum casts submitted to microwaves for 20 minutes exhibited less growth than the samples irradiated for shorter times (Davis et al., 1989).

There are also two available protocols for the microwave disinfection of disposable materials, such as swabs and gauze. Cardoso et al. (2007) used stock cultures of *Escherichia coli*, *S. aureus*, *S. epidermidis*, *P. aeruginosa*, and *C. albicans* to contaminate gauze and swabs. Thereafter, the materials were placed into autoclave bags, sealed, and irradiated for 30 seconds at 1000W. This time/power regime was efficient in the sterilization of the disposable materials tested. When a lower power was used (650W), 30 seconds of microwave irradiation were also sufficient to sterilize gauze contaminated with the same pathogenic bacteria and fungi (Border & Rice-Spearman, 1999). With the regard of oral cleaning devices, disinfection of toothbrushes contaminated with a suspension containing *S. mutans* was obtained after exposure to microwaves for 7 minutes at 1100W (Nelson-Filho et al., 2011).

However, the results from another study demonstrated that toothbrushes and tongue scrapers contaminated with *C. albicans*, *S. aureus*, and *S. mutans* were effectively disinfected after 1 minute of microwaving at 650W (Spolidorio et al., 2011).

Given the efficacy of microwave disinfection, much attention has been focused on the use of this method for denture decontamination. Various studies have been conducted to determine the most suitable time/power protocol when using microwave irradiation to disinfect dentures. Some studies have used home microwave ovens for the inactivation of pathogenic microorganisms, such as those recommended by the *Handbook of Disinfectants and Antiseptics* (Cole & Robison, 1996). Among these microorganisms, there are those considered indicators of sterilization, such as *S. aureus* (gram-positive bacteria), *P. aeruginosa* (gram-negative bacteria), and *B. subtilis* (sporulated aerobic microorganisms). In this context, Rohrer & Bulard (1985) investigated the possibility of using microwave irradiation to sterilize dentures and concluded that 8 minutes of irradiation at 720W were sufficient to sterilize the dentures contaminated with a mixture of five bacteria and a fungus. To obtain these results, the authors attached the dentures to a three dimensional rotating device. However, such device is not commercially available or practical for use by dentists or health care facilities. Ten years later, Thomas & Webb (1995) observed that an unmodified domestic microwave oven could be used in the disinfection of dentures. They also demonstrated that sterilization of dentures inoculated with *C. albicans* and *Streptococcus gordonii* could be achieved at 2, 4, 6, 8, and 10 minutes exposure times at high setting (650W) (Webb et al., 1998). Likewise, using an unmodified domestic microwave oven with a rotating table, Baysan et al. (1998) observed that microwave irradiation at 650W for 5 minutes promoted a reduction in the quantity of *C. albicans* and *S. aureus* present in resilient relining materials. The study of Meşe & Meşe (2007) also investigated the effect of microwave energy on the growth of *C. albicans* in resilient relining material and verified the reduction in the colony counts of the fungus after dry exposure to microwaves for 5 minutes (650W). The results obtained with the aforementioned microbiological studies, in which the test specimens or dentures were irradiated in a dry state, are variable with regard to the effectiveness of disinfection by microwaves. Since irradiation in water provides uniform heating of the materials, Dixon et al. (1999) suggested immersing the samples in water during exposure to microwaves. This procedure was considered adequate for eliminating microorganisms, including those located inside the pores. The authors inoculated *C. albicans* in resilient reliners and a heat-polymerized resin and showed that the specimens immersed in water during irradiation were completely disinfected after 5 minutes at maximum power, while those not immersed in water were only partially disinfected after the same time of exposure. Bearing in mind these results,

Neppelenbroek et al. (2003) evaluated the effectiveness of a home microwave oven for the inactivation of *S. aureus*, *P. aeruginosa*, *B. subtilis*, and *C. albicans* present in three reline resins. The contaminated samples were immersed in 200 mL of sterile distilled water and irradiated for 6 minutes at 650W. It was observed that all the test specimens were sterilized after irradiation, as no microbiological growth was noticed after the irradiated specimens had remained incubated for 7 days. Silva et al. (2006) evaluated the same protocol for disinfecting simulated complete dentures and observed that the lethal action varied according to the microorganisms tested. While complete disinfection was achieved for the dentures contaminated with *S. aureus* and *C. albicans*, those contaminated with *P. aeruginosa* and *B. subtilis* showed little, but detectable, microbial growth. The different results from those of Neppelenbroek et al. (2003) are probably related to the larger surface area of the complete dentures, given that the number of microorganism colonies on the acrylic resin surface is proportional to the total area involved. In addition, this difference could be related to the microorganisms tested. A greater resistance of *B. subtilis* to microwaves has been reported (Davis et al., 1989; Najdovski et al., 1991) and these results are probably related to the sporulation capacity of *B. subtilis.* Bacterial spores are metabolically inactive and particularly resistant to situations of stress, such as heating and radiation. Microwaves promote heating of the test specimens and water in which they are immersed, thus there is the possibility of spore formation during this procedure (Najdovski et al., 1991).

A more recent study (Mima et al., 2008) showed that the experimental protocols advocated by Dixon et al. (1999), Neppelenbroek et al. (2003), and Silva et al. (2006) could be used with lower exposure times. Test specimens contaminated with four microorganisms (*S. aureus*, *P. aeruginosa*, *B. subtilis,* and *C. albicans*) were immersed in water and submitted to microwave irradiation (650W) at different exposure times (5, 4, 3, 2, and 1 minutes). It was observed that all the test specimens irradiated for 3, 4, and 5 minutes were completely disinfected after microwave irradiation. When the time of irradiation was reduced to 2 minutes, the samples contaminated with *C. albicans* were completely disinfected while those inoculated with *S. aureus*, *P. aeruginosa*, and *B. subtilis* demonstrated microbial growth. When submitted to sterilization by humid heat, bacterial cells are inactivated at higher temperatures than fungal cells (Pelczar et al., 1993). Therefore, irradiation for at least 2 minutes promoted sufficient water heating to inactivate *C. albicans* but not the bacteria. Moreover, the yeast cells are larger than those of the bacteria (Verran & Maryan, 1997). Therefore, one could suppose that the *C. albicans* cells contained more water in their composition than did the other microorganisms tested, and therefore, they had been more susceptible to microwave irradiation. In spite of this in vitro study has evaluated small size test specimens that had surfaces with vitreous characteristics, differently from those observed clinically, its results indicated that the protocol should be evaluated in other experimental conditions. Therefore, laboratorial investigations were performed to evaluate the effectiveness of this protocol in the disinfection of dentures contaminated by several microorganisms. Sanitá et al. (2009) demonstrated that simulated complete dentures inoculated with different *Candida* spp. (*C. albicans*, *Candida tropicalis*, and the intrinsically resistant *Candida glabrata*, *Candida dubliniensis*, and *Candida krusei*) were completely disinfected by microwave irradiation for 3 minutes at 650W. Similar results have been reported for complete dentures contaminated with *S. aureus* and *P. aeruginosa* (Dovigo et al., 2009).

Given the promising results, some protocols were tested in clinical trials to evaluate the effectiveness of microwave irradiation in disinfecting patients'dentures (Glass et al., 2001b; D.G. Ribeiro et al., 2009). In one study, fragments of dentures that had been worn for periods ranging from 12 days to 48 years were immersed in a chemical solution (MicroDent Sanitizing and Cleaning System®) and exposed to microwaves for 2 minutes (Glass et al., 2001b). This method showed positive results for denture decontamination, considering that no microbial growth was observed on the fragments. In addition, the study of D.G. Ribeiro et al. (2009) showed that 3 minutes of irradiation at 650W resulted in complete inactivation of the biofilm present on dentures of 30 patients. It emerged also from this study that a lower reduction in the count of microorganisms (*C. albicans*, *S. aureus*, and *S. mutans*) was observed when a lower time of exposure was used (2 minutes).

In terms of treating denture stomatitis, Banting & Hill (2001) conducted the first clinical study that evaluated the effectiveness of microwave energy for denture disinfection as a co-adjuvant treatment. This study was performed in 2001, when the effective protocol of 3 minutes at 650W had not yet established. Patients received topical antifungal medication (nystatin/ three times a day) for 14 days and had their dentures irradiated (850W for 1 minute) on three different days (1st, 5th and 10th day). The authors observed that the method was effective in the reduction of the clinical signs of infection and of the invasive forms of *C. albicans* (pseudohyphas) adhered on the surfaces of the dentures. In another clinical study, microwave disinfection of dentures (10 minutes at 350W) in a daily basis during one week reduced the palatal inflammation and the numbers of *Candida* on cultures from the palates and dentures of patients (Webb et al., 2005). A more recent study (Neppelenbroek et al., 2008) also evaluated the effectiveness of denture microwave disinfection in the treatment of patients with denture stomatitis. The microwave treatment protocol adopted was 6 minutes of irradiation of the complete dentures at 650W, three times a week for 30 days. It was verified that disinfection of the dentures by microwaves was effective for the treatment of denture stomatitis and for the reduction of the mycelial forms of *Candida* spp. In agreement to

Banting & Hill (2001), these authors also observed that the levels of recurrence of *C. albicans* on the internal surfaces of the dentures and the supporting mucosa were significantly reduced in the patients whose dentures were microwaved (Neppelenbroek et al., 2008). Other clinical investigations have been carried out to evaluate the effectiveness of microwave irradiation in the treatment of denture stomatitis. A modification on the protocol proposed by Neppelenbroek et al. (2008) was evaluated by Vergani et al. (2008) and Silva et al. (2008). These authors demonstrated that denture microwave irradiation for 3 minutes at 650W, three times weekly for 14 days, is able to reduce the clinical signs of denture stomatitis on the palatal mucosa and the *Candida* colonization on complete dentures. Further clinical evaluations have also been conducted in order to evaluate the effectiveness of this protocol in the treatment of diabetic denture wearer patients with denture stomatitis (Sanitá et al., 2010, 2011). It was observed that microwave disinfection of complete dentures, by itself, was as effective as nystatin, the more conventional topical antifungal medication, in reducing the *Candida* counts and the clinical signs of denture stomatitis infection in patients with diabetes mellitus.

Based on the aforementioned studies, it can be seen that several regimes of microwave irradiation in relation to time/power are available. The protocol must be selected in accordance to the specific application of the microwave energy. Regardless these parameters, microwave irradiation is a potentially effective method for inactivating various microbial species present on dental materials, many of which are related to oral pathologies.

4. MECHANISMS OF ACTION ON THE MICROORGANISMS

While the inhibitory effect of microwave irradiation on microorganisms is being researched extensively, how microwave brings about this effect has been a matter of discussion. Some authors believe that microorganism inactivation by microwave irradiation is explained by a thermal effect (Fitzpatrick et al., 1978; Jeng et al., 1987; Yeo et al., 1999). Nevertheless, several studies suggest that, in addition to the heat generated around the microorganisms, there are other mechanisms resulting directly from the interaction of the electromagnetic field (Carrol & Lopez, 1969; Culkin & Fung, 1975; Olsen, 1965; Rohrer et al., 1986). Various mechanisms have been suggested to explain the nature of the so called non-thermal theory. Depending on the chemical composition of the microorganisms and the surrounding medium, the microbial cells may be selectively heated by the microwaves (Carrol & Lopez, 1969; Hiti et al., 2001; Yeo et al., 1999). Therefore, a certain frequency of microwave energy may be absorbed by certain fundamental biological molecules, such as the nucleic acids (Rohrer et al., 1986). Moreover, the level of molecular response from the biological system to the quantity of thermal energy may also explain the non-thermal effect of microwaves (Carrol & Lopez, 1969). The structural changes in the more peripheral layer around the biological macromolecules may alter their stability and function, and, consequently, these molecules may be denatured in an irreversible manner (Culkin & Fung, 1975). Studies have also demonstrated that the exposition of bacterial suspensions to microwave irradiation caused reduction on viable cell counts and increased the leaching of DNA and protein (Woo et al., 2000). These results suggest that microwaves caused changes in structural integrity and permeability of cell membrane and cell wall that have detrimentally affected the cell metabolism and lead to cell death (Campanha et al., 2007; Carrol & Lopez, 1969; Culkin & Fung, 1975; Olsen, 1965). Campanha et al. (2007) verified that leveduriform suspensions submitted to microwave irradiation at 650W for 6 minutes presented significantly lower cell count values and a larger number of substances released in comparison with the non-irradiated suspensions. The distinction between integral and non-integral cells was made based on the entry of methylene blue coloring into the cells, which is an indirect form of evaluating the cell membrane and wall integrity. Disintegrated cells were found in the irradiated suspension, indicating an alteration in the permeability or integrity of these structures. Moreover, the cells of this suspension lost their characteristic refringence, in spite of preserving their ellipsoidal leveduriform morphology. It was also demonstrated that after irradiation by microwaves, the release of electrolytes (K^+, Ca^{++}) and nucleic acids was significantly higher in the irradiated suspensions than that from the non-irradiated (Campanha et al., 2007). However, despite of cell inactivation, the optic density and cell concentration were not altered in comparison with those of the control suspensions, indicating that cells were not completely lysed. Irrespective of the mechanism of microwaves on pathogenic microorganisms being thermal or non-thermal, it is known that the effect of inactivation occurs mainly in the presence of water, this being an important factor for sterilization in microwave ovens. Freeze-dried or dry organisms are unlikely to be affected, even when submitted to prolonged exposures, indicating that humidity plays an important role in microwave energy absorption (Dixon et al., 1999; Watanabe et al., 2000). The water molecules present in the medium or inside the cells, being diploid, interact with the electromagnetic field of the microwaves. Consequently, numerous intermolecular collisions may occur and this vibration produces heating (Najdovski et al., 1991). This increase in temperature may cause protein and DNA denaturation (Ponne & Bartels, 1995). The consistent results from several studies, in which strains were completely inactivated when microwave irradiation was carried out with the specimens immersed in water, confirm this hypothesis.

According to scanning electron microscopic (SEM) studies (Neppelenbroek et al., 2003; Mima et al., 2008), microwave irradiation not only inactivated *C. albicans*, but also removed the nonviable yeast cells from resin surface. In this case, the irradiation was performed with the resin specimens immersed in water and, since water started to boil after approximately 1.5 minutes of irradiation, the movement of the water particles probably removed microbial cells from the resins. Verran & Maryan (1997) reported that the larger yeasts cells are more easily dislodged from acrylic resin surfaces compared with smaller bacteria. Considering the information discussed above, it can be concluded that, the nature of the lethality of the microwave irradiation for microorganisms may be a combination of thermal and non-thermal effects.

5. IMPACT ON THE PHYSICAL AND MECHANICAL CHARACTERISTICS OF DENTAL MATERIALS AND DEVICES

Several investigations have focused on finding the adequate microwave parameters for microbial inactivation and cross-infection control. Different irradiation protocols have proved to be remarkably effective for disinfection of dental prostheses and other materials frequently used in dental practice. However, to enable this method of disinfection to be safely recommended, it is important to clearly demonstrate that it does not exert deleterious effects on the physical and mechanical properties of the materials submitted to microwaves. For this reason, studies have been conducted to examine the effect of microwave disinfection on dental instruments and burs, impressions, gypsum products, acrylic resins, denture lining materials, and artificial teeth.

Rohrer & Bulard (1985) exposed dental air turbine handpieces to microwaves for 2, 4, 6, 8, 10, and 15 minutes (720W). After 25 cycles of 10 minutes, the dental handpiece tested showed no decrease in the pressure reading and no apparent alteration in sound or cutting power. Another study (Tate et al., 1996) evaluated instrument performance of two composite finishing and polishing systems before and after three cycles of microwave irradiation (6 minutes at 750W). The sample surfaces were examined with a profilometer after the finishing procedure and the results demonstrated that the systems tested can be submitted to microwave irradiation at least three times without affecting performance. Questions have also been raised about the effects of microwave regimens on the microscopic characteristics, durability, and strength of dental burs, which can have their sharpness and ability to effectively cut tooth structure altered. The effect of sterilization with microwaves on diamond burs was evaluated by Rizzo (1993). The author evaluated the dental burs by viewing them under stereomicroscope before and after sterilization cycles. It was found that no damage was present after 15 cycles. The possible influence of microwave irradiation on the cutting capacity of carbide burs was also investigated (Fais et al., 2009). The burs were used to cut glass plates in a cutting machine set for 12 cycles of 2.5 minutes each and, after each cycle, they were exposed to microwave irradiation for 5 minutes at 600W. The cutting capacity of the burs was determined by a weight-loss method. Compared to the control conditions, the microwaved burs showed a statistically significant decrease in their cutting capacity. Thus, microwave irradiation requires further investigations before final recommendations can be made for disinfection of carbide burs.

Microwave disinfection of rubber impressions was also suggested by some authors as an alternative approach of controlling microbial transmission. In this context, the reproducibility of rubber impressions after microwave irradiation (10 minutes/720W) has been evaluated and the results compared with other disinfection methods (Abdelaziz et al., 2004). Microwave sterilization had a small effect on accuracy of impressions and this procedure has been recommended as a suitable technique for sterilizing rubber impressions. Another application of microwaves in dentistry is the disinfection of gypsum casts. Although there are no studies evaluating the disinfection protocols on the properties of gypsum materials, the effect of drying casts by microwaves has been described. In this context, microwave irradiation of gypsum casts has been tested as to its effect on the resistance to fracture (Hersek et al., 2002; Luebke & Schneider, 1985; Tuncer et al., 1993) and hardness (Luebke & Chan, 1985). In general, the results indicated an improvement in these properties. However, there was some concern that a decrease in the compressive strength and the appearance of cracks or porosities in the surface might occur when gypsum casts were exposed to irradiation with a very high power (1450W). Other physical and mechanical properties, such as abrasion resistance and dimensional stability, should be performed to confirm the clinical applicability of this procedure.

One of the main applications of microwaves in dentistry is to disinfect dental prosthesis. A large number of investigations have been published in the past and recent years concerning its effectiveness and limitations. Laboratorial investigations aimed at identifying if microwave exposure affects the surface hardness of the denture base acrylic resins, relining materials, and artificial denture teeth. The hardness of a material is the result of the interaction of several properties, such as ductility, malleability, and resistance to cutting, and, therefore, hardness tests may be used as an indicator of these properties (Anusavice, 1996). Also, the hardness of materials is related to its resistance to local plastic deformation. Two universal types of microhardness test, Vickers and Knoop, are standard methods for measuring hard surfaces, while the Shore A measures hardness in terms of the elasticity of the material. It has been demonstrated that the Vickers hardness of a heat-polymerizable acrylic resin was not changed after different times of exposure to microwaves (6, 5, 4, 3, 2, and 1 minutes) at 650W (Machado et al., 2009; D.G. Ribeiro et al., 2008). The Knoop hardness values of a denture base resin were also not changed after two cycles of microwave disinfection for 6 minutes at 690W

(Sartori et al., 2008). In addition, no alterations in the Shore A hardness values were detected after microwave irradiation (3 minutes/500W) of resilient relining materials (Pavan et al., 2007). Other studies showed, however, increased hardness of denture base materials associated with microwave irradiation. Polyzois et al. (1995) evaluated two microwave disinfection protocols on the hardness of test specimens of a heat-polymerizable resin (3 or 15 minutes at 500W). Both protocols provided an increase in microhardness values. Similar findings were described by Machado et al. (2005) for two resilient relining materials submitted to seven irradiation cycles for 6 minutes (650W). D.G. Ribeiro et al. (2008) evaluated the effect of different times of exposure to microwaves (5, 4, 3, 2, and 1 minutes/650W) on the hardness of four reline materials and the findings suggested that the disinfectant method promoted an increase in the hardness of the reline resins. The increase in microhardness values after microwave disinfection may be related to the increase in temperature during the irradiation procedure. Arab et al. (1989) reported an increase in hardness values when heat-polymerizable resins were immersed in water heated to 100ºC. Similarly, an increase in Vickers hardness of a reline resin after heat treatment in a water bath at 55ºC for 10 minutes has been reported (Seó et al., 2007a). These results may be attributed to the reduction in the level of residual monomer, as a result of the complementary processes of polymerization (Lamb et al., 1983; Sideridou et al., 2004) and diffusion of residual monomer, both favored by the increase in water temperature during microwave irradiation.

Besides the acrylic denture base resins, dentures also comprise artificial teeth. For this reason, the effect of microwave disinfection on the surface hardness of artificial teeth commonly used for denture construction was also evaluated (Campanha et al., 2005). Two microwave disinfection cycles of 6 minutes each (650W), preceded or not by immersion in distilled water for 90 days, were tested. From these experiments, a reduction in surface hardness of the artificial denture teeth was observed after microwave irradiation. It seems that the high temperature associated with the movements of the molecules probably caused an increase in the speed of diffusion of the water molecules into the polymer, facilitating the movement of the polymeric chains during performance of the hardness (Takahashi et al., 1998). Thus, it is feasible that the reduction in hardness after irradiation is related to water sorption rather than to microwave irradiation. In fact, the teeth from the group that was microwaved after 90 days of water saturation presented no significant alteration in hardness after disinfection. Since microwave disinfection involves the exposition of dentures to water at high temperature, it has been hypothesized that this may affect the bond strength between the artificial teeth and the acrylic resin from which dentures are made. Results from a study evaluating the effect of microwave disinfection (6 minutes/650W) on the bond strength of two types of denture teeth to three acrylic resins showed that microwave disinfection did not adversely affect the bond strength of all tested materials, with the exception of one tooth/resin combination (R.C. Ribeiro et al., 2008). In another study microwave irradiation for 3 minutes at 650W promoted a reduction in the impact strength of the tooth/ acrylic resin interface, which could be explained by the increase in the degree of conversion of self-polymerizing resin, reducing the cohesion at the interface of the samples (Consani et al., 2008a).

The effects of microwave irradiation on other surface properties, such as roughness and porosity, have also been investigated. According to Allison & Douglas (1973), smoother surface retain a smaller quantity of biofilm, thus avoiding the proliferation of microorganisms on the acrylic surface of dentures. Surface roughness is an important characteristic of dental materials and, therefore, there is a direct correlation between the values of roughness and bacterial adherence. Moreover, according to Yannikakis et al. (2002),the presence of pores may reduce the mechanical properties of acrylic resin, as well as interfere in denture hygiene. It seems that the effect of microwave irradiation on a denture's surface roughness and porosity depends on the time of exposure, number of cycles, and the type of denture resin used. Novais et al. (2009) investigated the occurrence of porosity on the surface of four self-polymerizable acrylic resins and one heat-polymerizable resin, after two or seven cycles of microwave disinfection (6 minutes/ 650W). The number of pores found in two out of five resins remained similar after microwaving, while a reduction in porosity was observed in two resins after seven disinfection cycles. In these cases, it was suggested that the high temperatures may have been attained during exposure to microwaves, which led to a greater degree of conversion and continuation of polymerization. Seven cycles of microwave disinfection increased the number of pores in one material. According to the authors, this material presented a high level of residual monomer and, therefore, a probable explanation for the increase in the number of pores in this resin was related to monomer vaporization. Another investigation also showed that the use of microwave energy can modify the surface texture of acrylic resins (Sartori et al., 2006). It was reported that two microwave disinfection cycles (6 minutes at 690W) promoted an increase in surface roughness of an acrylic resin. However, only one material was evaluated and the effect of reduced exposure times on surface roughness was not investigated.

From the literature, it seems that microwave disinfection can play a role in promoting changes in denture materials. The flexural strength of five chairside reline resins and one denture base resin were evaluated after two and seven cycles of microwave disinfection at 650W for 6 minutes (Pavarina et al., 2005). The flexure strength of three resins presented significant increase in strength values.

In contrast, two reline resins presented reduced flexure values after microwave irradiation. The heating provided by each of the seven cycles of microwave disinfection may have increased the absorption of water of some of the evaluated materials, resulting in a reduction in the flexural strength values. In view of these results, shorter times of exposure to microwaves, and their effects on the flexural strength of acrylic resins, were investigated and the findings showed that the flexural strength of four reline materials and one heat-polymerizable resin was not detrimentally affected after 5, 4, 3, 2, and 1 minutes of microwave irradiation at 650W (D.G. Ribeiro et al., 2008). Indeed, the disinfection method was capable of significantly increase the flexural strength of one reline resin. A similar result was described elsewhere after a disinfection cycle for 3 minutes at 650W (Consani et al., 2008b). In addition, another investigation (Polyzois et al., 1995) observed that the flexural properties of a heat-polymerizable resin remained unaltered after the use of a low power (500W) associated with a long exposure time (15 minutes). However, in this study the samples were irradiated in a dry condition, a procedure that has been shown to be less effective for microbiological inactivation. In a more clinically relevant approach, the effect of denture microwave disinfection on the maximum fracture load, deflection at fracture, and fracture energy of intact and relined denture bases was evaluated. After exposed to microwave irradiation for 7 days (6 minutes/650W), the strength of the denture bases was similar to the strength of those immersed in water for 7 days (Seó et al., 2008).

An aspect that has also been investigated is the influence of microwave disinfection on bond strength between resilient liners and denture base acrylic resin. Test specimens made of resilient resins bonded to a denture base resin were submitted to microwave disinfection for two and seven irradiation cycles of 6 minutes at 650W (Machado et al., 2005). Microwave disinfection did not compromise the adhesion of resilient liners to the denture base resin. In a subsequent study, seven microwave disinfection cycles (6 minutes/650W) did not decrease the torsional bond strength between two hard reline resins and a denture base resin (Machado et al., 2006). Therefore, in general, the use of microwaves for denture disinfection does not appear to have any negative effect on the bond strength of reline materials frequently used in dental practice. Recently, a clinical study (R.C. Ribeiro et al., 2009) also reported the color stability of a hard chairside reline resin after microwave disinfection.

Another important aspect that should be considered prior to the selection of a disinfection procedure is the maintenance of adequate adaptation between the denture base and residual ridge. Clinically, this characteristic is fundamental both for denture retention and the preservation of the supporting tissues. Changes in denture base adaptation could act as one of the causes of alveolar bone loss and be indirectly responsible for decreasing the retention and stability of the denture (De Gee et al., 1979). A denture that exactly reproduces the supporting tissue may assure a uniform distribution of forces on the largest possible area of surface. Thus, analyses of the effect of microwave disinfection on the dimensional stability of denture bases resins have been conducted. Burns et al. (1990) aimed to determine the possible influence of microwaving on the dimensional stability of heat-polymerizable, self-polymerizable, and light-polymerizable resins. Test specimens were fabricated and submitted to measurements of mass and length before and after microwave irradiation (15 minutes/ 650W). The results showed that all the materials maintained dimensional stability after the disinfection procedure. In another study, microwave disinfection (3 or 5 minutes at 650W) promoted small, clinically insignificant dimensional changes on test specimens of a heat-polymerizable resin (Polyzois et al., 1995). Contrasting results were obtained in the study of Gonçalves et al. (2006), who evaluated the effect of two or seven cycles of microwave disinfection (6 minutes/ 650W) on the linear dimensional change of four reline resins and one denture base resin. Three of the evaluated resins presented significant alteration in the linear dimensional after disinfection. In spite of the positive and negative findings found by these different studies, all of the presented results cannot be extrapolated directly to a clinical situation, since the test specimens used did not simulate the dimensions and shape of denture bases. Standardized dentures were fabricated in the study of Thomas & Webb (1995) in order to evaluate the effect of microwaves on their dimensional stability. After 10 minutes of exposure to microwaves at 604W, some of the areas measured in the dentures presented significant shrinkage or expansion. Similarly, Sartori et al. (2008) observed that the disinfection cycles of 6 minutes (690W) could compromise the dimensional stability of denture base resins. In another study, however, a lower power setting (331W) decreased the occurrence of dimensional changes when the dentures were microwaved for 6 minutes (Thomas & Webb, 1995). The protocol of 6 minutes/ 650W of microwave irradiation was also tested for the dimensional stability of denture bases (Seó et al., 2007b). Repeated disinfection cycles were performed (two and seven) and an increase in shrinkage was observed both in intact bases and bases relined with heat-polymerizable resins. As the bases were immersed in water during irradiation, the results could be related to the possible occurrence of complementary polymerization as a result of water heating. In another investigation, the use of the same microwave power (650W), but with irradiation for 3 minutes, promoted no deleterious effects on the adaptation of denture bases (Consani et al., 2007). In fact, the authors observed that microwave irradiation improved the adaptation of bases in some experimental conditions. Pavan et al. (2005) also suggested that use of shorter irradiation times preserves the dimensional stability of dentures. The authors fabricated

30 denture bases that were submitted to irradiation for 3 minutes/500W or 10 minutes/ 604W. No dimensional alteration was observed in the bases irradiated for 3 minutes. The collective findings from the referenced studies indicate that brief microwave irradiation durations should be employed to preserve the adaptation of denture bases to the underlying tissue. Recently, Basso et al. (2010) conducted a clinical assessment of the impact of 3 minutes of microwave irradiation at 650W administered once or thrice weekly on the linear dimensional stability of complete dentures. . The collective findings from the referenced studies indicate that brief microwave irradiation durations should be employed to preserve the adaptation of denture bases to the underlying tissue. Recently, Basso et al. (2010) conducted a clinical assessment of the impact of 3 minutes of microwave irradiation at 650W administered once or thrice weekly on the linear dimensional stability of complete dentures. Measurements were conducted prior to the initial microwave disinfection (baseline) and subsequent to each week of disinfection. Additionally, the dentures were subjected to clinical monitoring. Irradiation three times per week demonstrated markedly greater shrinkage throughout all assessed weeks. This outcome may be ascribed to the thermal energy produced by microwave irradiation in a pre-polymerized substance. Despite three microwave disinfections demonstrating statistically significant shrinkage, the clinical assessment indicated no observable change. Consequently, the authors proposed that microwave irradiation may be utilized clinically for the disinfection of dentures and the treatment of denture stomatitis.

The aforementioned information indicates that conflicting results have been published concerning the hazards associated with microwave disinfection of dentures. The manifestation of adverse effects on the physical and mechanical properties of dentures is contingent upon the tested microwave methodology and the type of material assessed. The utilization of brief exposure durations may reduce the incidence of detrimental effects on dental prostheses. Considering all the data, it appears that a microwave duration of 3 minutes at 650W is sufficient for denture disinfection without inflicting considerable harm on the denture components.

6. CONCISE SUMMARY

This chapter presents scientific evidence supporting the effectiveness of microwave irradiation in reducing cross-infection and curing denture stomatitis. Multiple irradiation techniques were delineated and examined, and the microbiological efficacy of microwave energy was unequivocally established. In light of all the factors considered, we conclude that microwave disinfection is an effective, rapid, straightforward, and cost-efficient method that can be utilized by dentists, assistants, technicians, patients, and/or their caretakers to inactivate bacteria.

Furthermore, the utilization of a microwave oven necessitates no specialized storage and does not promote resistance in fungus or other microbes. This technology may have significant potential applications in dentistry offices, laboratories, and healthcare institutions where patients, particularly those with removable dentures, are treated.

7. REFERENCES

AbdelazizK. M.HassanA. M.HodgesJ. S. 2004 Reproducibility of sterilized rubber impressions. Brazilian Dental Journal, 15 3 (December 2004), 209213 , 0103-6440

AllisonR. J.DouglasW. H. 1973 Microcolonization of the denture-ftting surface by Candida albicans. Journal of Dentistry, 1 5 (June 1973), 198201 , 0300-5712

AnkolaA. V.HebbalM.EshwarS. 2009 How clean is the toothbrush that cleans your tooth? International Journal of Dental Hygiene, 7 4 (November 2009), 237240 , 1601-5029

AnusaviceK. J. 1996 Phillips' science of dental materials. (10th Ed), W.B. Saunders, 0-72165-741-9 United States

ArabJ.NewtonJ. P.LloydC. H. 1989 The effect of an elevated level of residual monomer on the whitening of denture base and its physical properties. Journal of Dentistry, 17 4 (August 1989), 189194 , 0300-5712

BantingD. W.HillS. A. 2001 Microwave disinfection of dentures for the treatment of oral candidiasis. Special Care in Dentistry, 21 1 (January 2001), 48 , 0275-1879

BassoM. F.GiampaoloE. T.VerganiC. E.MachadoA. L.PavarinaA. C.CompagnoniM. A. 2010 Influence of microwave disinfection on the linear dimensional stability of complete dentures: a clinical study. The International Journal of Prosthodontics, 23 4 (August 2010), 318320 , 0893-2174

BaysanA.WhileyR.WrightP. S. 1998 Use of microwave energy a long-term soft lining material contaminated with Candida albicans or Staphylococcus aureus. The Journal of Prosthetic Dentistry, 79 4 (April 1998), 454458 , 0022-3913

BergE.NielsenO.SkaugN. 2005 High-level microwave disinfection of dental gypsum casts. The International Journal of Prosthodontics, 18 6 (December 2005), 520525 , 0893-2174

BorderB. G.Rice-SpearmanL. 1999 Microwaves in the laboratory: effective decontamination. Clinical Laboratory Science, 12 3 (June 1999), 156160 , 0089-4959X

BrownG. H.MorrisonW. C. 1954 An exploration of the effects of strong radio-frequency fields on micro-organism in aqueous solutions. Food Technology, 8 361366 , 0015-6639

Budtz-JörgensenE.TheiladeE. 1983 Regional variations in viable bacterial and yeast counts of 1-week-old denture plaque in denture-induced stomatitis. Scandinavian Journal of Dental Research, 91 4 (August 1983), 288295 , 0002-9845X

Budtz-JörgensenE. 1990 Etiology, pathogenesis, therapy, and prophylaxis of oral yeast infections. Acta odontologica Scandinavica, 48 1 (February 1990), 6169 , 0001-6357

BurnsD. R.KazanogluA.MoonP. C.GunsolleyJ. C. 1990 Dimensional stability of acrylic resin materials after microwave sterilization. The International Journal of Prosthodontics, 3 5 (October 1990), 489493 , 0893-2174

CampanhaN. H.PavarinaA. C.VerganiC. E.MachadoA. L. 2005 Effect of microwave sterilization and water storage on the Vickers hardness of acrylic resin denture teeth. The Journal of Prosthetic Dentistry, 93 5 (May 2005), 483487 , 0022-3913

CampanhaN. H.PavarinaA. C.BrunettiI. L.VerganiC. E.MachadoA. L.SpolidorioD. M. P. 2007 Candida albicans inactivation and cell membrane integrity damage by microwave irradiation. Mycoses, 50 2 (March 2007), 140147 , 1439-0507

CardosoV. H.GoncalvesD. L.AngiolettoE.Dal-PizzolF.StreckE. L. 2007 Microwave disinfection of gauze contaminated with bacteria and fungi. Indian Journal of Medical Microbiology, 25 4 (October 2007), 428429 , 0255-0857

CarrolD. E.LopezA. 1969 Lethality of radio-frequency energy upon microorganisms in liquid, buffered, and alcoholic food systems. Journal of Food Science, 34 4 (July 1969), 320324 , 1750-3841

ChandraJ.MukherjeeP.K.LeidichS.D.FaddoulF.F.HoyerL.L.DouglasL.J.GhannoumM. A. 2001 Antifungal resistance of candidal biofilms formed on denture acrylic in vitro. Journal of Dental Research, 80 3 (March 2001), 903908 , 0022-3913

ChauV. B.SaundersT. R.PimslerM.ElfringD. R. 1995 In-depth disinfection of acrylic resins. The Journal of Prosthetic Dentistry, 74 3 (September 1995), 309313 , 0022-3913

CliffordT. J.BurnettC. A. 1995 The practice of consultants in restorative dentistry (UK) in routine infection control for impressions and laboratory work. The European Journal of Prosthodontics and Restorative Dentistry, 3 4 (June 1995), 175177 , 0965-7452

CodinoR. J.MarshallW. E. 1976 Control of infection in the dental operatory. Dental Survey, 52 5 (May 1976), 4250 , 0011-8788

ColeE. C.RobisonR. 1996 Test methodology for evaluation of germicides. In: Handbook of Disinfectants and Antiseptics. J.M. Ascenzi, 113 , Marcel Dekker, 0-82479-524-5 York, United States

ColomboA. L.NucciM.SalomãoR.BranchiniM. L.RichtmannR.DerossiA.WeyS. B. 1999 High rate of non-albicans candidemia in Brazilian tertiary care hospitals. Diagnostic Microbiology and Infectious Disease, 34 4 (August 1999), 281286 , 0732-8893

ConsaniR. L.MesquitaM. F.de ArrudaNobilo. M. A.HenriquesG. E. 2007 Influence of simulated microwave disinfection on complete denture base adaptation using different flask closure methods. The Journal of Prosthetic Dentistry, 97 3 (March 2007), 173178 , 0022-3913

ConsaniR. L.MesquitaM. F.ZampieriM. H.MendesW. B.ConsaniS. 2008 Effect of the simulated disinfection by microwave energy on the impact strength of the tooth/acrylic resin adhesion. The Open Dentistry Journal, 2 (January 2008), 37 , 1874-2106

ConsaniR. L.VieiraE. B.MesquitaM. F.MendesW. B.Arioli-FilhoJ. N. 2008 Effect of microwave disinfection on physical and mechanical properties of acrylic resins. Brazilian Dental Journal, 19 4 348353 , 0103-6440

CottoneJ. A.TererhalmyG. T.MolinariJ. A. 1991 Practical Infection Control in Dentistry (1st **Edition), Lea &** Febiger, 0-68302-138-9 PA

CulkinK. A.FungD. Y. C. 1975 Destruction of Escherichia coli and Salmonella typhimurium in microwave-cooked soups. Journal of Milk and Food Technology, 38 1 815 , 0022-2747

DavisD. R.CurtisD. A.WhiteJ. M. 1989 Microwave irradiation of contaminated dental casts. Quintessence International, 20 8 (August 1989), 583585 , 0033-6572

De GeeA. J.TenHarkel. E. C.DavidsonC. L. 1979 Measuring procedure for the determination of the three-dimensional shape of dentures. The Journal of Prosthetic Dentistry, 42 2 (August 1979), 149153 , 0022-3913

DixonD. L.BreedingL. C.FalerT. A. 1999 Microwave disinfection of denture base materials colonized with Candida albicans. The Journal of Prosthetic Dentistry, 81 2 (February 1999), 207214 , 0022-3913

DovigoL. N.PavarinaA. C.RibeiroD. G.de OliveiraJ. A.VerganiC. E.MachadoA. L. 2009 Microwave disinfection of complete dentures contaminated in vitro with selected bacteria. Journal of Prosthodontics, 18 7 (October 2009), 611617 , 0105-9941X

EgusaH.WatamotoT.AbeK.KobayashiM.KanedaY.AshidaS.MatsumotoT.YataniH. 2008 An analysis of the persistent presence of opportunistic pathogens on patient-derived dental impressions and gypsum casts. The International Journal of Prosthodontics, 21 1 (February 2008), 6268 , 0893-2174

FaisL. M.PinelliL. A.AdaboG. L.SilvaR. H.MarceloC. C.GuaglianoniD. G. 2009 Influence of microwave sterilization on the cutting capacity of carbide burs. Journal of Applied Oral Science, 17 6 (December 2009), 584589 , 1678-7757

FitzpatrickJ. A.Kwao-PaulJ.MasseyJ. 1978 Sterilization of bacteria by means of microwave heating. Journal of Clinical Engineering, 3 1 (March 1978), 4447 , 0363-8855

Friedrich JrE. G.PhillipsL. E. 1988 Microwave sterilization of Candida on underwear fabric. A preliminary report. The Journal of Reproductive Medicine, 33 5 (May 1988), 421422 , 0024-7758

GlassR. T.BullardJ. W.HadleyC. S.MixE. W.ConradR. S. 2001a Partial spectrum of microorganisms found in dentures and possible disease implications. The Journal of the American Osteopathic Association, 101 2 (February 2001), 9294 , 1945-1997

GlassR. T.GoodsonL. B.BullardJ. W.ConradR. S. 2001b Comparison of the effectiveness of several denture sanitizing systems: a clinical study. Compendium of Continuing Education in Dentistry, 22 12 (December 2001), 10931096 , 1098, 1100-2 passim; quiz 1108, 1548-8578

GonçalvesA. R.MachadoA. L.GiampaoloE. T.PavarinaA. C.VerganiC. E. 2006 Linear dimensional changes of denture base and hard chair-side reline resins after disinfection. Journal of Applied Polymer Science, 102 2 (October 2006), 18211826 , 0021-8995

HendersonC. W.SchwartzR. S.HerboldE. T.MayhewR. B. 1987 Evaluation of the barrier system, an infection control system for the dental laboratory. The Journal of prosthetic dentistry, 58 4 (October 1987), 517521 , 1097-6841

HersekN.CanayS.AkçaK.CiftçiY. 2002 Tensile strength of type IV dental stones dried in a microwave oven. The Journal of Prosthetic Dentistry, 87 5 (May 2002), 499502 , 1097-6841

HitiK.WalochnikJ.FaschingerC.Haller-SchoberE. M.AspockH. 2001 Microwave treatment of contact lens cases contaminated with Acanthamoeba. Cornea, 20 5 (July 2001), 467470 , 1536-4798

HoffmanP. N.HanleyM. J. 1994 Assessment of a microwave-based clinical waste decontamination unit. The Journal of Applied Bacteriology, 77 6 (December 1994), 607612 , 0021-8847

IlbayS. G.GüvenerS.AlkumruH. N. 1994 Processing dentures using a microwave technique. Journal of Oral Rehabilitation, 21 1 (January 1994), 103109 , 1365-2842

Infection control in the dental office. Council on Dental Materials and Devices. Council on Dental Therapeutics. 1978 Journal of the American Dental Association, 97 4 (October 1978), 673677 , 0002-8177

JasnowS. B.SmithJ. L. 1975 Microwave sanitization of color additives used cosmetics: feasibility study. Applied Microbiology, 30 2 (August 1975), 205211 , 0003-6919

JengD. K.KaczmarekK. A.WoodworthA. G.BalaskyG. 1987 Mechanism of microwave sterilization in the dry state. Applied and Environmental Microbiology, 53 9 (September 1987), 21332137 , 1098-5336

JorgeJ.H.GiampaoloE.T.VerganiC.E.PavarinaA.C.MachadoA.L.CarlosI.Z.2009 Effect of microwave postpolymerization treatment and of storage time in water on the cytotoxicity of denture base and reline acrylic resins. Quintessence International, 40 10 (November-December 2009), e93100 , 0033-6572

KahlerH. 1929 The nature of the effect of a high-frequency electric field upon Paramecium. Public Health Reports, 44 7 (February 1929), 339347 , 0033-3549

KahnR. C.LancasterM. V.Kate JrW. 1982 The microbiologic crosscontamination of dental prostheses. The Journal of Prosthetic Dentistry, 47 5 (May 1982), 556559 , 1097-6841

KohnW. G.HarteJ. A.MalvitzD. M.CollinsA. S.ClevelandJ. L.EklundK. J. 2004 Centers for Disease Control and Prevention. Guidelines for infection control in dental health care settings-2003. The Journal of the American Dental Association, 135 1 (January 2004), 3347 , 0002-8177

LambD.J.EllisB.PriestleyD. 1983 The effects of process variables on levels of residual monomer in autopolymerizing dental acrylic resin. Journal of Dentistry, 11 1 (March 1983), 8088 , 0300-5712

LatimerJ. M. 1977 Microwave oven irradiation as a method for bacterial decontamination in a clinical microbiology laboratory. Journal of Clinical Microbiology, 6 4 (October 1977), 340342 , 0095-1137

LeungR. L.SchonfeldS. E. 1983 Gypsum casts as a potential source of microbial cross-contamination. The Journal of Prosthetic Dentistry, 49 2 (February 1983), 210211 , 1097-6841

LuebkeR. J.SchneiderR. L. 1985 Microwave oven drying of artificial stone. The Journal of Prosthetic Dentistry, 53 2 (February 1985), 261265 , 1097-6841

LuebkeR. J.ChanK. C. 1985 Effect of microwave oven drying on surface hardness of dental gypsum products. The Journal of Prosthetic Dentistry, 54 3 (September 1985), 431435 , 1097-6841

MachadoA. L.BreedingL. C.PuckettA. D. 2005 Effect of microwave disinfection on the hardness and adhesion of two resilient liners. The Journal of Prosthetic Dentistry, 94 2 (August 2005), 183189 , 1097-6841

MachadoA. L.BreedingL. C.PuckettA. D. 2006 Effect of microwave disinfection procedures on torsional bond strengths of two hard chairside denture reline materials. Journal of Prosthodontics, 15 6 (November-December 2006), 337344 , 0153-2849X

MachadoA. L.BreedingL. C.VerganiC. E.daCruz.PerezL. E. 2009 Hardness and surface roughness of reline and denture base acrylic resins after repeated disinfection procedures. The Journal of Prosthetic Dentistry, 102 2 (August 2009), 115122 , 1097-6841

MeşeA.MeşeS. 2007 Effect of microwave energy on fungal growth of resilient denture liner material. Biotechnology & Biotechnological Equipment, 21 1 (January 2007), 9193 , 1310-2818

Meunier-CarpentierF.KiehnT. E.ArmstrongD. 1981 Fungemia in the immunocompromised host. The American Journal of Medicine, 71 3 (September 1981), 363370 , 0002-9343

MimaE. G.PavarinaA. C.NeppelenbroekK. H.VerganiC. E.SpolidorioD. M.MachadoA. L. 2008 Effect of different exposure times of microwave irradiation on the disinfection of a hard chairside reline resin. Journal of Prosthodontics, 17 4 (June 2008), 312317 , 0153-2849X

MonsenegoP. 2000 Presence of microorganisms on the fitting denture complete surface: study in vivo. Journal of Oral Rehabilitation, 27 8 (August 2000), 708713 , 1365-2842

NajdovskiL.DragasA. Z.KotnikV. 1991 The killing activity of microwaves on some non-sporogenic and sporogenic medically important bacterial strains. The Journal of hospital infection, 19 4 (December 1991), 239247 , 0195-6701

Nelson-FilhoP.FariaG.daSilva. R. A.RossiM. A.ItoI. Y. 2006 Evaluation of the contamination and disinfection methods of toothbrushes used by 24- to 48-month-old children. Journal of Dentistry for Children, 73 3 (September-December 2006), 152158 , 1935-5068

Nelson-FilhoP.daSilva. L. A.daSilva. R. A.daSilva. L. L.FerreiraP. D.ItoI. Y. 2011 Efficacy of microwaves and chlorhexidine on the disinfection of pacifiers and toothbrushes: an in vitro study. Pediatric Dentistry, 33 1 (January-February 2011), 1013 , 0164-1263

NeppelenbroekK. H.PavarinaA. C.SpolidorioD. M.VerganiC. E.MimaE. G.MachadoA. L. 2003 Effectiveness of microwave sterilization on three hard chairside reline resins. The International Journal of Prosthodontics, 16 6 (November-December 2003), 616620 , 0893-2174

NeppelenbroekK. H.PavarinaA. C.PalomariSpolidorio. D. M.SgavioliMassucato. E. M.SpolidorioL. C.VerganiC. E. 2008 Effectiveness of microwave disinfection of complete dentures on the treatment of Candida-related denture stomatitis. Journal of Oral Rehabilitation, 35 11 (November 2008), 836846 , 1365-2842

NewtonA. V. 1962 Denture sore mouth. A possible etiology. British Dental Journal, 112 1 (May 1962), 357360 , 0007-0610

NikawaH.IwanagaH.KamedaM.HamadaT. 1992 In vitro evaluation of C. albicans adherence to soft denture-lining materials. The Journal of Prosthetic Dentistry, 68 5 (November 1992), 804808 , 1097-6841

NovaisP. M.GiampaoloE. T.VerganiC. E.MachadoA. L.PavarinaA. C.JorgeJ. H. 2009 The occurrence of porosity in reline acrylic resins. Effect of microwave disinfection. Gerodontology, 26 1 (March 2009), 6571 , 1741-2358

Nunes deMello. J. A.BraunK. O.RachedR. N.Del BelCury. A. A. 2003 Reducing the negative effects of chemical polishing in acrylic resins by use of an additional cycle of polymerization. The Journal of Prosthetic Dentistry, 89 6 (June 2003), 598602 , 1097-6841

OlsenC. M. 1965 Microwaves inhibit bread mold. Food Engineering, 37 7 (1965), 5153 , 0260-8774

PavanS.ArioliFilho. J. N.DosSantos. P. H.MolloF.de. A.Jr 2005 Effect of microwave treatments on dimensional accuracy of maxillary acrylic resin denture base. Brazilian Dental Journal, 16 2 (May-August 2005), 119123 , 1806-4760

PavanS.ArioliFilho. J. N.DosSantos. P. H.NogueiraS. S.BatistaA. U. 2007 Effect of disinfection treatments on the hardness of soft denture liner materials. Journal of prosthodontics, 16 2 (March-April 2007), 101106 , 0153-2849X

PavarinaA. C.NeppelenbroekK. H.GuinesiA. S.VerganiC. E.MachadoA. L.GiampaoloE. T. 2005 Effect of microwave disinfection on the flexural strength of hard chairside reline resins. Journal of Dentistry, 33 9 (October 2005), 741748 , 0187-9176X

Pelczar JrM. J.ChanE. C. S.KriegN. R. 1993 Control of 1 microorganisms: principles 2 and physical agents. In: Microbiology: concepts and applications, McGraw-Hill, 200220 , 0-07112-914-6 York

PolyzoisG. L.ZissisA. J.YannikakisS. A. 1995 The effect of glutaraldehyde and microwave disinfection on some properties of acrylic denture resin. The International Journal of Prosthodontics, 8 2 (March-April 1995), 150154 , 0893-2174

PonneC. T.BartelsP. V. 1995 Interaction of electromagnetic energy with biological material- relation to food processing. Radiation Physics and Chemistry, 45 4 (April 1995), 591607 , 0096-9806X

PowellG. L.RunnellsR. D.SaxonB. A.WhisenantB. K. 1990 The presence and identification of organisms transmitted to dental laboratories. The Journal of Prosthetic Dentistry, 64 2 (August 1990), 235237 , 0022-3913

RayK. C.FullerM. L. 1970 Isolation of Mycobacterium from dental impression material. The Journal of Prosthetic Dentistry, 24 3 (September 1970), 335338 , 0022-3913

RibeiroD. G.PavarinaA. C.MachadoA. L.GiampaoloE. T.VerganiC. E. 2008 Flexural strength and hardness of reline and denture base acrylic resins after different exposure times of microwave disinfection. Quintessence International, 39 10 (November 2008), 833840 , 0033-6572

RibeiroD. G.PavarinaA. C.DovigoL. N.PalomariSpolidorio. D. M.GiampaoloE. T.VerganiC. E. 2009 Denture disinfection by microwave irradiation: a randomized clinical study. Journal of Dentistry, 37 9 (September 2009), 666672 , 0300-5712

RibeiroR. C.GiampaoloE. T.MachadoA. L.VerganiC. E.PavarinaA. C. 2008 Effect of microwave disinfection on the bond strength of denture teeth to acrylic resins. The International Journal of Adhesion and Adhesives, 28 6 (September 2008), 296301 , 0143-7496

RibeiroR. C.IzumidaF. E.MoffaE. B.BassoM. F. M.GiampaoloE. T.VerganiC. E. 2009 Color stability of reline resin after microwave disinfection: clinical evaluation. Journal of Dental Research, 88 Special Issue Letter A, Abstract 1755 0022-0345

RizzoR. 1993 The effects of sterilization with microwaves on diamond burs. Minerva Stomatologica, 42 3 (March 1993), 9396 , 0026-4970

RohrerM. D.BulardR. A. 1985 Microwave sterilization. Journal of the American Dental Association, 110 2 (February 1985), 194198 , 0002-8177

RohrerM. D.TerryM. A.BulardR. A.GravesD. C.TaylorE. M. 1986 Microwave sterilization of hydrophilic contact lenses. American Journal of Ophthalmology, 101 5 (January 1986), 4957 , 0002-9394

RosaspinaS.SalvatorelliG.AnzanelD.BovolentaR. 1994 Effect of microwave radiation on Candida albicans. Microbios, 78 314 5559 , 0026-2633

RossiT.PeltonenR.LaineJ.EerolaE.Vuopio-VarkilaJ.KotilainenP. 1996 Eradication of the long-term carriage of methicillin-resistant Staphylococcus aureus in patients wearing dentures: a follow-up of 10 patients. The Journal of Hospital Infection, 34 4 (December 1996), 311320 , 0195-6701

SandeM. A.GadotF.WenzelR. P. 1975 Point source epidemic of Mycoplasma pneumoniae infection in a prosthodontics laboratory. The American Review of Respiratory Disease, 112 2 (August 1975), 213217 , 0003-0805

SanitáP. V.VerganiC. E.GiampaoloE. T.PavarinaA. C.MachadoA. L. 2009 Growth of Candida species on complete dentures: effect of microwave disinfection. Mycoses, 52 2 (March 2009), 154160 , 0933-7407

SanitáP. V.VerganiC. E.MachadoA. L.GiampaoloE. T.PavarinaA. C. 2010 Denture microwave disinfection for reducing Candida infection in diabetic patients. Journal of Dental Research, 89 Special Issue Letter B, Abstract 1798 0022-0345

SanitáP. V.MachadoA. L.DovigoL. N.GiampaoloE. T.PavarinaA. C.VerganiC. E. 2011 Denture microwave disinfection in treating diabetics with denture stomatitis. Journal of Dental Research, 90 Special Issue Letter A, Abstract 2416 0022-0345

SartoriE. A.SchmidtC. B.WalberL. F.ShinkaiR. S. 2006 Effect of microwave disinfection on denture base adaptation and resin surface roughness. Brazilian Dental Journal, 17 3 (2006), 195200 , 0103-6440

SartoriE. A.SchmidtC. B.MotaE. G.HirakataL. M.ShinkaiR. S. 2008 Cumulative effect of disinfection procedures on microhardness and tridimensional stability of a poly(methyl methacrylate) denture base resin. Journal of Biomedical Materials Research. Part B, Applied Biomaterials, 86B 2 (August 2008), 360364 , 1552-4973

SeóR. S.VerganiC. E.GiampaoloE. T.PavarinaA. C.MachadoA. L. 2007a Effect of a post-polymerization treatment on the flexural strength and Vickers hardness of reline and acrylic denture base resins. Journal of Applied Oral Science, 15 6 (December 2007), 506511 , 1678-7757

SeóR. S.VerganiC. E.PavarinaA. C.CompagnoniM. A.MachadoA. L. 2007b Influence of microwave disinfection on the dimensional stability of intact and relined acrylic resin denture bases. The Journal of Prosthetic Dentistry, 98 3 (September 2007), 216223 , 0022-3913

SeóR. S.VerganiC. E.GiampaoloE. T.PavarinaA. C.dosSantos.NunesReis. J. & Machado, A.L. 2008 Effect of disinfection by microwave irradiation on the strength of intact and relined denture bases and the water sorption and solubility of denture base and reline materials. Journal of Applied Polymer Science, 107 1 (January 2008), 300308 , 1097-4628

SideridouI.DSAchiliasKyrikouE. 2004 Thermal expansion characteristics of light-cured dental resins and resin composites. Biomaterials, 25 15 (July 2004), 30873097 , 0142-9612

SilvaM. M.VerganiC. E.GiampaoloE. T.NeppelenbroekK. H.SpolidorioD. M. P.MachadoA. L. 2006 Effectiviness of microwave irradiation on the disinfection of complete dentures. The International Journal of Prosthodontics, 19 3 (May-June 2006), 151156 , 0893-2174

SilvaM. M.VerganiC. E.MimaE. G. O.PavarinaA. C.MachadoA. L.GiampaoloE. T. 2008 Effect of microwave disinfection in the treatment of denture stomatitis. Journal of Dental Research, 87 Special Issue Letter B, Abstract 1276 0022-0345

SofouA.LarsenT.FiehnN. E.OwallB. 2002 Contamination level of alginate impressions arriving at a dental laboratory. Clinical Oral Investigations, 6 3 (September 2002), 161165 , 1432-6981

SpolidorioD.TardivoT.dosReis.DerceliJ.NeppelenbroekK.DuqueC.SpolidorioL. PiresJ. 2011 Evaluation of two alternative methods for disinfection of toothbrushes and tongue scrapers. International Journal of Dental Hygiene, (February 2011), doi: 10.1111/j.16015037 .2011.00503.x. [Epub ahead of print], 1601-5029

TakahashiY.ChaiJ.KawaguchiM. 1998 Effect of water sorption on the resistance to plastic deformation of a denture base material relined with four different denture reline materials. The International Journal of Prosthodontics, 11 1 (January-February 1998), 4954 , 0893-2174

TarantinoL.TomassiniE.PettiS.SimonettiD'Arca. A. 1997 Use of a microwave device for dental instrument sterilization: possibilities and limitations. Minerva Stomatologica, 46 10 (October 1997), 561566 , 0026-4970

TateW. H.GoldschmidtM. C.WardM. T.GrantR. L. 1995 Disinfection and sterilization of composite polishing instruments. American Journal of Dentistry, 8 5 (October 1995), 270272 , 0894-8275

TateW. H.GoldschmidtM. C.PowersJ. M. 1996 Performance of composite finishing and polishing instruments after sterilization. American Journal of Dentistry, 9 2 (April 1996), 6164 , 0894-8275

ThomasC. J.WebbB. C. 1995 Microwaving of acrylic resin dentures. The European Journal of Prosthodontics and Restorative Dentistry, 3 4 (June 1995), 179182 , 0965-7452

TuncerN.TufekçiogluH. B.CalikkocaogluS. 1993 Investigation on the compressive strength of several gypsum products dried by microwave oven with different programs. The Journal of Prosthetic Dentistry, 69 3 (March 1993), 333339 , 0022-3913

VerganiC. E.GiampaoloE. T.PavarinaA. C.SilvaM. M.MimaE. G. O.MachadoA. L. 2008 Efficacy of denture microwave disinfection in treating denture stomatitis. Journal of Dental Research, 87 Special Issue Letter B, Abstract 1289 0022-0345

VerranJ.KossarS.Mc CordJ. F. 1996 Microbiological study of selected risk areas in dental technology laboratories. Journal of Dentistry, 24 1-2 , (January-March 1996), 7780 , 0300-5712

VerranJ.MaryanC. J. 1997 Retention of Candida albicans on acrylic resin and silicone of different surface topography. The Journal of Prosthetic Dentistry, 77 5 (May 1997), 535539 , 0022-3913

WakefieldC. W. 1980 Laboratory contamination of dental prostheses. The Journal of Prosthetic Dentistry, 44 2 (August 1980), 143146 , 0022-3913

WatanabeK.KakitaY.KashigeN.MiakeF.TsukijiT. 2000 Effect of ionic strength on the inactivation of micro-organisms by microwave irradiation. Letters in Applied Microbiology, 31 1 (July 2000), 5256 , 0266-8254

WebbB. C.ThomasC. J.HartyD. W.WillcoxM. D. 1998 Effectiveness of two methods of denture sterilization. Journal of Oral Rehabilitation, 25 6 (June 1998), 416423 , 0030-5182X

WebbB. C.ThomasC. J.WhittleT. 2005 A 2-year study of Candida-associated denture stomatitis treatment in aged care subjects. Gerodontology, 22 3 (September 2005), 168176 , 0734-0664

WilliamsH. N.FalklerW. A.Jr HaslerJ. F.LibonatiJ. P. 1985 The recovery and significance of nonoral opportunistic pathogenic bacteria in dental laboratory pumice. The Journal of Prosthetic Dentistry, 54 5 (November 1985), 725730 , 0022-3913

WooI.RheeI.ParkH. 2000 Differential damage in bacterial cells by microwave radiation on the basis of cell wall structure. Applied and Environmental Microbiology, 66 5 (May 2000), 22432237 , 0099-2240

WrightP. S.ClarkP.HardieJ. M. 1985 The prevalence and significance of yeasts in persons wearing complete dentures with soft-lining materials. Journal of Dental Research, 64 2 (February 1985), 122125 , 0022-0345

YannikakisS.ZissisA.PolyzoisG.AndreopoulosA. 2002 Evaluation of porosity in microwave-processed acrylic resin using a photographic method. The Journal of Prosthetic Dentistry, 87 6 (June 2002), 613619 , 0022-3913

YeoC. B.WatsonI. A.Stewart-TullD. E.KohV. H. 1999 Heat transfer analysis of Staphylococcus aureus on stainless steel with microwave radiation. Journal of Applied Microbiology, 87 3 (September 1999), 396401 , 1364-5072

PART 3
HEALTH AND ENVIRONMENT

5 Chapter

MICROWAVE ENHANCED ADVANCED OXIDATION PROCESS APPLICATION TO TREATMENT OF DAIRY MANURE

1. INTRODUCTION: AN OVERVIEW

Microwave heating is a dielectric heating method in which heat is produced by the interaction of dielectric materials with electromagnetic radiation. The dielectric constant quantifies a material's ability to impede microwave energy during transmission, whereas the loss factor assesses the material's capability to dissipate that energy. Materials with a high loss factor are readily heated with microwave irradiation. The mechanism of energy dissipation in materials occurs through ionic conduction and dipolar rotation. The dielectric characteristics of a material are generally influenced by temperature, moisture content, density, and material shape. Microwave heating offers superior heating rates, eliminates direct contact between the heating source and the material, enables selective heating, minimizes equipment space, and enhances process control compared to conventional heating methods. Furthermore, microwave irradiation can cause damage to both DNA and bacterial cell membranes. The impact on the treatment process, in addition to heating, is termed the athermal effect (Hong et al. 2006).

Microwave heating has been utilized in several operations and manufacturing sectors, including food processing, wood drying, and waste treatment (Jones et al., 2002). Microwave heating has numerous applications in the sewage wastewater treatment sector, specifically for sludge treatment, aimed at volume reduction, dewaterability enhancement, digestibility improvement, nutrient release augmentation, pathogen eradication, and heavy metal stabilization (Menendez et al., 2002; Liao et al., 2005; Wojciechowaska, 2005; Hong et al., 2006; Eskicioglu et al., 2007; Hsieh et al., 2007; Yu et al., 2010). Reports indicate that the total treatment efficacy may be enhanced through the integration of microwave heating with chemical processes (thermo-chemical) for sludge treatment (Chan, et al., 2007; Qiao, et al., 2008). The prevalent compounds utilized in thermo-chemical processes include acids, bases, and oxidants.

The microwave-assisted advanced oxidation process (MW/H2O2-AOP) employs microwave irradiation alongside hydrogen peroxide to produce hydroxyl radicals that interact with organic molecules. The structure of solid materials is modified, allowing nutrients to be released into the solution throughout the process. The treatment also decreases the suspended solids (SS) concentration in the slurries. The mechanism is presumed to resemble the wet-air oxidation process, wherein the initial phase entails the degradation of substantial particulate organic materials, such as carbohydrates and proteins, into smaller, more soluble organic constituents, followed by a subsequent phase involving the further oxidation or gasification of certain resultant organic products (Shanableh & Shimizu, 2000; Liao et al., 2007). The determinants influencing the efficacy of the MW/H2O2-AOP have been delineated; these include microwave temperature, hydrogen peroxide dosage, microwave intensity, reaction duration, and acid addition. The consequences differed based on the application, contingent upon the soluble elements of interest and the diverse organic waste slurries utilized in the procedure (Wong et al., 2007). Dairy manure comprises lipids, proteins, lignin, carbohydrates, and inorganic residue, and is abundant in various nutrients such as nitrogen, phosphorus, and minerals (Rico et al., 2007); it should be seen as a useful biological resource rather than a waste product. Lignocellulosic constituents from dairy manure can be turned into fermentable saccharides through enzymatic or acid hydrolysis, which can then be changed into ethanol and other useful products (Wen, et al., 2004). Nutrients, including phosphate and ammonia-nitrogen, can be extracted by a struvite crystallization method. Carbonaceous materials can undergo anaerobic digestion to produce methane, an important bioenergy resource. Nonetheless, the majority of phosphorus in dairy manure (about 65%) exists in a form that is not readily soluble, particularly in its structural composition (Barnet, 1994). Moreover, elevated suspended particles are characteristic of dairy manure; therefore, a pre-treatment phase is necessary to solubilize the nutrients and dissolve the solids to facilitate resource recovery.

The MW/H_2O_2-AOP, which can disintegrate solids and release nutrients into solution, can be a potential effective pre-treatment method for nutrient and energy recovery from dairy manure. The effectiveness of the MW/H_2O_2-AOP under both batch operation and a continuous mode of operation for solubilisation of nutrients and organic matters from dairy manure are therefore reported in this chapter.

2. METHODS

2.1. Apparatus

A lab-scale Milestone Ethos TC microwave oven digestion system (Milestone Inc., USA) was used for a batch mode operation. The system operates at a frequency of 2,450 MHz with a maximum power output of 1,000 W. The system has the capacity of handling up to 12 vessels in a single run: one reference and 11 sample vessels, each with a volume of 100 mL. A thermocouple is inserted into the reference vessel, providing real time temperature monitoring during the runs. The maximum operating temperature and pressure are 220º C and 435 psig, respectively.

A household microwave oven (Panasonic Genius Prestige Countertop) was modified for use in a continuous mode operation. The system operates at a frequency of 2,450 MHz with a maximum power output of 1,000 W. A silicon tube with 6.35 mm diameter was wound into a continuous horizontal coil and held by a custom made perforated polypropylene shelf. The total length of the silicon tube inside the MW is 40 m and the volume is approximately 1.3 L. The dairy manure was pumped into the MW oven through the silicon tube and was collected from the outlet. An appropriate amount of concentrated sulfuric acid was added to the liquid manure to the desired acid concentration before the mixture was pumped into the MW oven. The desired H_2O_2 dosage in the process was controlled by the flow rate and the concentration of the prepared H_2O_2. Different exit temperatures after the MW treatment can be achieved by controlling the flow rate of both the liquid manure and H_2O_2; a higher flow rate yields a lower exit temperature. The overall heating rate of 5 °C/min was used in this study. The unit was not pressurized; the exit temperatures were therefore lower than the boiling point of water.

2.2. Substrate

Dairy manure was obtained from the UBC Dairy Education & Research Centre in Agassiz, British Columbia, Canada. The dairy manure used for the batch operation was the solid portion obtained after the solid-liquid separation. Once collected from the farm the dairy manure was stored in a closed container at 4ºC. The dairy manure sample had total solids (TS) of 33%, which contained large amounts of sand, the bedding material, as well as undigested lignocellulosic materials. Distilled water was added to the dairy manure samples and subsequently decanted to remove the sand. The resulting dairy manure had a TS of 5.6% for the first part of the batch operation (Part 1), while 3.7 % TS for the second part of the study at pH 4 (Part 2).

For a continuous operation, the liquid manure after the solid-liquid separation was again passed through a U.S. Standard No. 18 sieve (1 mm openings) to remove large fibres and particles to avoid solids deposition in the silicon tube in the MW unit. It was then diluted four times with distilled water to have TS of 0.83%.

2.3. Experimental design

2.3.1. Batch operation

For Part 1, few variables were held constant, including the dosage of hydrogen peroxide (1.5 mL of 30% concentration, or expressed as 0.28 g H_2O_2/g TS), treatment time (20 minutes), and treatment temperature (120 °C). In an effort to isolate the effects of pH and hydrogen peroxide on the microwave process, experiments under acidic, neutral, and basic conditions, as well as in the absence and presence of hydrogen peroxide were tested. The experimental conditions included microwave treatment only (MW), microwave treatment with hydrogen peroxide (MW/H_2O_2), microwave treatment in acidic condition (MW/H^+), microwave treatment with hydrogen peroxide in acidic condition (MW/H_2O_2/H^+), microwave treatment in basic condition (MW/OH^-), and microwave treatment with hydrogen peroxide in basic condition (MW/H_2O_2/OH^-). Four replicates were conducted for each experimental condition. In adjusting to acidic conditions, dilute H_2SO_4 was dripped into the dairy manure solution to reach a pH of 4; whereas for basic conditions, dilute NaOH solution was added to reach a pH of 10.

For Part 2, the acidified dairy manure was chosen. Variables including treatment temperature, treatment time, and hydrogen peroxide dosage at pH 4 were tested for their effects on solids disintegration and nutrient solubization. Based on a computer statistical program, the Box-Benken design for response surface plots was chosen, and a series of 15 trials was required (Sall, et al., 2005). The temperature range was between 80 and 160 °C, treatment time ranged from 10 to 20 min, while hydrogen peroxide dosage was from 0 to 0.14 g H_2O_2/g TS. The rate of temperature increase for all the experiments were set at 20 °C per minute up to the designated treatment temperature and subsequently held for the specified time period. The experimental design is listed in Table 1.

2.3.2. Continuous operation

The MW/H_2O_2–AOP experiments were operated at temperatures of 60, 70, 80 and 90º C. The H_2O_2 **dosage was fixed at 0.1% (v/v) and the acid dosages were 0.2, 0.5 and 1% (v/v) for each temperature. A set of experiments without H_2O_2** was performed to serve as control. Overall, sixteen individual experiments each with three replicates were conducted.

Table 1. JMP design for Part 2 of batch operation

Set #	Temp ºC	Heat time (min)	Dosage (mL)	g H_2O_2/g TS
1	120	15	0.25	0.068
2	160	20	0.25	0.068
3	120	15	0.25	0.068

Set #	Temp ºC	Heat time (min)	Dosage (mL)	g H_2O_2/g TS
4	80	10	0.25	0.068
5	80	15	0	0.000
6	160	15	0.5	0.135
7	160	15	0	0.000
8	80	20	0.25	0.068
9	160	10	0.25	0.068
10	120	10	0	0.000
11	120	15	0.25	0.068
12	80	15	0.5	0.135
13	120	20	0	0.000
14	120	20	0.5	0.135
15	120	10	0.5	0.135

2.4. Chemical analysis

The samples were centrifuged at 3500 rpm for 10 minutes first, and then their supernatants were extracted for analysis for soluble chemical oxygen demand (SCOD), orthophosphate, soluble ammonia, volatile fatty acids (VFA) and reducing sugar. For dairy manure, the orthophosphate analysis could be affected by dilution due to interference, resulting in erroneous values; in order to obtain correct results, the initial dairy manure sample was first diluted to 0.5% TS before proceeding with the analysis (Wolf et al., 2005). It was also proven in a previous study that after microwave treatment the results obtained are reliable (Kenge, et al., 2009). The initial dairy manure samples were also analyzed for TS, total chemical oxygen demand (TCOD), total phosphate (TP) and total Kjeldahl nitrogen (TKN). All of the chemical analyses, except that of reducing sugar, followed the procedures outlined in Standard Methods (APHA, 1998). The colorimetric method was employed to determine the total reducing sugar content using anthrone reagent (Raunkjer, et al., 1994).

All chemical analyses were determined by a flow injection system, except determinations of TS and COD (Lachat Quik-Chem 8000 Automatic Ion Analyzer, Lachat Instruments, USA). A Hewlett Packard 5890 Series II gas chromatograph, equipped with a flame ionization detector (FID), was used to measure VFA. Volatile separation was accomplished with an HP FFAP (free fatty acid phase) column (0.25 m × 0.31 mm with 0.52 μ film thickness). The injection temperature was set at 175 ºC and the flame ionization detector was at 250 ºC. Helium gas was used as the carrier at a head pressure of 69 kPa.

3. RESULTS AND DISCUSSION

3.1. Batch operation

3.1.1. Part 1

The effects of pH and H_2O_2 on the nutrient release were evaluated in Part 1. The concentration of ortho-P and VFA increased with an addition of acid without the MW treatment, while concentration of both SCOD and reducing sugar decreased. It was suggested that the agglomeration of fine suspended particles of dairy manure occurred under the acidic condition, resulting in a decrease of the SCOD concentration. The reduction of reducing sugar concentration might be due to chemical reactions. There was no significant change of any initial concentration with the addition of NaOH (Table 3). Upon the addition of acid, the SCOD concentration decreased by about half. This may be attributed to chemical reactions between various soluble constituents in the samples. As a result, increased particle size caused the precipitation, and it was subsequently filtered out prior to this test. It appeared that acid addition aided the solubization of ortho-P, and the VFA production. However, it was not clear the role acid played in the reaction mechanism.

Reducing sugar was produced from all treatments. Microwave heating enhanced its yield, regardless of acidic or basic conditions. A significant amount of reducing sugar was produced in either the MW or MW/OH^- treatment. The best yield was obtained with the MW/H^+/H_2O_2-AOP. The addition of hydrogen peroxide did not significantly help in producing more reducing sugar (Table 2). For the microwave pretreatment of dairy manure for anaerobic digestion, sugar (glucan/xylan) production was also reported (Jin et al., 2009). In their study, sugar yield was not affected by the different types of acid applied, sulfuric acid or hydrochloric acid, but was rather affected by the concentration of acid. The best yield in their study was produced either in 0.5% (v/v) of sulfuric acid or in 0.185% (v/v) of hydrochloric acid, but the yield did not increase further with a higher acid concentration. Sugar production was the worst in the given conditions with 2% H_2SO_4 addition. The comparison between the study by Jin et al. (2009) and this study should be made with the note that higher concentrations of acid, or base, and a lower hydrogen peroxide dosage were used in their study.

The low yield of reducing sugar was due to a diluted acid used in this study. Concentrated acid used for sugar production was more effective than diluted acid hydrolysis (Sun and Cheng, 2002). As high as 75% concentrated acid was used for treating dairy manure to produce more than 84% of glucose (g/100g cellulose) at 120 °C and 30 min of reaction time. The yield of glucose decreased with an increase of reaction time (Liao et al., 2006). Their results pointed out that the reaction period was also critical for the sugar yield.

When treated with MW alone, SCOD values increased due to the disintegration of organic particles. However, for the MW/H^+ and $MW/H^+/H_2O_2$ treatment, the SCOD value fell to below that of the MW treatment alone. On the other hand, it appeared that the disintegration of organic maters was enhanced in basic conditions. The high concentrations of SCOD were obtained for both MW/OH-/H2O2 and MW/OH-. These results were not quite similar to those of Jin et al. (2009). In their study, the high acid concentration (2% of acid) yielded the highest SCOD, about 35% of TCOD whereas the low acid concentration (0.5%) produced less SCOD than that of either NaOH or calcium oxide addition. In this study, high SCOD was obtained with the basic treatment. Despite the high SCOD yield with NaOH treatment, ortho-P, ammonia, and VFA release appeared to be lacking.

VFA was produced in a few of the trials. One common variable in the VFA production trials was the H_2SO_4 addition. It was clear that when acid was added, VFA was produced with or without MW. As hydrogen peroxide was added, VFA was reduced slightly. At the same time, TCOD was reduced for the acid trials with hydrogen peroxide, which indicated that instead of retaining carbon to form products such as VFA, carbon left the system as CO2 as an end oxidation product. The presence of acid was also stabilizing the hydroxyl radicals to be more potent in oxidizing organics.

Phosphates have been known to be soluble in solution after treatment with the MW/H_2O_2-AOP. Orthophosphate was released in large amounts upon the addition of acid, at almost half the total phosphorus in solution. With the MW/H_2O_2-AOP treatment on top of acid addition, the release was even greater, at 62% of total phosphorus from the initial substrate. Soluble phosphate release was not significant for the basic treatment, with or without hydrogen peroxide addition. Dairy manure contained various ions, such as carbonate, ammonia, magnesium, calcium, sodium and sulfate, as well as phosphate. At a high pH (basic) condition, various phosphate precipitates can be formed in the sample, such as struvite, and many forms of calcium phosphate (Wang and Nancollas, 2008; Jin, et al., 2009); therefore, orthophosphate release after the MW treatment would be easily precipitated out of solution again. Basic condition would not be suitable for dairy manure treatment in terms of orthophosphate release. As mentioned earlier, basic conditions would be suitable for the purpose of solid disintegration, as a very high SCOD concentration was obtained in base treatment. The results indicated that in order to release considerable amounts of orthophosphate into solution, acid addition is required for the MW treatment. Hydrogen peroxide would also help an increase of phosphate release, but is not necessarily required in the process.

Ammonia was not produced in large quantities after the MW/H_2O_2-AOP treatment (Table 2). The initial ammonia concentration in solution barely increased after treatment with any chemical addition.

The results indicated that reducing sugar can be obtained from dairy manure with the MW treatment, regardless of treatment conditions. In order to solubilize phosphorus, dairy manure should be treated in acidic condition with an addition of hydrogen peroxide to maximize its yield. If the main goal is to disintegrate manure solids, the basic condition shall be chosen, which would yield the highest SCOD concentration.

Table 2. Results from Part 1 of batch operation

	o-PO_4(mg P/L)	NH_4^+ (mg N/L)	SCOD (g/L)	Reducing sugar (mg/L)
initial concentration	6.3 + 0.2	600 + 25	15 + 0.6	672 + 16
initial acid addition	130 + 1.4	600 + 28	6.0 + 0.1	216 + 31
initial base addition	4.6 + 0.1	480 + 0	17 + 1.5	869 + 69
MW	8.0 + 0.8	830 + 88	22 + 1.9	1190 + 95
MW/H_2O_2	7.0 + 0.3	830 + 33	19 + 2.0	1130 + 91
MW/H^+	240 + 16	740 + 220	14 + 2.3	788 + 133
MW/H^+/H_2O_2	240 + 4.5	790 + 180	14 + 2.3	1460 + 62
MW/OH^-	30 + 1.9	670 + 200	27 + 1.5	1360 + 126
MW/OH^-/H_2O_2	13 + 3.5	610 + 110	25 + 3.8	1180 + 217

3.1.2. Part 2

Based on the results from Part 1, an acidic condition was chosen for Part 2 of the study. The design was modeled after a study by Kenge et al. (2010), where a control without the addition of hydrogen peroxide was included. In that study,hydrogen peroxide (0.30-0.59 g H_2O_2/g TS) was in excess; its incremental effects on the solubilization of nutrients could not be fully evaluated. Large amounts of hydrogen peroxide were also not helpful in producing reducing sugar from dairy manure (Jin et al., 2009; Kenge et al., 2010). Therefore, lower dosages of hydrogen peroxide, ranging from 0 to 0.14 g H_2O_2/g TS were used in Part 2, instead of 0.28 g H_2O_2/g TS of hydrogen peroxide used in Part 1.

3.1.2.1. Reducing sugar

The surface plot depicting the trend of reducing sugar production is showed in Figure 1. With the addition of hydrogen peroxide, reducing sugar release increased. It can be expected that with the addition of hydrogen peroxide, reducing sugar release increases due to its aid in the breakdown of lignocellulosic materials in dairy manure.

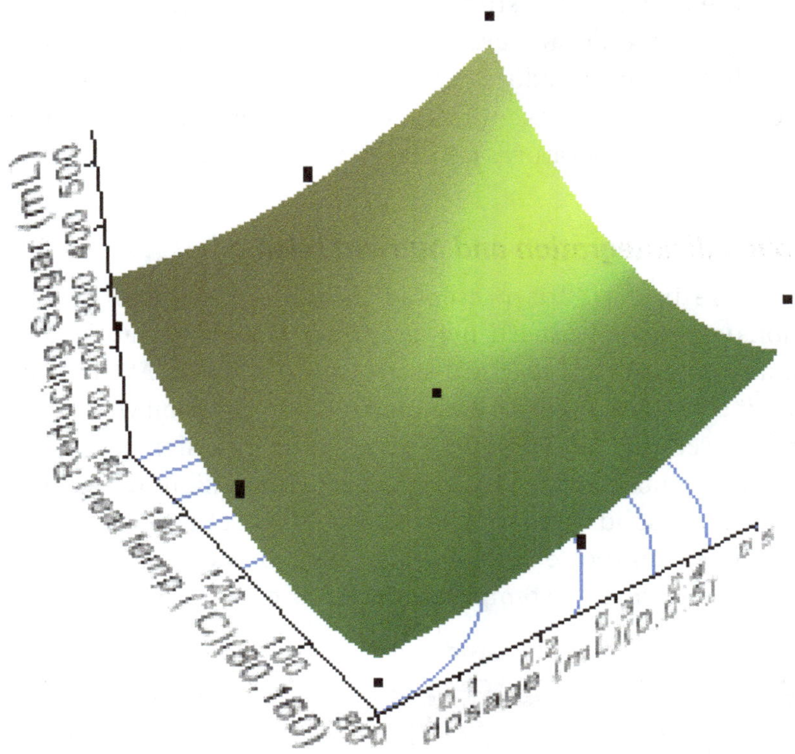

Figure 1. Surface plot for Part 2 – reducing sugars

Reducing sugar concentration also increased with an increase of temperature. In the study by Kenge et al., (2010), temperature was the most significant factor affecting reducing sugar production; an increase of microwave temperature increased reducing sugar production. An increase in hydrogen peroxide dosage also increased reducing sugar yield at lower temperatures; however, its yield remained relatively constant at higher microwave temperatures. A much higher dosage of hydrogen peroxide was applied in the earlier study, causing solubilized reducing sugars to be further oxidized to form other oxidation products. Due to a balance between reducing sugars being solubilized and oxidized at the same time, reducing sugar concentration in solution remained relatively constant. The rates of reducing sugar production and sugar oxidation to form other compounds have not been measured, but it appeared that equilibrium between the two rates was reached in this case. However, results of this study showed that reducing sugar increased slightly with an increase of hydrogen peroxide at high operating temperatures. This might be due to the fact that there was no excess hydrogen peroxide in solution to engage and aid in further oxidation reactions.

The results obtained from this study and previously indicated that temperature and hydrogen peroxide dosage are factors affecting reducing sugar production (Kenge et al., 2010). The procedure can be conducted with or without H2O2 for the generation of reducing sugars; significant quantities of reducing sugars can also be achieved at elevated temperatures and extended heating durations in the absence of H2O2.

3.1.2.2. Solids disintegration and nutrient release

The SCOD release in Part 2 is presented in Table 3. Generally, elevated SCOD concentrations were achieved with increased operating temperatures. Temperature was the primary determinant for the disintegration of solids. At the same operating temperature, it could be run for an extended heating duration without the addition of H2O2, or for a reduced heating duration with the inclusion of H2O2 to achieve a comparable SCOD concentration. Hydrogen peroxide exhibited a minimal effect on the disintegration of solids at low dosages. The temperature of the microwave was a crucial component in the release of phosphates. Orthophosphate concentration rose with elevated microwave temperatures (Figure 2).

Table 3. Results for Part 2 of batch operation

Set #	Reducing sugar (mg/L)	o-PO_4 (mg P/L)	NH_4+ (mg N/L)	SCOD (g/L)	TCOD (g/L)	VFA (mg/L)
1	106 ± 19	109 ± 4.8	43 ± 0.5	2.9 ± 0.2	20 ± 3.7	127 ± 6
2	344 ± 58	367 ± 4.0	71 ± 2.6	4.6 ± 0.2	35 ± 3.6	252 ± 7
3	126 ± 169	167 ± 3.9	57 ± 1.0	3.0 ± 0.4	33 ± 2.1	203 ± 5
4	65 ± 95	113 ± 4.0	40 ± 3.4	4.1 ± 0.3	19 ± 4.9	133 ± 14
5	60 ± 69	333 ± 5.0	50 ± 1.7	3.9 ± 0.1	37 ± 5.3	236 ± 26
6	517 ± 53	356 ± 4.3	73 ± 1.4	5.0 ± 0.1	36 ± 0.2	238 ± 9
7	230 ± 79	121 ± 0.5	51 ± 0.6	3.7 ± 0.4	19 ± 2.3	153 ± 20
8	104 ± 50	162 ± 8.4	51 ± 3.3	2.8 ± 0.6	38 ± 6.6	188 ± 14
9	296 ± 47	172 ± 3.4	66 ± 0.6	4.0 ± 0.2	37 ± 3.3	248 ± 12
10	110 ± 40	16 ± 6.7	54 ± 1.3	3.4 ± 0	31 ± 0.9	213 ± 19
11	172 ± 29	331 ± 6.0	56 ± 1.8	3.2 ± 0.1	34 ± 3.0	254 ± 10
12	387 ± 47	106 ± 3.8	42 ± 1.0	3.6 ± 0	20 ± 1.3	140 ± 5
13	111 ± 35	334 ± 10.8	55 ± 1.6	3.7 ± 0.5	37 ± 1.4	220 ± 22
14	183 ± 32	159 ± 6.5	56 ± 0.7	3.2 ± 0	32 ± 2.6	188 ± 16
15	149 ± 94	163 ± 3.6	55 ± 3.0	3.3 ± 0.3	28 ± 2.5	194 ± 17
Initial	322 ± 48	60 ± 3.2	16 ± 0.4	5.9 ± 0.1	36 ± 4.1	188 ± 19

The polyphosphates in dairy manure can decompose into ortho-P at high temperatures, with the rate and amount of decomposition being temperature-dependent (Kuroda et al., 2002). Acids also facilitate the solubilization of organic phosphorus and polyphosphate (Chan et al., 2007). Figure 2 illustrates a trend indicating that treatment duration significantly enhances the release of ortho-P, accompanied by a marginal increase in hydrogen peroxide. Regardless of hydrogen peroxide concentration and treatment temperature, ortho-P elevated with extended treatment duration (Figure 2). The findings from both segments of this investigation indicate that the addition of acid is essential for the liberation of ortho-P from dairy manure into solution. The orth-P results obtained in this study were consistent with results from previous studies (Kenge, et al., 2010; Yu, et al., 2010). It was reported by Jin, et al. (2010) that the optimal condition for orthophosphate release was at 135º C and 26 min in microwave treatment of dairy manure. It seemed that the MW/H_2O_2-AOP was operated near the optimal condition for the release of orthophosphate and other soluble components in this study. In view of the results, the process should be operated at a high temperature region (120-160º C), a long reaction time (15-20 min) and dosage of 0.07-0.14 g H_2O_2/g TS for maximizing soluble components from dairy manure.

Ammonia is one of the soluble products from the MW/H_2O_2-AOP, and it is a constituent of a potent fertilizer, struvite. No dramatic increases in ammonia production were found.

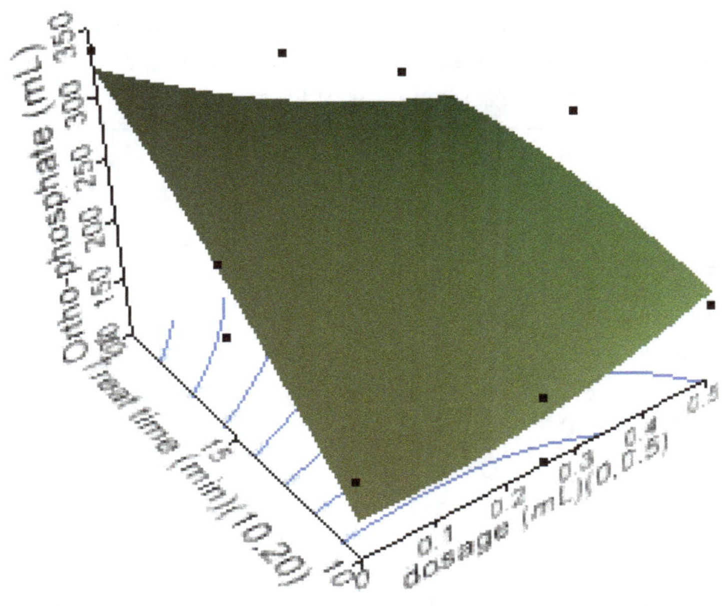

Figure 2. Surface plot for Part 2 – orthophosphate

3.2. Continuous operation

The disadvantages of a batch operation are: (1) the materials must be transported twice from the reaction vessel, and (2) the amount of sample that can be treated at one time is limited by the batch reaction time and the sample vessel volume. Besides these shortcomings, hydrogen peroxide was also introduced prior to microwave irradiation, limiting the synergistic effects between H_2O_2 and microwave irradiation. The alternative to batch operation is a continuous process, which is more suitable for an industrial application.

The results are presented in Figures 3 to 7. The most pronounced effect on the solubilization of solids was due to microwave heating temperature (Fig. 3).

The SCOD solubilization remained relatively steady at a range of 60 to 70º C, regardless of the acid concentration. Acid concentration aids the process to maintain or even increase the hydroxyl radical concentration by stabilizing them and thereby inhibiting their degradation. The narrow range of treatment temperature involved may mask the synergetic effects of adding acid to the process as the continuous mode of operation is not pressurized to exceed boiling point. The SCOD increased with an increase of microwave temperature to a higher range of 80 to 90º C. This was due to the fact that heating would also increase the decomposition of H_2O_2 into hydroxyl radicals and therefore enhance the oxidation process when H_2O_2 was applied simultaneously with microwave heating (Eskicioglu, et al., 2008). Heating alone can also add to the degradation of the substrate's structural integrity. The results showed that SCOD solublization from dairy manure was most affected by MW temperature. To a lesser extent, it was also affected by acid concentration, especially at the higher temperature levels of 80 to 90º C.

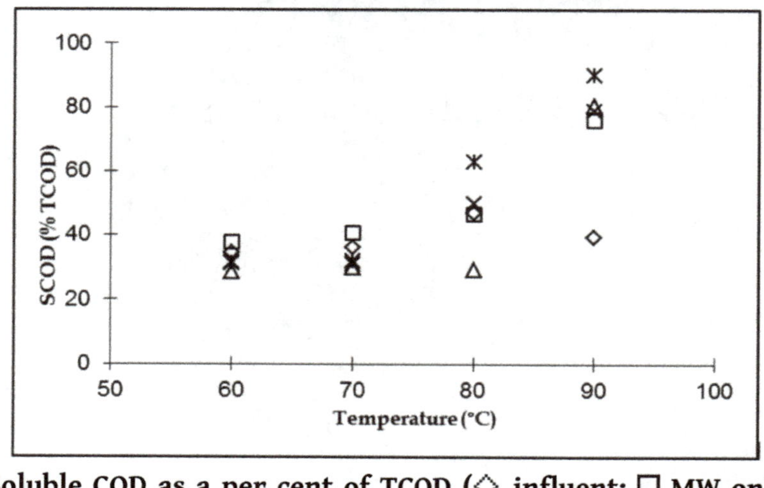

Figure 3. Soluble COD as a per cent of TCOD (◇ influent; □ MW only; △ MW/H2O2-AOP with 0.2% acid; ✕ MW/H2O2-AOP with 0.5% acid; ✳ MW/H2O2-AOP with 1.0% acid)

It was evident that an acidic condition was critical for phosphorus solubilization from dairy manure (Fig. 4). Without the acid addition, as microwave temperature increased from 60 to 90º C, the ortho-P decreased from 34% to 11% of TP, which was lower than in the untreated substrate. Phosphorus solubilization was greatly improved with an addition of acid. However, acid dosages did not display significant effects on the solubilization of phosphorus; an acid addition of 0.5% appeared to be sufficient.

There was no significant release of ammonia over the range of 60 to 80º C, regardless of the acid dosages (Fig. 5). However, up to 70% of TKN was solubilised at 90º C. The effect of acid dosage on NH_4-N solubilization was similar to that of phosphorus solubilization; the MW/H_2O_2-AOP needed to be operated in an acidic condition; 0.5% of acid concentration was sufficient.

The VFA comprised of various organic acids, such as acetic, propionic, i-butyric, i-valeric, valeric, hexanoic, and heptanoic acids. However, acetic acid is the major component of the VFA, ranged from 46 to 53%. A considerable amount of VFA was released from the dairy manure after the MW/H_2O_2-AOP treatment. The component of acetic acid increased in the treated solutions, ranging from 53 to 84% of total VFA. Overall, the best VFA production of approximately 330 to 420 mg/L was solubilized with an acid dosage of 0.5%. Acid in combination with the MW/H_2O_2-AOP produced more VFA with increasing temperature, except for the highest acid dosage of 1%, where a slight decrease was found at the higher temperatures. This trend may not be conclusive to say higher acid dosage is detrimental to VFA production, but it indicates that higher acid concentrations may have an effect in reducing VFA in dairy manure (Fig. 6).

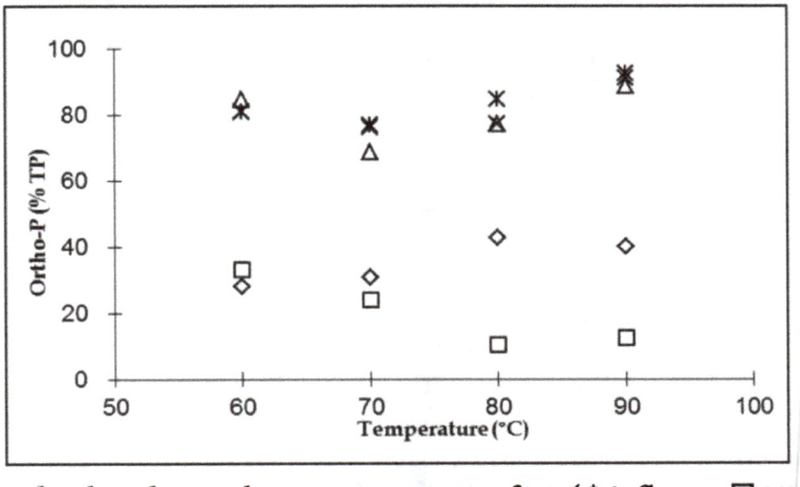

Figure 4. Orthophosphate release as a per cent of TP (◇ influent; □ MW only; △ MW/H2O2- AOP with 0.2% acid; ✕ MW/H2O2-AOP with 0.5% acid; ✱ MW/H2O2-AOP with 1.0% acid)

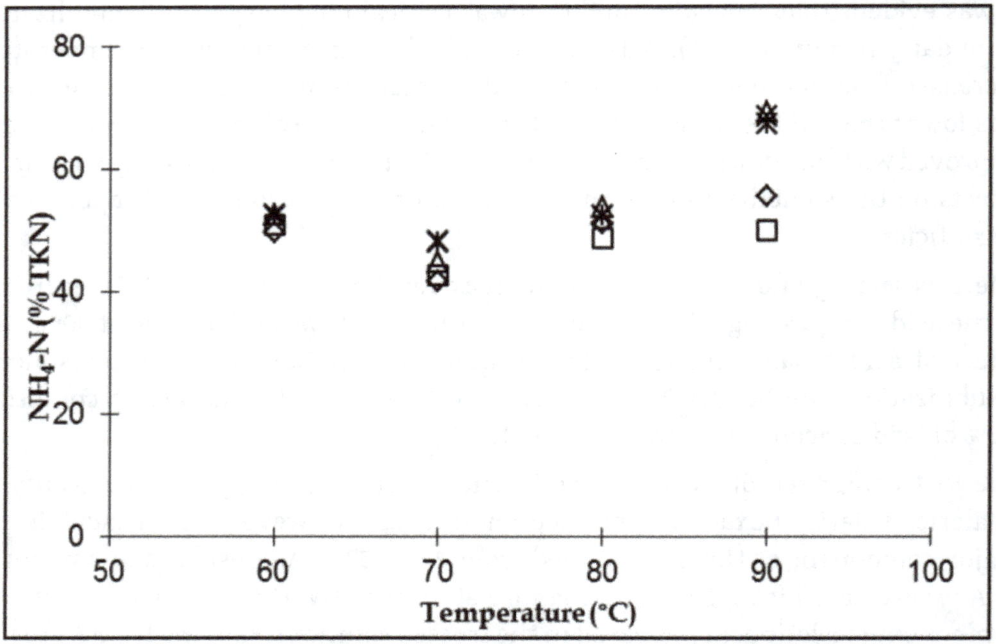

Figure 5. Ammonia release as a per cent of (◇ influent; □ MW only; △ MW/H2O2-AOP with 0.2% acid; ✕ MW/H2O2-AOP with 0.5% acid; ✱ MW/H2O2-AOP with 1.0% acid)

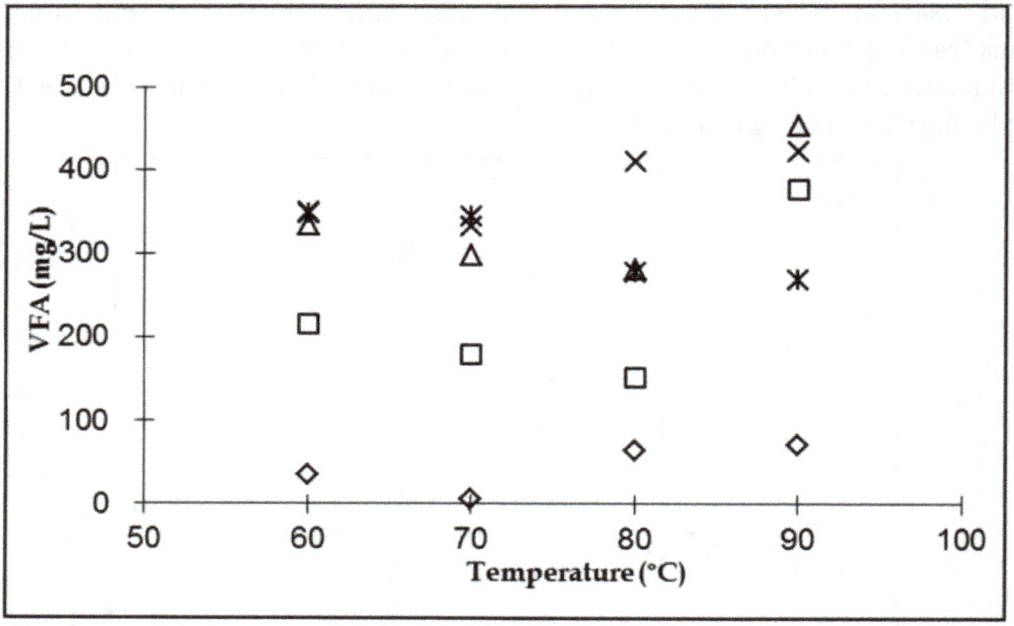

Figure 6. Productions of volatile fatty acids(◇ influent; □ MW only; △ MW/H2O2-AOP with 0.2% acid; ✕ MW/H2O2-AOP with 0.5% acid; ✱ MW/H2O2-AOP with 1.0% acid)

For the MW only treatment, the reducing sugar concentrations remained similar to those in the influents; however, reducing sugar concentrations decreased after treatment with the MW/H_2O_2-AOP. For the MW/H_2O_2-AOP, the reducing sugar concentrations increased gradually with an increase of microwave temperature. With an increase in acid concentrations for every treatment temperature, reducing sugar concentrations were further decreased (Figure 7). The reducing sugars produced from dairy manure are from the hydrolysis of hemicellulose and cellulose, the resulting solution most likely contains a mixture of pentose and hexose. Indeed, reducing sugars of arabinose, xylose, galactose and glucose have been identified from the hydrolysis of dairy manure using an ion chromatograph (Wen, et al., 2004).

The decrease in reducing sugars could be explained as the further oxidation of soluble reducing sugars into VFA and CO_2 during the MW/H_2O_2-AOP. The oxidation process is more favourable at a lower pH, in other words, a higher acid dosage. Very low concentrations (less than 90 mg/L of reducing sugars) were obtained from different treatments at low microwave temperatures.

This was due to the manure being sieved and diluted; therefore, the lignocellulosic materials, a precursor to sugar production, were low in the manure used in this study. The other reason was that temperatures and acid concentration applied were much lower than that of optimal conditions. As indicated in conventional acid hydrolysis of lignocellulosic materials in dairy manure for the production of simple sugars, samples required pre-treatment with 75% acid concentration and 30 min of reaction time for the de-crystallization of fibre, and then were further hydrolyzed in 12.5% of acid at 135º C for 10 minutes (Liao, et al., 2006). For the MW/H_2O_2-AOP treatment of dairy manure, a higher reducing sugar production was obtained at above 120º C (Kenge, et al., 2010). The results show that the important factors affecting reducing sugar yield were temperature and acid concentration. The presence of small amounts of hydrogen peroxide during the treatment process would decrease the degrees of polymerization and cellulose crystallinity, at the same time, igt would increase the accessible surface area and pore size of lignocellulosics. However, too much of hydrogen peroxide would decrease reducing sugar yield. It would favor the further oxidative reaction to form VFA or CO_2 (Kenge, et al., 2010).

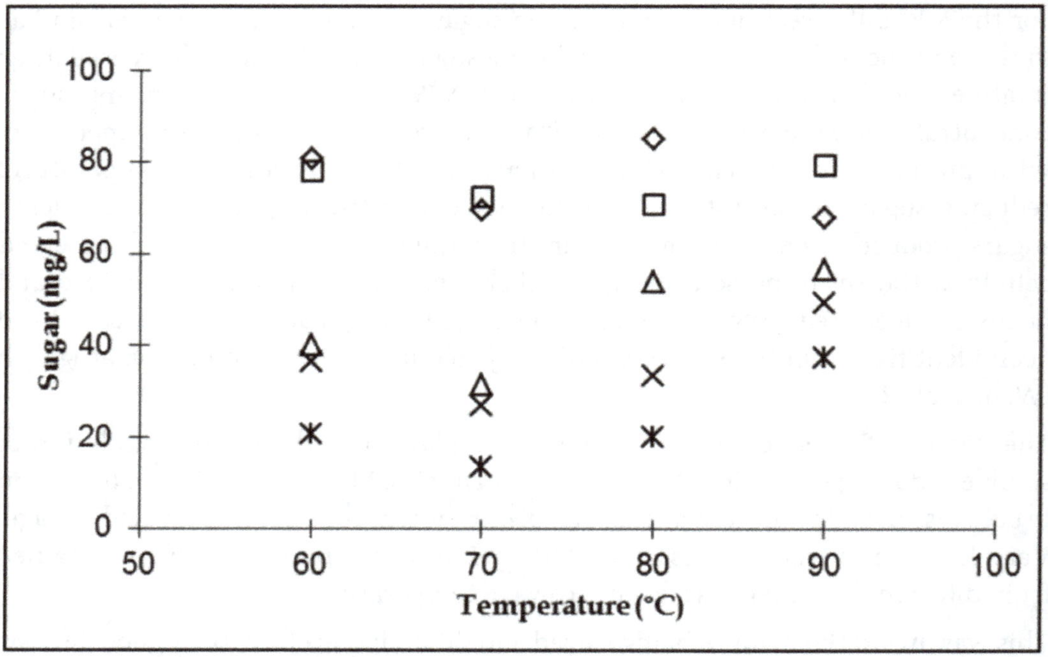

Figure 7. Production of reducing sugars (◇ influent; □ MW only; △ MW/H2O2-AOP with 0.2% acid; ╳ MW/H2O2-AOP with 0.5% acid; ⋇ MW/H2O2-AOP with 1.0% acid)

3.2.1. Comparison of batch and continuous operations

Four experimental sets using a batch microwave unit were also conducted for the purpose of comparison to the continuous process. Three replicates were used at two testing temperatures of 60 and 90º C. The rate of temperature increase was set at 5º C/minute to attain the target temperature, at which time, the reaction was terminated. Two sets of experiments at each temperature were operated: one with both acid (0.2% v/v) and H_2O_2 (0.1% v/v), the other without.

The results of SCOD and ortho-P for both batch and continuous systems are presented in Table 4. Without the additions of H_2O_2 and acid, a simple decomposition process occurred under thermal treatment. As a result, there was little solids disintegration and little release for both nutrients. For the MW/H_2O_2-AOP, it was the thermal oxidation process that involved not only the breakdown of large particulate organic matters, but also, the further oxidation or gasification of some of the resulting organic products. Both SCOD and ortho-P concentrations for the MW/H_2O_2-AOP were higher than those of the MW treatment only (Table 4). A higher microwave temperature also increased phosphorus solubilization and solids disintegration. Higher ortho-P and SCOD concentrations were obtained from the continuous mode operation than from a batch system. The superior performance of a continuous system could be attributed to the synergistic effects of MW irradiation and H_2O_2. Since H_2O_2 **was injected into the influent stream immediately prior to MW irradiation in a continuous mode unit, the synergistic effects between H_2O_2 and MW irradiation became manifested.**

Table 4. Comparison of batch and continuous operation

Temperature (°C)	H_2O_2 and H_2SO_4	Ortho-P Increase (%)		SCOD Increase (%)	
		Batch	Continuous	Batch	Continuous
60	No	-3.2	5.3	-0.6	1.8
	Yes	26	57	12	37
90	No	-7.1	-14	0.1	-6.6
	Yes	28	71	12	34

This study also demonstrates the feasibility to produce reducing sugar from dairy manure for the subsequent processes, such as ethanol production, and useful fermentation products. To obtain a large quantity of reducing sugars from dairy manure, the MW process should be operated at a small amount of H_2O_2 **or without** H_2O_2. Nutrient is also released in the MW/H_2O_2-AOP, which can also be used for energy production, or struvite as a fertilizer.

Nonetheless, the process has yet to be fully tuned; it was anticipated that continuous mode operating under pressured conditions could prove to be more successful. Further experiments should be undertaken to encompass a broader spectrum of treatment temperatures. A pilot-scale continuous mode MW/H2O2-AOP system is being implemented to further assess the technology's viability for solids reduction and nutrient solubilization. An assessment of economic feasibility will also be performed.

3.3. Economic implication

Applying the MW/H_2O_2-AOP to treat dairy manure, there will be significant increases of VFA and nutrients (N & P), and a decrease of solids content in the liquid filtrate. This will be very beneficial in that 1) with reduced solids content in the dairy manure, the more advanced anaerobic reactor, such as the fixe-film reactor, can be adopted for anaerobic digestion of the treated filtrate at much shorter HRTs, resulting in more methane production; and 2) with more VFA and other readily biodegradable substances in the treated dairy manure filtrate, the methane conversion efficiency will be much greater; the methane production rate from the manure filtrate can be as much as five to six times of that of the untreated manure filtrate (Lo, et al., 1985). The soluble phosphorus in the filtrate can be increased substantially by using the MW/H_2O_2-AOP. As a result, struvite production would also increase substantially. In addition, the MW/H_2O_2-AOP can destroy fecal coliforms in wastewater, making it possible for the final effluent available for barn cleaning and other purposes (Yu, et al., 2010). This not only reduces the water usage on farm, but also helps by reducing the pollution of fresh water resources.

4. CONCISE SUMMARY

The MW/H_2O_2-AOP could be an effective pre-treatment method to release nutrient, and disintegrate solids from dairy manure. It is recommended that the process shall be operated at high microwave temperatures (120-160º C), a low hydrogen peroxide dosage (0.07-0.14 g H_2O_2/g TS) and a long treatment period (15-20 min).

Microwave temperature and acid concentration are factors affecting reducing sugars production. Hydrogen peroxide also plays a role in a less extent. Higher microwave temperatures favor the production of reducing sugars. The process could be operated with or without H_2O_2 for reducing sugar production; the substantial amounts of reducing sugars could be obtained at a higher temperature and a longer heating period without H_2O_2.

Temperature was the dominant factor for solids disintegration., At the same operating temperature, the process could be operated either at a longer heating period without addition of H_2O_2, or at a shorter heating period with an addition of H_2O_2 to yield the similar SCOD concentration. Heating time also affected orthophosphate release. Heating time and temperature were significant factors for ammonia release and VFA.

The MW/H_2O_2-AOP can be operated in a continuous operation, with equal or better results than a batch process. When the MW/H_2O_2-AOP is operated in a continuous mode, it maximizes the synergistic effects between H_2O_2 and MW irradiation, thereby promoting more nutrient solubilization and solid disintegration.

5. REFERENCES

APHA. 1998 Standard Methods for the Examination of Water and Wastewater, 20th Ed. Am. Public Health Association, 0-87553-223-3 D.C., USA.

Barnett G. 1994 "Phosphorus forms in animal manure". Bioresource Technology, 4 139 147 , 0960-8524

Chan W. Wong W. Liao P. Lo K. 2007 Sewage sludge nutrient solubilization using a single-stage microwave treatment. Journal of Environ. Science & Health A, 42 59 63 , 1093-4529

Eskicioglu C. Terzian N. Kennedy K. Droste R. 2007 Athermal microwave effects for enhancing digestibility of waste activated sludge. Water Research, 41, 11, 2457 2466 , 0043-1354

Eskicioglu C. Prorotb A. Marinc J. Drostec R. Kennedy K. 2008 Synergetic pretreatment of sewage sludge by microwave irradiation in presence of H_2O_2 for enhanced anaerobic digestion. Water Research, 42, 18, 4674 4682.

Hong S. Park J. Teeradej N. Lee Y. Cho T. Park C. 2006 Pretreatment of sludge with microwaves for pathogen destruction and improved anaerobic digestion performance. Water Environmental Research, 78, 76 81 , 1061-4303

Hsieh C. Lo S. Chiuch P. Kuan W. Chen C. 2007 Microwave-enhanced stabilization of heavy metal sludge. Journal of Hazardous Materials, B139, 160 166 , 0043-1354

Jin Y. Hu Z. Wen Z. 2009 Enhancing anaerobic digestibility and phosphorus recovery of dairy manure through microwave-based thermochemical pretreatment. Water Research, 43 3493 3502 , 0043-1354

Jones D. Lelyveld T. Mavrofidis S. Kingman S. Miles N. 2002 Resource, Conservation and recycling, 34, 75 90 , 0921-3449

Kenge A. Liao P. Lo K. 2009 Solubilization of municipal anaerobic sludge using microwave-enhanced advanced oxidation process. Journal of Environ Science & Health A, 44, 502 506, 1093-4529

Kenge A. Liao P. Lo K. 2010 Nutrient release from solid dairy manure using the microwave advanced oxidation process. Biomass & Bioenergy, 0961-9534 (Submitted)

Kuroda A. Takiguchi N. Gotanda T. Normura K. Kato J. Ikeda T. Ohtake H. 2002 A simple method to release to release polyphosphate from activated sludge for phosphorus reuse and recycling. Biotechnology and Bioengineering, 78, 333 338 , 0006-3592

Liao P. Wong W. Lo K. 2005 Release of phosphorus from sewage sludge using microwave technology, Journal of Environmental Engineering and Science, 4, 77 81 , 0149-6256X

Liao P. Lo K.. Chan W. Wong W. 2007 Sludge Reduction and Volatile Fatty Acid Recovery Using Microwave Advanced Oxidation Process. Journal of Environ. Science & Health A, 42, 633 639 , 1093-4529

Liao W. Liu Y. Liu C. Wen Z. Chen S. 2006 Acid hydrolysis of fibers from dairy manure. Bioresource Technology, 97, 1687 1695 , 0960-8524

Lo K. Liao P. 1985 High-rate anaerobic digestion of screened dairy manure. Journal of Agricultural Engineering Research, 32, 349 358 , 0021-8634

Menendez J. Inguanzo M. Pis J. 2002 Microwave-induced pyrolysis of sewage sludge. Water research, 36, 3261 3264 , 0043-1354.

Qiao W. Wang W. Xun R. Lu W. Yin K. 2008 Sewage sludge hydrothermal treatment by microwave irradiation combined with alkali addition 43, Journal Mater Science, 43, 2431 2436 , 0022-2461

Raunkjer K.. Hvitved-Jacobsen T. Nielsen P. 1994 Measurement of pools of protein, carbohydrate and lipid in domestic wastewater. Water Research, 28, 251 262 , 0043-1354

Rico J. Garcia H. Rico C. Tejero I. 2007 Characterisation of solid and liquid fractions of dairy manure with regard to their component distribution and methane production. Bioresource Technology, 98, 971 979 , 0960-8524

Sall J. Creighton L. Lehman 2005 2005. JMP Start Statistics: A guide to statistics and data analysis using JMP and JMP IN software. 3rd edition. Brooks/Cole-Thomason Learning. 0-534-99749-X Belmont, California.

Shanableh A. Shimizu Y. 2000 Treatment of sewage using hydrothermal oxidation technology application challenge. Water Science and Technology, 41, 85 92 , 0273-1223

Sun Y. Cheng J. 2002 Hydrolysis of lignocellulosic materials for ethanol production: a review. Bioresource Technolog, 83, 1 11 , 0960-8524

Wang L. Nancollas G. 2008 Calcium orthophosphates: crystallization and dissolution. Chemical Reviews, 108, 4628 4669 , 0009-2665.

Wojiechowaska E. 2005 Application of microwaves for ssewage sludge conditioning. Water Research, 39, 4749 4574 , 0043-1354

Wolf A. Kleinman P. Sharpley A. Beegle D. 2005 Development of a water-extractable phosphorus test for manure: An interlaboratory study. Soil Sci. Soc. Am. J., 69, 695 700 , 0361-5995

Wen Z. Liao W. Chen S. 2004 Hydrolysis of animal manure lignocellulosics for reducing sugar production. Bioresource Technology, 91 31 39 . 0960-8524

Wong W. Lo K. Liao P. 2007 Factors affecting nutrient solubilization from sewage sludge in microwave advanced oxidation process. Journal of Environ. Science & Health A, 42 825 829 , 1093-4529

Yu Y. Chan W. Liao P. Lo K. 2010 Disinfection and solubilization of sewage sludge using the microwave enhanced advanced oxidation process. Journal of Hazardous Materials 181, 1143 1147 , 0304-3894

Yu Y. Lo I.. Liao P. Lo K. 2010 Treatment of dairy manure using the microwave enhanced advanced oxidation process under a continuous mode operation. Journal of Environmental Science and Health B 45, 838 843 , 0360-1234

6 Chapter MICROWAVE TREATMENT OF MEAT

1. INTRODUCTION: AN OVERVIEW

The recent expansion of the chilled and frozen food business can be attributed to the swift advancement of reheating appliances, including both domestic and commercial microwave ovens, as well as traditional ovens. The objective of these product producers is to heat the products to a temperature that guarantees the decrease of pathogenic germs to a safe level. It is standard practice to cook chilled items in a microwave oven to a minimum temperature of 70°C for at least 2 minutes, or to an equivalent temperature/time condition, such as 75°C for 30 seconds or 65°C for 10 minutes. A minor alteration in microwave heating duration significantly impacts the viability of Salmonella spp. In infected beef loaves, Salmonella persisted after 3.5 minutes of heating but was absent when the heating duration was extended to 4 minutes (European Chilled Food Federation, 1996; Farber et al., 1998; Levre & Valentini, 1998).

The expanding variety of processed goods is accompanied by microwave warming instructions globally, alongside a rise in microwave usage. Microwave systems have been developed for industrial purposes such as drying, pre-cooking bacon and pork, pasteurizing ready meals, and tempering meat and fish. Nonetheless, there is an emerging use including blanching, baking, and microwave phyto-extraction. Microwave heating was mostly utilized in agriculture for grain drying and pest control. In addition, it was used for food drying, blanching, pasteurization and cooking (Chen et al., 1971; Lin & Li, 1971; Maurer et al., 1971; Bhartia et al., 1973; Nelson, 1973; Jaynes, 1975; Avisse & Varaquaux, 1977; Nykvist & Decareau, 1976; Sobiech, 1980; Shivhare et al., 1994; Tulasidas et al., 1995; Begum & Brewer, 2001; Hong et al., 2001; Sumnu, 2001; Ramesh et al., 2002; Beaudry et al., 2003; Severini et al., 2004; Williams et al., 2004; Zhang et al., 2004).

Frozen materials are usually thawed or tempered before further processing in food industry. Thawing is usually regarded as complete when all the material has reached 0 °C and no free ice is present. This is the minimum temperature at which the meat can be boned or other products cut or separated by hand. Lower temperatures (e.g. -5 to -2 °C) are acceptable for mechanical chopping of product, but such material is `tempered' rather than thawed. The two mentioned processes are not same with together because tempering only constitutes the initial phase of a complete thawing process while thawing is often considered as simply the reversal of the freezing process. However, inherent in thawing is a major problem that it does not occur in freezing operation. Thawing operation has some steps, at first the surface areas of food to rise in temperature and bacterial multiplication can restart. On large objects surface spoilage can occur before the centre regions have fully thawed. It is interesting to know that, there is basic difference between conventional thawing and tempering systems with microwave. Conventional thawing and tempering systems supply heat to the surface and then rely on conduction to transfer that heat into the centre of product but microwave systems use electromagnetic radiation to generate heat within the food. The main detrimental effect of freezing and thawing on meat is the large increasing in amount of proteinaceous fluid (drip) released on final cutting. Several studies showed that fast thawing rates would increase amount of drip. Microwave thawing has some advantages: in this way, the ice formation during freezing breaks up cell structure and fluids are reduced during thawing but microwave thawing has a better quality product than thawing at 4 or 21 °C. It is necessary to more investigate the microwave processing because of large applications of microwave in meat processing (tempering, thawing and cooking) as well as quality control of meat and meat products (Riihonen & Linko, 1990).

2. DEFINITION

High-frequency heating is considered a thermal process which causes oscillation of water molecules, friction, and resultant heat generation. In several studies it was permitted the radiofrequency heating include 13.56, 27.12, and for microwave applications is 40.68 MHz and 433, 915, 2450, and 5800 MHz. Radiation is a mechanism for heating meat by electromagnetic energy. When the electrons in atom move from a higher to a lower energy state, it sends energy as waves. These waves are not in the same energy level and frequencies. Lower energy electro-magnetic radiation (microwave, radio, TV) occurs as very long waves with frequencies ranging from 300MHz to 300 GHz. Unlike gamma and X-rays, non-ionizing MW energy is sufficient to move the atoms of a molecule, but can not change its chemical bounds. Also MW move at the same speed as light waves essentially. Metallic objects reflect them; some dielectric materials absorb them while others transmit them. Water, carbon, and foods which are high in water are good MW absorbers, while thermoplastics, glass and ceramics allow them to pass through and they can not absorb it completely. When the polar molecules such as H_2O place in a magnetic field, they line up with the field. If the field alternates, the polar molecules alternate at the MW frequency to maintain this alignment. As they rotate, they disrupt H-bonds between adjacent water molecules, generating heat. Freedom of the polar molecules plays an important role in the rate of heat generation. Because the movement of water molecules in ice is restricted, it is a poor MW energy absorber. In an electric field kinetic energy accelerates the ions in solution (Na^+, Cl^- and Ca^{++}). They crash with other ions and give up heat. The more ions in solution, the higher the kinetic energy release (Decareau and Peterson, 1986; Hugas et al. 2002; Aymerich et al. 2008).

All microwave ovens have a similar design that includes a magnetron device as a power source and a waveguide to bring radiation to a heating chamber. A radiofrequency oven which is known as RF applicator is equipped with a generator coupled with a pair of electrodes. Microwaves can penetrate to a depth of 5 to 7 cm, which results in faster cooking and fewer nutrient changes compared with conventional ovens. It was showed 2 $\log_{10}$ CFU/g reductions in total microbial flora after microwaving meat balls at 2450 MHz frequency, in an 800 W oven, for 300 seconds. Conveyor zed systems have been applied to thawing of meat, in some cases using water surrounding the material to aid temperature uniformity. Electromagnetic (900-3000 MHz) waves are directed at the product through waveguides without electrodes. Potentially very rapid, the application is limited by thermal instability and penetration depth. Instability results from preferential absorption of energy by warmer sections and by different ingredients, such as fat. Warmer sections may be present at the start of the process; for example, the surface temperature may be warmer than the middle, or they may be produced during the process. In the extreme, such warming can lead to some parts of the food being cooked while others remain frozen. In addition to these problems, the high capital costs of equipments have greatly limited commercial applications. Attempts to avoid runaway heating have involved low-power (and hence longer duration) microwaving, cycling of power on and off to allow equalization periods, and cooling of surfaces with air or liquid nitrogen. Penetration depth depends upon temperature and frequency, being generally much greater at frozen temperatures and greater at lower frequencies (Murano 2003; Yilmaz et al. 2005; James & James, 2010).

As mentioned before, microwaves include waves with wavelengths from 3mm to 3m. In the electromagnetic spectrum, microwaves are located between radio waves at low frequencies and infrared at higher frequencies (Fig. 1). In addition to industrial applications of microwaves in meat tempering, thawing and cooking, microwave was used to detect the fraudulent addition of water in meat products. By using microwave, a nondestructive method was developed for measuring fat in fish and meat, measuring water activity in protein gels and also sensory controlling meat and fish freshness based on the change in dielectric properties (Kent et al.

1993; 2000, 2001& 2002; Kent 1990; Boggaard et al. 2003; Clerjon et al. 2003; Tejada et al. 2007; Clerjon & Damez 2007).

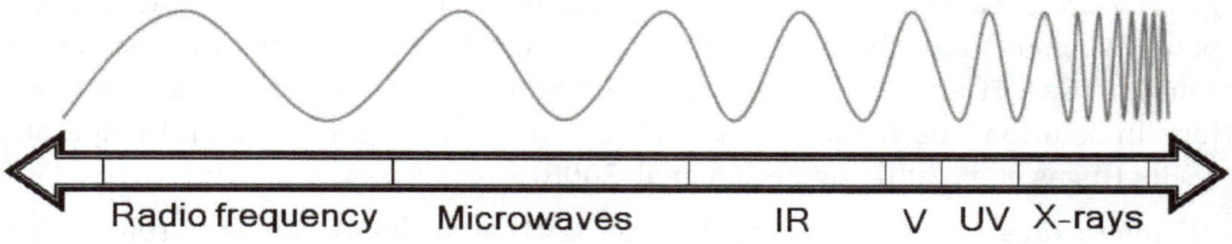

Figure 1. The electromagnetic spectrum.

2.1. Dielectric properties of foods

Microwaves operating at 915 MHz are utilized for industrial heating, whereas those at 2450 MHz are employed in residential microwave ovens globally.Ice, possessing a minimal loss factor, is nearly transparent to microwaves. Oils are esters of long-chain fatty acids that exhibit significantly reduced mobility relative to water molecules when subjected to fluctuating electromagnetic fields. The dielectric constant and loss factor of oil are therefore very small compared to free water. The air voids in some foods reduce the loss factor and increase the penetration depth of microwaves at 915MHz and 2450 MHz. Variation in dielectric properties among high protein products can be as large as among other groups of food at 915 MHz and 2450 MHz. Temperature and salt are important factors on dielectric properties of meat products such as cooked ham and beef. The loss factor of cooked ham is much larger than cooked beef. The penetration depth of microwaves at 915MHz and 2450 MHz in ham is less than 0.5 cm. The dielectric constants and penetration depth of microwaves in these foods decrease, while the loss factors increase, along with increase in temperatures at 915 MHz and 1800 MHz. In general, predicting the dielectric properties of complex food products is difficult and its direct measurement needs to be made over specific composition, temperature and frequency ranges (Wang et al., 2003; Guan et al., 2004).

2.2. Water migration adjustment as temperature controller

Modifying the meat composition in a microwave burger and bun combination can enhance temperature control during microwave heating. It is essential to modify the food using emulsifiers, gums, proteins, and sugars to regulate the water content of the meat and the bun independently, ensuring that, when treated simultaneously in a microwave, the bun does not incinerate while the meat cooks. The incorporation of vegetable fiber results in fiber swelling and fat absorption, which interacts with beef protein to create a complex matrix that inhibits the release of fat and moisture. (Anderson & Berry, 2001; Lyng et al., 2002; Parker & Vollmer, 2004).

2.3. Microwave interactions with food components

Fluctuations in electric fields, food ingredients, and the positioning of food within a microwave oven can result in non-uniform heating, which may hinder optimal interaction among food components and permit the survival of microbes. Various techniques to enhance the uniformity of microwave heating include rotating and oscillating food items, surrounding the product with an absorbing medium (such as water), employing pulsed power, and adjusting frequency and phase. Nevertheless, understanding the dielectric properties of the food is essential for developing effective processes. (Yang & Gunasekaran, 2001; Guan et al., 2004).

3. COMMERCIAL MICROWAVE SYSTEMS

Since the 1970s, a diverse array of batch and continuous commercial microwave systems for thawing and tempering has been developed, and microwave drying and cooking are currently employed in food processing. The devices were utilized to cure meat, fish, butter, and berries. Several batch systems were employed either as a component of a hybrid microwave/conduction system or to enhance extensive conduction systems by rapidly tempering small quantities of critically needed material. Industrial systems may be categorized as batch or continuous, varying in scale from small systems capable of processing a single block (25 kg) of food to extensive continuous tunnels that can handle up to 8 tons per hour. A solitary 25 kg block of frozen meat can be defrosted in 65 seconds. The throughput is highly contingent upon product composition and the final temperature attained post-tempering. The systems are constructed from stainless steel for food applications, ensuring uniform load distribution throughout the oven; a multi-mode cavity applicator generates consistent heat dispersion throughout the oven. The selection of belt material and configuration is determined by the characteristics of the product undergoing heating. Both ends of the conveyor are equipped with specialized vestibules to mitigate any microwave energy leakage into the surrounding environment. To ensure consistent heat distribution across diverse load configurations, each oven section is equipped with a waveguide splitter with dual microwave feed points and mode stirrers (Swain & James, 2005).

4. UTILIZATION OF MICROWAVE HEATING IN MEAT PROCESSING

4.1. Microwave tempering

Tempering of frozen food, can make food easier to slice and reduce drip loss. Frozen meat is heated to just below the freezing point (-2 to -4 °C) can be tempered by MW energy in tempering process then allowed to thaw fully at refrigeration temperature. At temperatures slightly below 0 °C, the outer layer of the meat can absorb significant amounts of energy, resulting in overheating near the surface. Frozen foods thawed in the MW may experience `runaway heating' due to selective heating of the liquid phase (opposed to the crystalline phase of the ice). MW power penetration is greater at 915MHz than at 2450MHz so it is more suitable for heating thick masses of materials. Tempering frozen food has been more successful at 915MHz than at 2450MHz, partly because of surface overheating at the latter frequency. This can be balanced, to some degree, by circulating cold air around the product; however, this may increase drip losses by up to 10%. MW tempering requires less time and space, produces little or no weight loss, increases juice retention, and reduces bacterial growth (Decareau & Peterson, 1986; Rosenberg & Boe, 1987).

There are widespread applications of industrial microwave systems for tempering meat. The blocks processed directly from frozen storage can be acceptably tempered in a batch microwave unit to a mean temperature of approximately -3 °C (range -5 to 0 °C) with no hot spots. In general, microwave tempering of blocks for 8 h in ambient temperatures is unsatisfactory. If it is not possible to produce a uniform power/time combination, surface temperatures, especially at the corners of the meat blocks, rose to unacceptable levels and there was substantial drip loss from thawed surfaces. Trials have to be carried out to determine the correct power and time setting for each type of block in optimal tempering. Blocks sorted into batches of similar type should be processed directly from frozen storage under the predetermined conditions. Successful tempering can be made in some minutes using a microwave system, compared with the 1 to 14 days required in industrial air tempering systems. Continuous conveyorized microwave tempering systems using either a single 60 kW magnetron or two 40 kW magnetrons can temper 2 to 2.5 tones per hour depending upon fat content. In large-throughput operations the continuous microwave tempering plant provides considerable flexibility, in that changes in raw material requirements, for the post-tempering processes, can be accommodated in minutes. Using air tempering systems, at least one day and up to eight days are required to accommodate equivalent changes. Microwave systems bring higher product yield due to reductions in evaporative and drip loss during tempering. Since the majority of conventional plants temper material in a wrapped form, evaporative losses are insignificant, while substantial periods at air temperatures above 0 °C would be required before thawing of surface tissues occurred and drip became apparent (Swain & James, 2005).

4.2. Microwave thawing

Microwave thawing employs electromagnetic waves transmitted to the product via waveguides, eliminating the need for electrodes. Although microwave energy is a rapid way for thawing frozen food, its use is constrained by thermal instability. During microwave heating, certain portions of the meal may be cooked while others stay frozen. The absorption of electromagnetic radiation by frozen food can increase the temperature, especially at -5 °C. During irradiation, if a region of the material is slightly hotter than its surroundings, proportionately more energy will be absorbed. Enthalpy increasing cause the absorption to increases and the unevenness of heating worsens at an ever-increasing rate. If irradiation is continued after reaching the hot spot to its initial freezing point, the temperature rises at appalling rate. Lowering the power density to allow thermal conduction to even out the enthalpy distribution through the food can reduce such runaway heating. Since the main instability tends to occur at the surface, attempts have been made to cool the surface during thawing using air or liquid nitrogen. A hybrid microwave/vacuum thawing system was developed in the 1980s in which boiling of surface water at low temperature was used to cool the surface. The system consisted of a cylindrical vacuum chamber approximately 1 meter in diameter and 1 meter long. The chamber could be evacuated to absolute pressures as low as 10 mbar by a water ring pump in series with a rotary pump. Microwaves at a frequency of 915 MHz were introduced into the chamber via two waveguides positioned near the top at the front and rear of the plant. The microwaves were produced by a 25kW generator, though power had to be limited to 2.5kW to avoid arcing problems in the chamber. A large circular twisted disc was rotated within the chamber during thawing to produce a more even microwave field. After thawing, the average meat temperature was 9.4 °C (ranging from -1 to 23.3 °C between blocks) and the maximum surface temperature measured on any block was 28.4 °C. Weight losses averaged 7.6%, which appeared large, but weight loss in commercial systems was stated to range from 2 to 10%. The overall energy efficiency of the plant was 49% with 24% of the energy being absorbed by the structure of the chamber. The combination of microwave energy and cold air with different ambient temperatures can reduce thawing time and avoid runaway heating during microwave-assisted thawing. The microwave power was cycled on and off using hot and cold points control schemes to a predetermined temperature gradient. When appropriate conditions were used, thawing time was reduced by a factor of seven compared to convective thawing at ambient temperature. The rate of sublimation with low microwave power at the surface resulted low weight losses during the process. The temperatures claimed at the end of thawing processes lasting between 24.5 and 34.25 minutes ranged from -1.1 to -2.0 °C. The shielded region was in fluid communication with the cavity so that thawed material may flow from the cavity into the shielded region. Even if microwave-thawing systems have not been commercially successful but microwave-tempering systems have found successful applications in the meat and dairy industry (Bialod et al., 1978; Virtanen et al. 1997; James, 1984; Yagi & Shibata, 2002; Swain & James, 2005).

4.2.1. Vacuum-heat thawing (VHT)

A vacuum-heat thawing system operates by transferring the heat of condensing steam under vacuum to the frozen product. Theoretically, a condensing vapor in the presence of a minimum amount of a non-condensable gas can achieve a surface film heat transfer coefficient far higher than that achieved in water thawing. The principle of operation is that when steam is generated under vacuum, the vapor temperature will correspond to its equivalent vapor pressure. For example, if the vapor pressure is maintained at 1106 Nm^{-2}, steam will be generated at 15 °C. The steam will condense onto any cooler surface such as a frozen product. The benefits of latent heat transfer cooking without any problems which would occur at atmospheric pressure. Thawing cycles are very rapid with thin materials, enabling high daily throughputs to be achieved. The advantage of a high h-value is increasing less marked as material thickness, and beef quarters or 25 kg meat blocks require thawing times permitting no more than one cycle per day. The cost of largest capacity units (10-12 tones) can restrict its application. The frozen product is continuously tumbled with vacuum tumble thawing systems while steam under vacuum condenses on the exposed surfaces of the food. Very fast thawing is claimed to be obtained with small bulk-frozen individual products. However, delicate products such as fish fillets cannot be thawed in such systems (Swain & James, 2005).

4.2.2. Radio frequency thawing

Heat is generated in frozen food during radio frequency thawing due to dielectric losses when the product is exposed to an alternating electric field. A uniform regular slab is positioned between parallel electrodes in an idealized scenario of radio frequency heating without any heat exchange. When an alternating electromotive force (emf) is provided during the thawing and tempering of electrodes via microwave processing, the resulting field within the slab is uniform, ensuring that the energy and associated temperature increase are consistent throughout all sections of the food. Food items typically do not conform to the geometry of perfect parallelepipeds; for instance, frozen meat comprises at least two components: fat and lean tissue. While loading, frozen meats absorb heat from their environment, causing an increase in surface temperature, which results in the dielectric system lacking the homogeneous temperature distribution necessary for consistent heating. By using a conveyorized system to keep the product moving past the electrodes and/or surrounding the material by water, commercial systems have been produced for blocks of oily fish and white fish. Successful thawing of 13 cm thick meat, and 14 cm thick offal, blocks has also been reported but the temperature range at the end of thawing (44 minutes) was stated to be -2 °C to 19 °C and -2 °C to 4 °C respectively, and the product may not have been fully thawed.

To overcome runaway heating with slabs of frozen pork bellies, workers have tried coating the electrodes with lard, placing the bellies in oil, water and saline baths and wrapping the meat in cheesecloth soaked in saline solution. Only the last treatment was successful but even that was not deemed practical. An industrial system was installed to thaw frozen blocks of boned-out poultry in the late 1980s. The continuous plant had a throughput of 450 kg per hour with four 12.5kW generators operating at a frequency of 27.12 MHz providing the heating. It was claimed that the process reduced thawing time from 3 to 4 days to less than 2 hours and cut weight loss from 7 to 0.5% (Satchell and Doty, 1951; Jason & Sanders, 1962; Sanders, 1966; Swain & James, 2005).

4.3. Microwave or radio frequency cooking

Microwave and radio frequency cooking are recent innovations in the meat sector. Microwave-cooked roasts achieved endpoint temperatures more rapidly than those prepared using conventional methods. Meat prepared in a microwave lacks the characteristic browned surface associated with other cooking methods due to the brief cooking duration. Radio frequency, as a volumetric heating method, is an expedient cooking alternative that generates heat within the product, hence decreasing cooking times and maybe facilitating more uniform heating. Microwave cooking preserves greater quantities of vitamins, including retinol, thiamin, and riboflavin, in comparison to meat cooked in an earth oven. The difference in vitamin retention could be due to a higher cooking temperature of earth-oven cooking compared with microwave cooking (Welke et al., 1986; Kumar & Aalbersberg, 2006; Zhang et al. 2006).

In cooking process, the contractile proteins of meat (myosin and actin) will denature and the connective tissue (collagen) under goes to solubilize. MW heating can solubilize more collagen (percentage of hydroxyproline) than does boiling. In general, MW-cooked meat and poultry have higher cooking losses than those cooked by conventional methods (Moody et al., 1978; Riffero & Holmes, 1983; Yarmand & Homayouni, 2009); however, species (beef, pork, and poultry) and cut have important role in losses and palatability. Cooking losses, evaporation and drip losses were greater for steaks cooked in MW convection ovens compared with those cooked in forced air convection or conventional ovens, though they found no difference in juiciness, tenderness, beef flavor, or external color of the steaks. Evaporative losses were higher for oven dry-roasted beef than for MW- convection roasts, which were higher than MW roasted samples. Drip losses were highest for MW, followed by oven roasted, then by MW-convection roasted samples. Although Instron shear values did not differ, oven dry-roasted samples were judged as tender by taste panelists. Lean meat cook yield was lower for MW-cooked rib eye roasts while it was the same for round and chuck roasts, whether cooked conventionally or in the MW. Tenderness, softness, natural flavor and tenderness (shear force) were unaffected. MW-cooked rib eye roasts were browner and less juicy. Visual color and tenderness improved for meat products cooked in the MW oven. There were no differences in drip loss of MW-cooked chops, though evaporative and total losses were lower than for broiling. Flavor scores were highest for broiled chops and did not differ due to MW power level. Chops cooked at low MW power were juiciest and tenderness varied inversely with cooking rate. Overall acceptability was lowest for chops cooked at high MW power level. Textural analysis showed greater peak forces for hot air oven-cooked products compared to MW-cooked products. MW cooking time was shorter and produced more uniform heating, though flavor was better in hot air oven-cooked products. Cooking time of chicken breasts increase with decreasing power level, but cooking losses are not affected. Both sensory and instrumental tenderness (Instron compression) were best at 60% power level, while juiciness, mealiness and flavor were unaffected by power level. It was reported that convection-MW-cooked chicken was more tender, juicy and acceptable than MW-cooked chicken even if its flavor intensity was similar (Hines et al., 1980; Fulton & Davis, 1983; Payton & Baldwin, 1985; Hammernick-Oltrogge & Prusa, 1987; Howat et al., 1987; Barbeau & Schnepf, 1989; Yarmand & Homayouni, 2009).

Thiamin retention ranged from 77% in conventionally cooked chicken breasts to 98% in MW-cooked chicken legs. Exposure of pre-rigor broiler muscle to MW energy can decrease glycogen metabolism but it has no effect on ATP retention and is not improve tenderness. When meat is cooked in the MW, addition of fluids and other ingredients, and cooking from the frozen state, appear to improve the overall outcome. Curing of buffalo meat using a polyphosphate solution may increase pH from 5.70 to 6.12 and reduce cooking loss from 17.5 to 10.3%. When cooking started from the frozen state, roasts cooked by MW on low power were comparable to those conventionally cooked in sensory quality. Roasts cooked from the frozen state by MW on high power had lower palatability scores (except flavor) and higher shear values. While MW cooking resulted in lower fat contents, and higher water-holding capacity, TBA value and cook yield than conventional oven cooking. Visually, MW-heated beef may appear to be unevenly cooked. MW-thawed and cooked roasts and steaks may be redder in the interior, and lighter and more yellow on the exterior, than their broiled counterparts. The irregular shape of many whole meat cuts results in significantly non-uniform heating. In addition, bone reflects MW, causing overheating in the regions adjacent to the bone (Moody et al., 1978; Dunn & Heath, 1979; Drew et al., 1980; Riffero & Holmes, 1983; Mendiratta et al., 1998; Hoda et al., 2002).

The temperatures experienced by a MW-heated beef slab can vary by as much as 50 °C. This variation is a function of slab thickness (between 1 and 3 cm) and time, or a function of time alone for slabs >4 cm thick. Thermal insulation at the faces of the slab increases this variation. MW is used primarily for reheating precooked meat products, making cook loss even more problematic. Because fluid losses appear to be the primary roadblock in producing precooked, reheatable meat products, adding fluid (enhancement) has been evaluated in an effort to produce a juicy product. Pumping fresh meat to 10% over original weight to produce 1% salt and 0.3% phosphate in the final product can produce an acceptably palatable, value-added, precooked product (boneless pork roast) designed for MW reheating. However, flavor may suffer during MW reheating. Reheating of precooked frozen turkey in MW and conventional ovens tends to create stale, aldehyde-like aroma. Meaty-brothy flavor and aroma were more intense in MW-reheated meat. MW-reheated light meat was flat or bland flavored, less juicy and higher in moisture content than conventionally reheated meat. MW cooking had no effect on chicken flavor or aroma, though MW reheating resulted in less warmed-over flavor than conventional reheating. TBA analyses showed that multiple reheating of dark turkey meat using MW energy retarded lipid oxidation, including that caused by NaCl (2.0%) which was a pro-oxidant compared to KCl (2.0%). MW reheating did not influence warmed-over aroma or flavor or TBA values of cooked stored roast beef. Development of off-flavors may be somewhat species-specific (Cipra & Bowers, 1971; Hsieh & Baldwin, 1984; Steiner et al., 1985; Boles & Parrish, 1990; King & Bosch, 1990).

Cooking fresh meat in a microwave can represent a challenge since the cool air surrounding the meat results in lack of browning. While the meat heats up inside, there is mass transfer from the interior to the outside, resulting in a tough, dry and flavourless product. Ingredient selection can target these problems, such as a salt-based coating to attract microwave energy to the surface of the meat along with a colouring agent, a water binding ingredient (such as starch) to reduce moisture loss and an enzyme to retain tenderness. Flavours can be added for overall improvements. In general it is recommended to keep the salt concentration as low as possible to improve the penetration depth and slow the heating at the surface while providing more uniform heating and avoiding thermal runaway. On the other hand, the addition of salt may be recommended in some cases as it can increase the heating rate at the surface for particular applications (Schiffmann, 1986; Taki, 1991; Schiffmann, 1997).

5. MICROSTRUCTURE OF MEATS TREATED BY MICROWAVE TECHNOLOGY

Quality control methods and microstructural evaluation were employed to study the qualitative parameters of meat and meat products. Among these methods, the environmental scanning electron microscopy (ESEM) has its own advantages for evaluation of microstructural changes in meat especially for comparing the effect of various heat treatments (Arvanitoyannis & van Houwelingen-Koukaliaroglou, 2003; Papadima, et al., 1999; Tzouros & Arvanitoyannis, 2001; Yarmand & Homayouni, 2009; 2010).

5.1. Optical microscopy

Microscopical techniques are used to characterize meat structure. Optical microscopy achieved the observation of fat globules distribution and protein gel in emulsion-type buffalo meat sausages. It revealed that caseinate and modified whey form distinct dairy protein gel regions within meat batters, and this could explain their ability to enhance the textural properties of the meat batters compared with the other dairy proteins (Krishnan and Sharma 1990; Barbut, 2006).

5.2. SEM and ESEM

Scanning electron microscopy (SEM) and environmental scanning electron microscopy (ESEM) have been used for the examination of living and fresh botanical samples including fungal mycelium and cross sections of stems from different plant sources. ESEM has been known as one of the most interesting new developments in the field of electron microscopy. A number of studies have demonstrated the use of ESEM for hydrated biological samples. Some research compared unprocessed ESEM specimens and samples prepared by conventional methods. Much research has been done on the study of wool fiber. Processes for industrial wool commence with sheep breeding and conclude with the study of finished fabric. Investigations in many of these processes can be largely enhanced using ESEM microscopes and techniques. Techniques of ESEM have been demonstrated by Baumgarten (1990) and Peters (1990). In this study ESEM has been applied to study the microstructure of goat semimembranosus muscle, using raw muscle as a control and comparing it to different heat treatments including conventional and microwave heating. Scanning electron microscopy was useful to show structure differences in low-fat sausages. The environmental scanning electron microscopy (ESEM) is capable of examine specimens in a gaseous environmental saturated with water vapor and higher resolution micrographs have been illustrated at the presence of gas. The environmental scanning microscope has been described as a scanning electron microscopy technique to retain a minimum water vapor pressure of at least 609 pa in the chamber specimen. ESEM creates the possibility of testing practically any sample which is wet (Fig. 2). The difference of ESEM from conventional SEM is its capacity to examine materials consisting of liquids and oils in their natural situation without any initial preparation for the samples (Danilatos, 1981; Danilatos & Postle, 1982; Danilatos, 1989; Danilatos, 1991; Klose, et al., 1992; O'Brien, et al., 1992; Wallace, et al., 1992; Uwins, 1994; Morin et al. 2004; Cáceres et al. 2008).

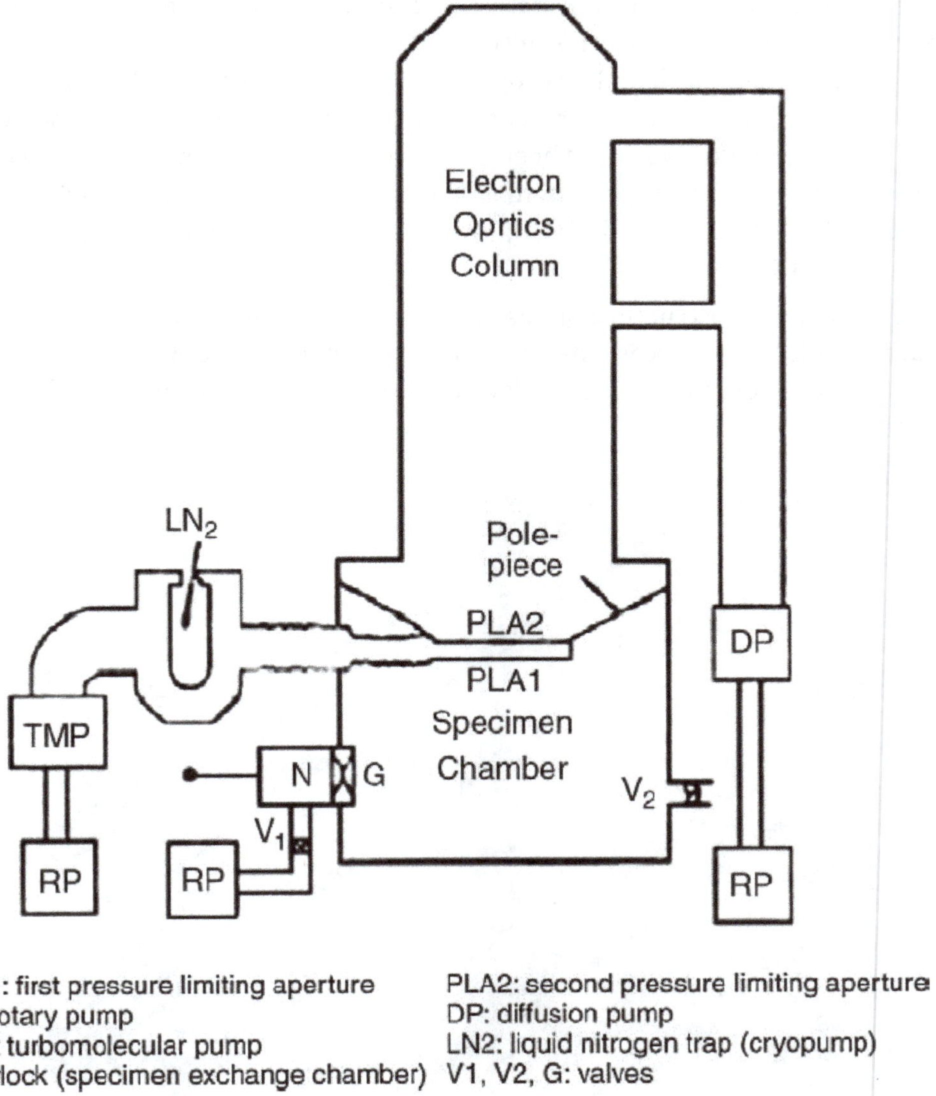

Figure 2. Diagrammatic representation of a two-stage differential pumping system for an ESEM. Source: (Danilatos, 1991).

The two main parts of the instrument include an electron gun chamber or electron opticsolumn and specimen chamber. The gun chamber is located at the top part of instrument and provides a flow of electrons by heating a tungsten filament, lanthanum hexaboride filament or field emission source. The specimen chamber is capable of working at a very poor vacuum unlike any other types of ESEM which require high vacuum. Therefore this specimen will not dry and ESEM is used to observe specimens in the fresh state. Some important advantages have been shown for the gas around the samples in the specimen chamber. Accumulation of charge on insulating samples can be recognized as a basic advantage for this technique. This phenomenon occurs by ionization of the gases inside the specimen chamber and nowadays much work has been done on it. The gas itself can be employed as a detector in the microscope system which is another advantage for ESEM (Parsons et al., 1974; Moncrieff et al., 1978; Crawford, 1979; DaniJatos, 1983, 1982, 1990; Uwins, 1994).

Study of the microstructure of goat SM muscle was carried out by Environmental Scanning Electron Microscopy (Yarmand & Homayouni, 2009; 2010). The result shows bundles of muscle fiber in raw SM muscle that is parallel to each other (Fig. 3 and Fig. 4).

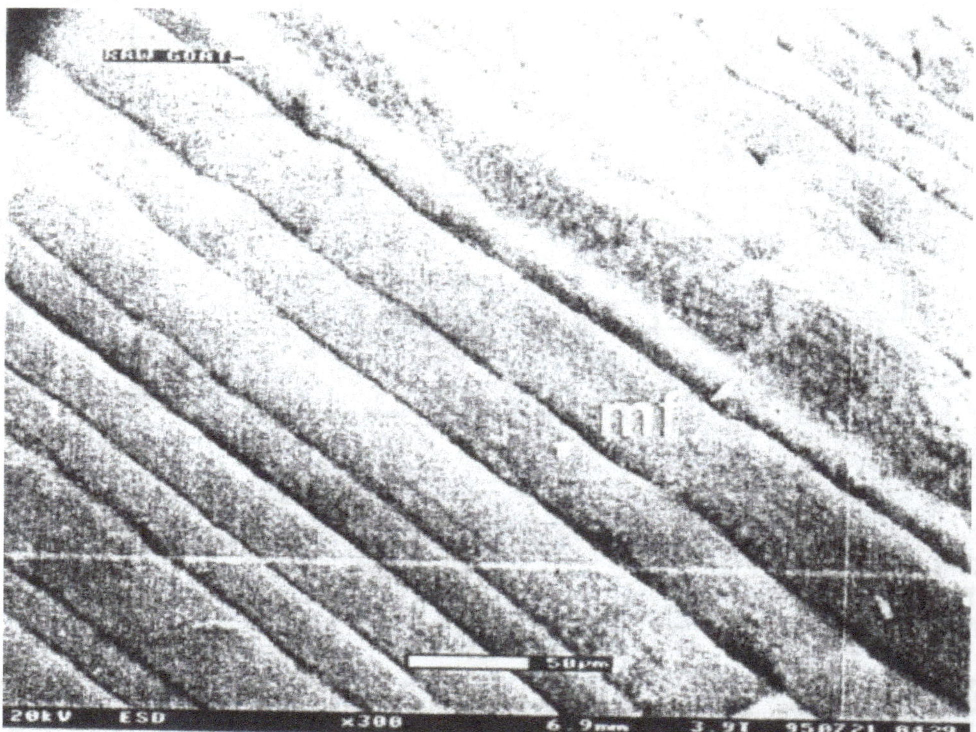

Figure 3. Illustration of structure of raw goat SM muscle by ESEM (mf: muscle fiber).

The collagen fibers surrounding the muscle fiber are not clear in some Figures, but in Fig. 5 the collagen fibers appear clearer and the myofibril surfaces seem normal. In order to observe collagen more clearly than before, Miller stain was used. This technique increased contrast by increasing the conductivity of the material (Miller, 1971).

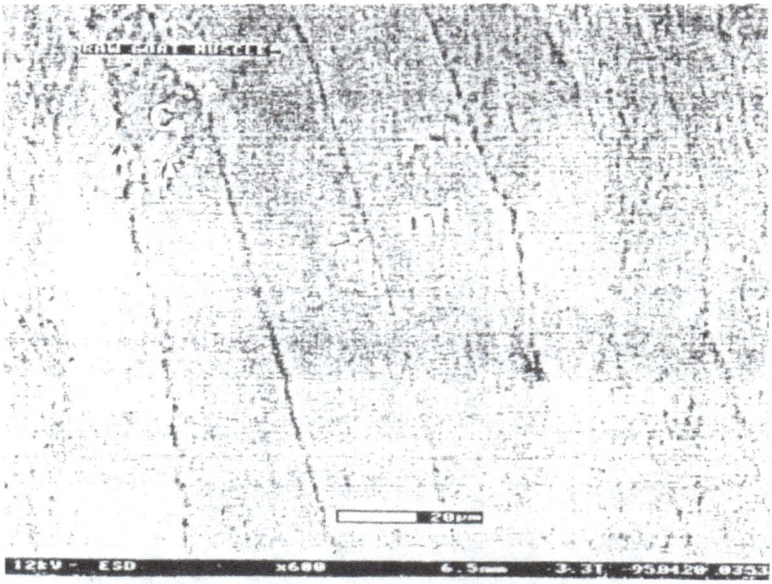

Figure 4. Another illustration of structure of raw goat SM muscle by ESEM (c: collagen, mf: muscle fiber).

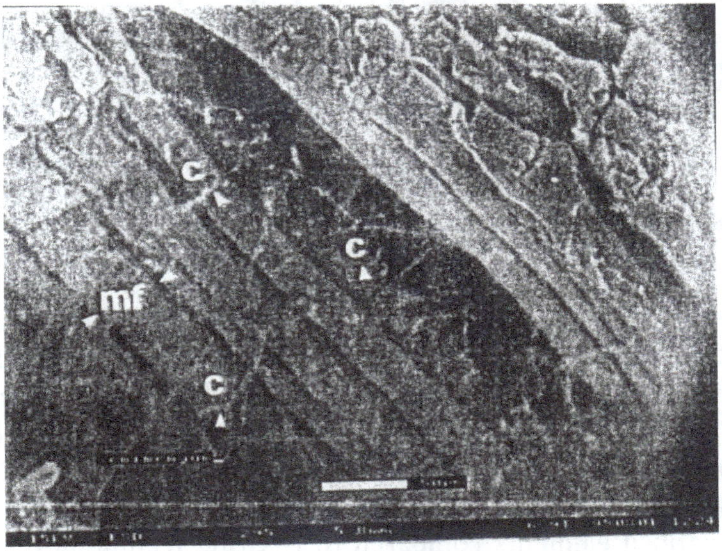

Figure 5. Structure of SM muscle using Miller stain. Collagen network appears bright (c: collagen, mf: muscle fiber).

As shown in Fig. 6 collagenous fibers are clear and appeared as a white network with covering bundles of SM muscle fiber. Cross section of SM muscle has been shown in Fig. 7 In application of ESEM, it is important to consider possibility of dehydration of the fresh tissues which might occur in the sample chamber. There is also the possibility which might occur from electron beam on the sample if the operating system is not clearly set and adjusted prior to imaging. To prevent dehydration and localized beam damage in correctly specimen, it can be maintained in semi-wet condition by monitoring the system at a lower voltage (2-5 kv) or at lower temperature by applying a cooling step (-20 to 30 °C) and at higher gas pressure up to 20 torr. (Uwins, 1994).

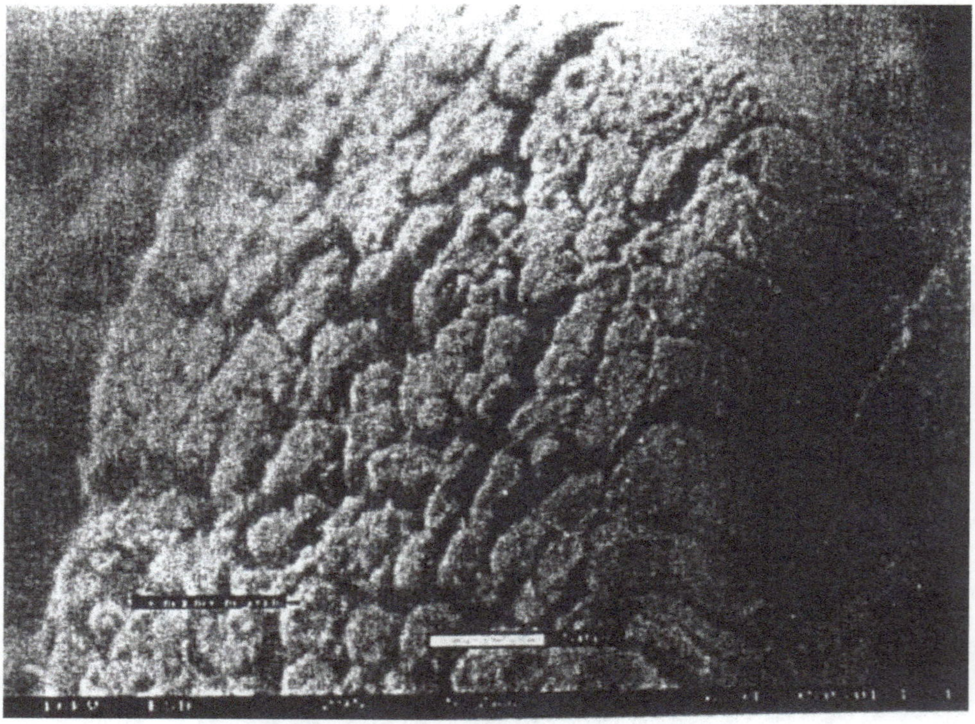

Figure 6. Cross section of SM muscle.

In the study of the microstructure of SM muscle, more damage was observed when domestic microwave heating was used. This resulted in more shrinkage and breakdown in the SM muscle. The myofibril surface did not seem to be normal but showed little evidence of tissue damage. The cross section of SM (Fig. 8) illustrated the damage at the surface of specimen, further application of the ESEM illustrated that domestic microwave heating causes more physical damage to connective tissue and myofibril elements compared to roasting technique (Fig. 9). Results showed that domestic microwave heating caused little hydrolysis in connective tissue.

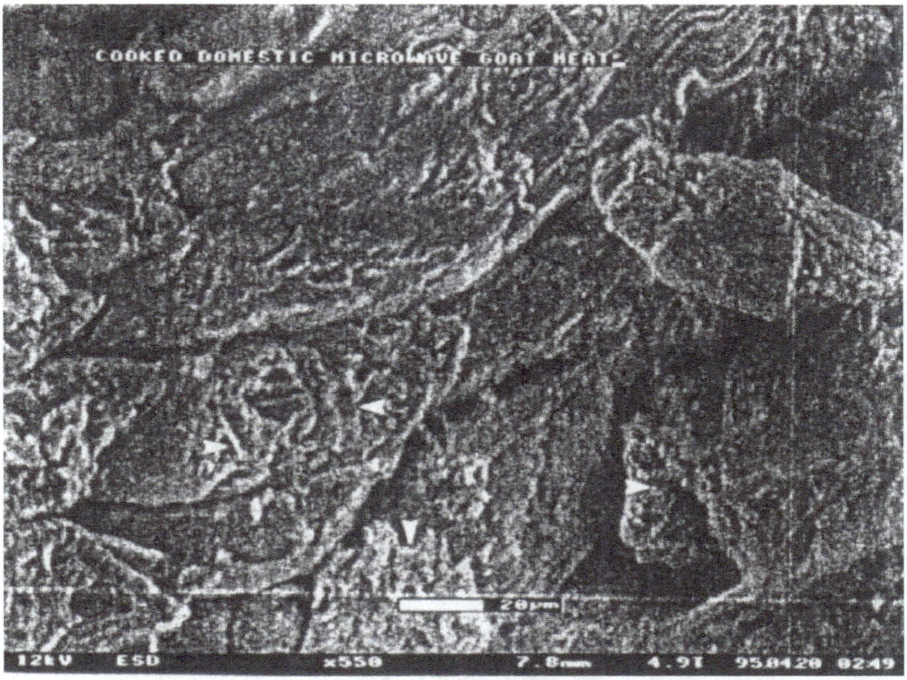

Figure 7. Environmental Scanning Electron Microscopy of domestic microwave heated SM muscle. Surface damages were shown by arrows in cross section view.

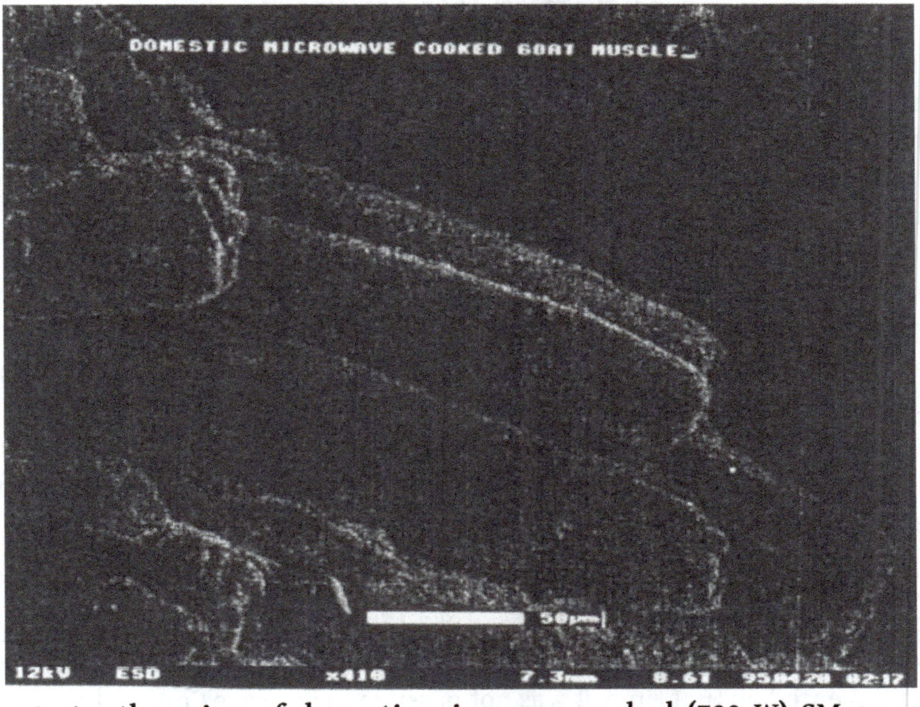

Figure 8. Another view of domestic microwave cooked (700 W) SM muscle by ESEM.

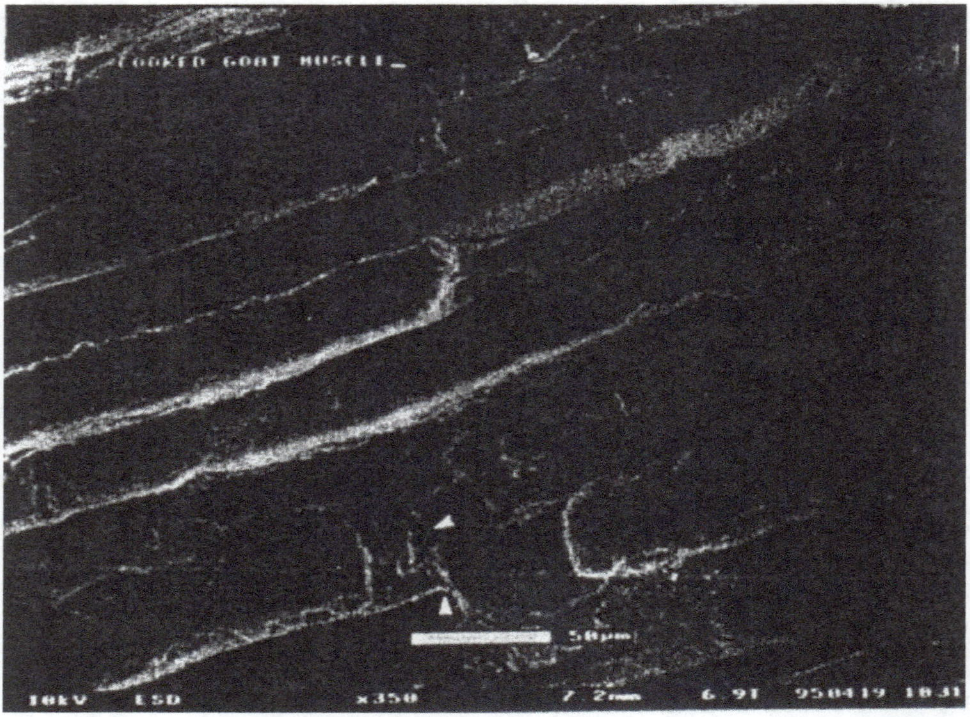

Figure 9. Effect of conventional heating on SM muscle by ESEM. Transverse breakage was shown by arrows.

As a result of these investigations, it is possible to identify and characterize the fine structure of lamb, veal and goat semimembranosus muscle. Moreover to study and compare the effect of various heat treatments such as conventional and microwaves at two wattages levels (700 and 12000 W) on structure of this muscle. In lamb and goat SM muscles results show that microwave heating cause more structural damage at both levels comparing to conventional heating and heat distribution in microwave heating is responsible for the surface damage to muscle fiber and separation of some parts and denaturation of collagen. While in veal SM muscle because of original texture of the veal, results are different. Domestic microwave heating caused less physical damage to the connective tissue network and myofibril elements comparing to the conventional cooking results. Further research is still required to study the fine structure of semimembranosus or other muscle from various animals (Yarmand & Homayouni, 2009; 2010).

5.3. CSLM

Confocal microscopy was described at 1957 (Fig. 10). It has some advantages in composition with conventional light microscopy. Since the laser as a light source was not invented on that time, it was not possible to improve Confocal Microscopy. In addition lack of a suitable computer due to collecting data of image presentation, also causes a problem in the progress of this instrument. Reflection Scanning Light Microscopy (TRSLM) was made as the second confocal system. For the first time the sectioning effect in confocal microscopy (CM) was presented. Generally the major construction of the second microscope was entirely different from the first microscope suggested by Minsky (1957). This instrument was not satisfactory because of poor image quality and resolution. In spite of the sectioning in TRSLM, the original could not be called as three-dimensional image (Sheppard and Wilson, 1978).

Early presentation of the imaging abilities and the initial improved confocal was displayed in publication by Brakenhoff (1979). Sheppard et al (1977) and Brakenhoff (1979) both employed laser as a light source. They also used a mechanical scanning that moves the specimens through the stationary confocal point. The nucleoid structure of the Echerichia Coli was studied by Brakenhoff microscope. Many experiments described the advantage of the combination of fluorescent techniques and high aperture confocal microscopy for the 3D imaging of biological structure. Fluorescence confocal microscopy was used for food emulsions and nerve cells. Fluorescence confocal microscopy also applied in the investigation of living cells. Calcium probes for the study of calcium spread in cells and searching SH groups (Brakenhoff, 1979; Grynkiewicz et al., 1985; Wijnaeadts Van Resandt et al., 1985).

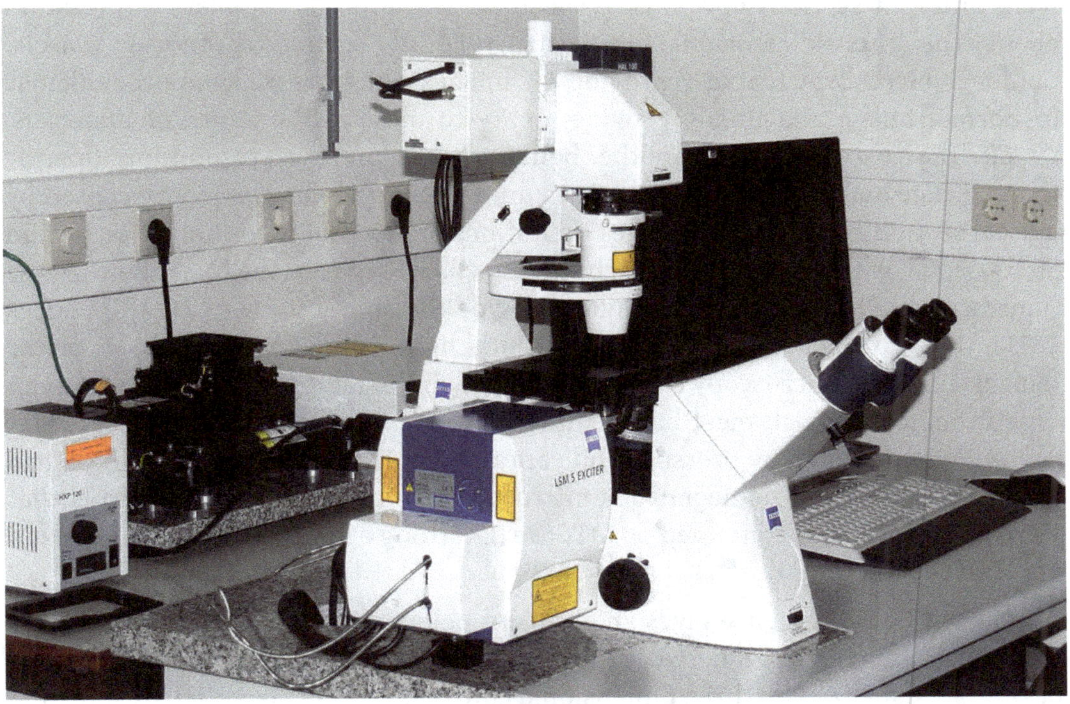

Figure 10. Confocal microscopy

Confocal scanning laser microscopy has been applied as a new technique in the study of microstructure of food during recent years. sheppard and Gu (1993) applied confocal microscopy for presentation of three dimensional fluorescence images of muscle fibers. According to them, a simple model of striated muscle fiber is quite useful for such image modeling. Fluorescence confocal scanning laser microscopy (FCSLM) model DMRBE was used for study of structure of goat SM muscle after various heat treatments. All specimens were stained with Picro Sirius Red technique. Confocal scanning laser microscopy has several advantages over conventional microscopes. According to Brakenhoff et al., (1988) Confocal scanning light microscopy forms a bridge between conventional light microscopy with its limited resolution (but capable of imaging hydrated, possibly live, specimens) on the one side, and electron microscopy with its higher resolution on the other. The principal of the basic confocal microscope has been explained by Sheppard and Choudhury (1977). This microscope might be operated in two techniques: reflection or in fluorescence. Optical sectioning and 3D reconstruction might be considered as a very important superior of confocal microscopy. Optical sectioning is non-invasive and allows for the imaging of thick objects and a three dimensional reconstruction of the sample. The Performance of confocal microscope suggests new possibilities in microstructure studies of food by the disturbance free visualizing of the three-dimensional internal structure. Such a set of observations from different sections of a muscle specimen will present complete information about the three-dimensional microstructure of semimembranosus muscle. Confocal microscopy is able to provide X-Z images which are images perpendicular to normal image and finally provide ability to observe the depth of materials. Despite that, using CSLM has another benefit that of improved resolution, superior fluorescence and improved quality of sample. Improved sample quality caused by optical sectioning which diminishes limitations of physical sectioning. The system is equipped with fluorescence techniques which exposes the specimen to the fluorescence. Fluorescence techniques provide the ability for the CSLM to choose CSLM the wavelength band of fluorescence light which contributes to better images. Confocal microscopy can show that the size of fat droplets varies with gum type in minced ostrich meat batter. In chicken meat gels, it showed that low-fat protein gels obtained by pressure and containing microbial transglutaminase had a more compact and homogeneous microstructure compared with controls that were pressurized but contained no MTGase (Chattong et al. 2007; Trespalacios and Pla, 2007).

CSLM used in fluorescence gives an excellent perspective of relationship between muscle fiber and collagen network including endomysial collagen fibers. An image of raw SM visualized by three dimensional construction of this muscle is shown in Fig. 11.

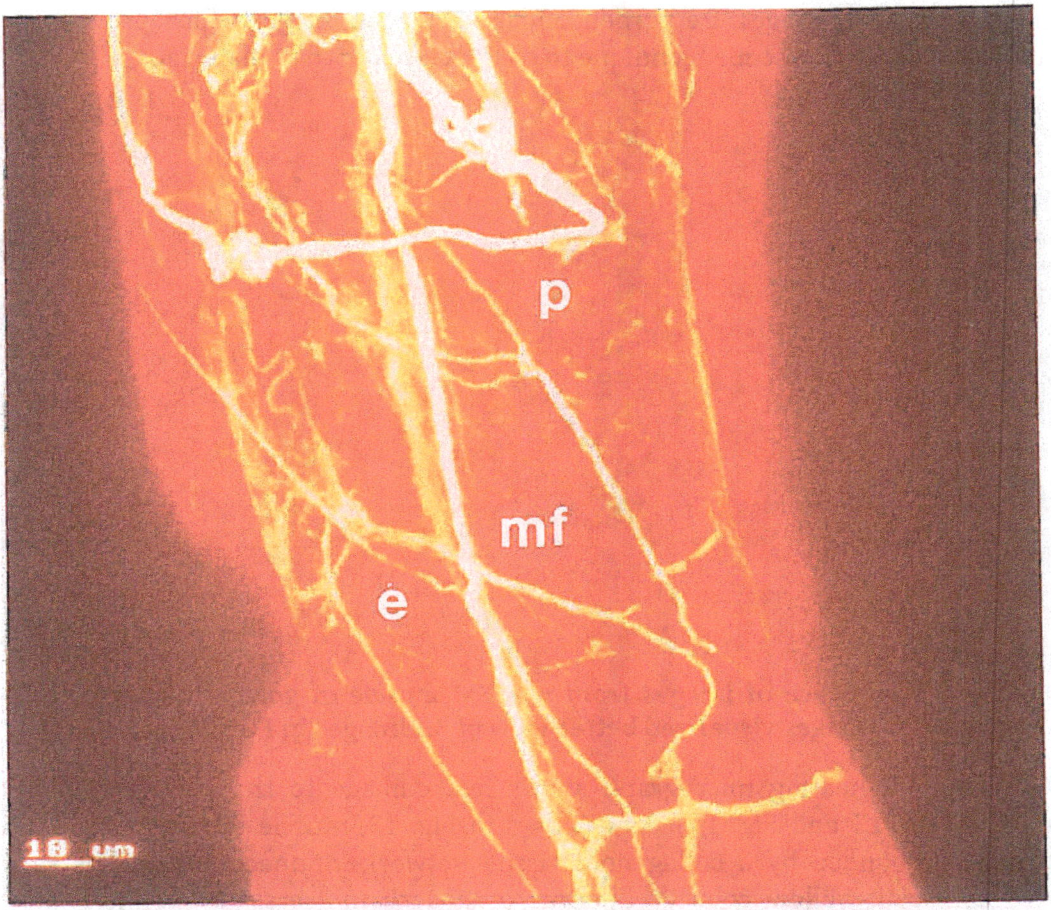

Figure 11. Illustration of raw SM muscle of goat by FCSLM. Perimysial (P) and endomysial (e) collagenous network were appeared. Mf: muscle fiber.

Optical sectioning through the specimen is responsible for the three-dimensional production of pictures, as previously stated. Fluorescence confocal scanning laser microscopy (FCSLM) has been employed in the examination of the semimembranosus muscle in goats. (Yarmand & Homayouni, 2009; 2010). Four treatments raw (control), conventional and microwave heating at two power levels 700 and 12000 W were examined. As mentioned previously the main advantage of CSLM is the capability of preparing three dimensional pictures which enables the relative distribution of connective tissue to be visualized. Because of the ability to 3D image muscle when heat treatment reduces the length or width of a muscle structure, it will be apparent in the images. Fig. 12 represents how FCSLM build this construction. Overlap of sequence images in Fig.12; give us better view of 3D images which illustrate the depth of material by providing X-Z images in final observation. As described earlier in the principle of confocal scanning electron microscopy, X-Z images are perpendicular to normal images which provide the ability for this microscope technique to observe the depth of material.

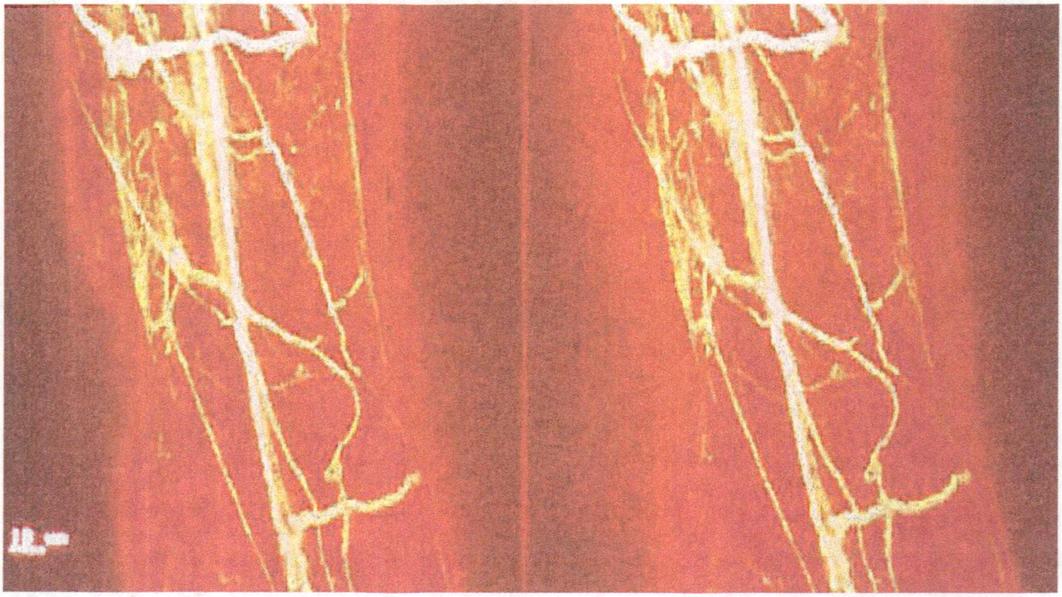

Figure 12. Sequence of images from raw SM muscle of goat. Overlap of these images visualized better view of 3D images in FCSLM.

Conventional heating micrograph shows less damage to the structure of SM muscle (Fig. 13 and Fig. 14). As it shows in Fig. 13, muscle fiber appeared red which surrounded by a collagenous tissue network. Connective tissue is seen as fluorescent yellow color. Sequences of images indicate the manner of 3D construction of conventional micrographs by FCSLM. In the confocal pictures the signal from collagen fluorescence stained with Picro Sirius Red give a yellow golden hue whereas the muscle fluorescence is illustrated as red. Using the fluorescence property with CSLM allowed us to select the wavelength band of fluorescence light contributing to the image. In this way, we are able to distinguish various structures inside the SM muscle and study their 3-D relationships. Conventional heating micrograph shows less damage to the structure of SM muscle (Fig. 13). As it shows in Fig. 13, muscle fiber appeared red which was surrounded by connective tissue network. Connective tissue seems fluorescent with yellow color.

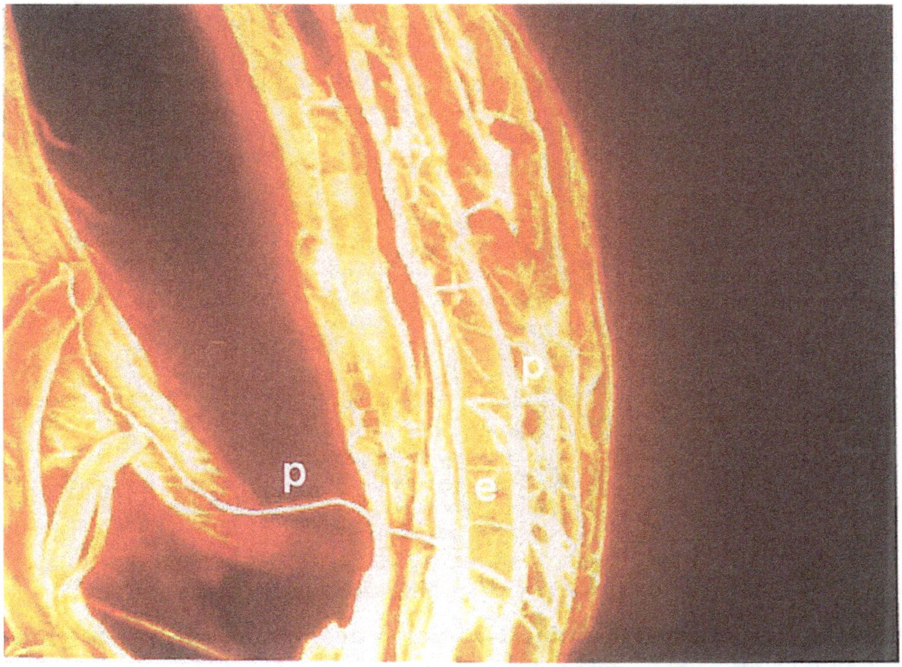

Figure 13. Conventional heating image of goat SM muscle by FCSLM Perimysial (p) and endomysial (e) connective tissue.

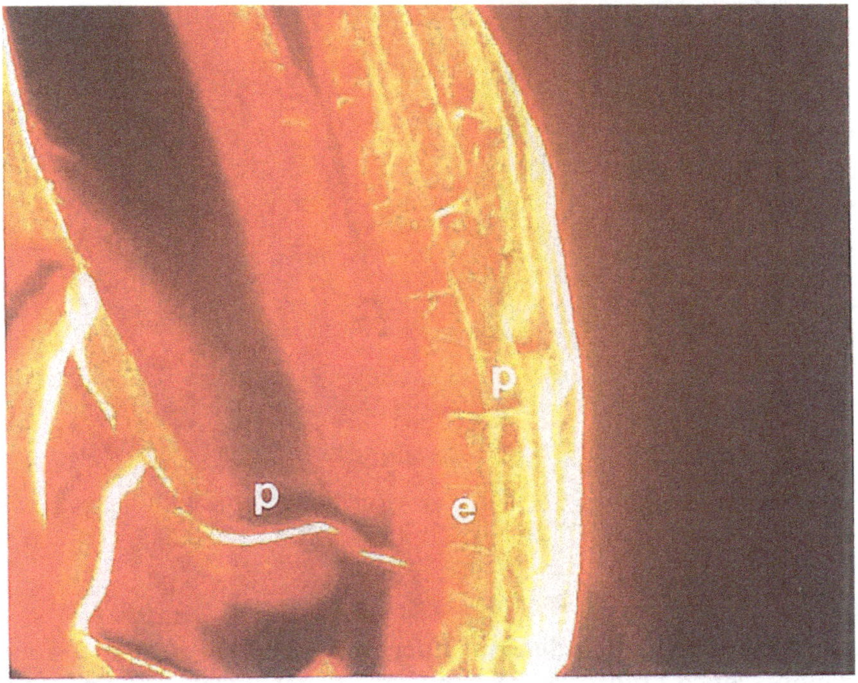

Figure 14. Another view of conventional heating of goat SM muscle. Perimysial (p) and endomysial (e) connective tissue have been shown by FCSLM.

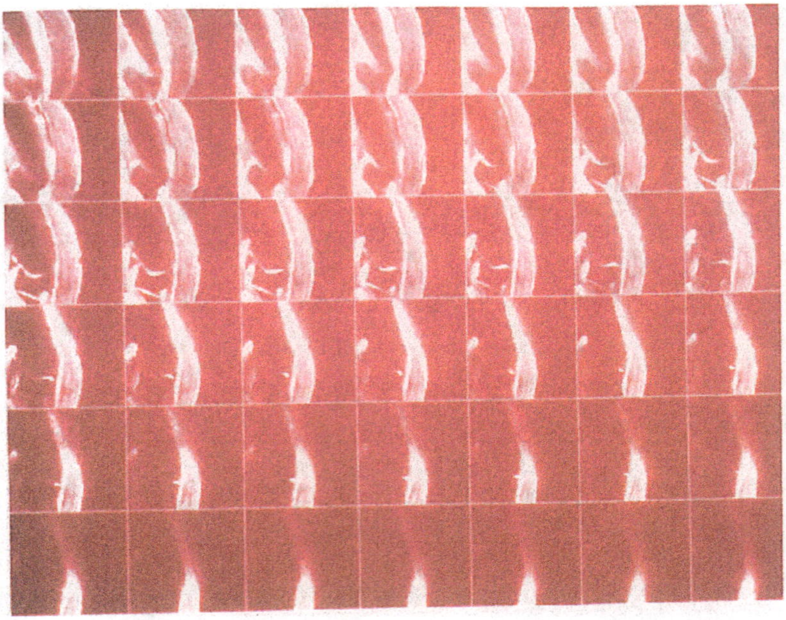

Figure 15. Demonstration of domestic microwave heated of goat SM muscle by FCSLM. Separation of connective tissue network has been shown in micrograph. Mf: muscle fiber, c: connective tissue.

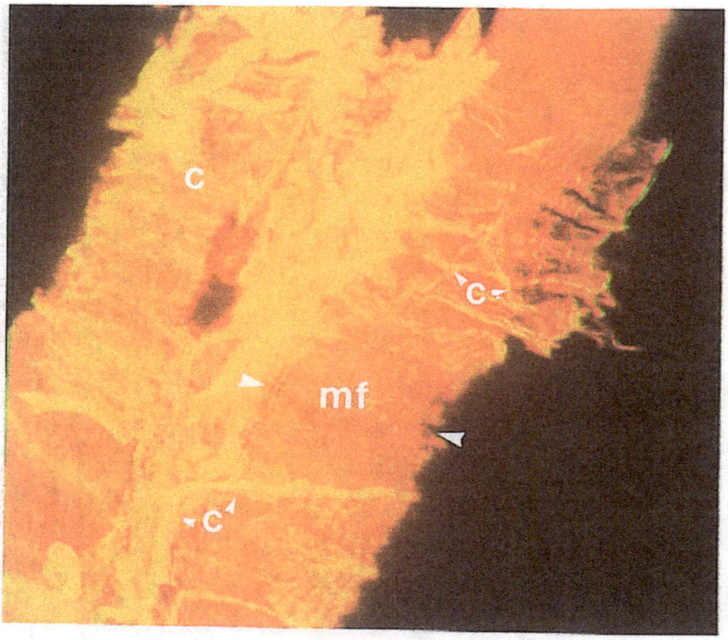

Figure 16. Presentation of industrial microwave heated of goat SM muscle by FCSLM. Fluorescence Fluorescence property of perimysial (p) and endomysial (e) connective tissue have been illustrated. Arrows indicate degradation and breakdown area of muscle fiber. mf: muscle fiber, scale bar: 10

More damage was observed in the structure of goat SM muscle after using industrial microwave heat treatment. Degradation and breakdown also appeared in microwave heating (Fig. 15 **and** 16). This result is in agreement with environmental scanning electron microscopy of goat SM muscle. Fig. 17 **illustrated another view of industrial microwave cooked. The same result is provided by FCSLM for domestic microwave muscle. The sequence of images provided as a set of observations by fluorescence confocal scanning laser microscope, is shown in** Fig. 12. It seems that in raw (control) specimen, fluorescence property of connective tissue is more apparent and it seems that it slightly decreases after heating by various techniques. This reduction was visualized in Fig. 13 **and appeared more in microwave heating** Fig. 15 **and** 16. This suggests that particular heat transfer occurred. However, still further researches are needed in order to understand the effect of heat on fluorescence property of collagen.

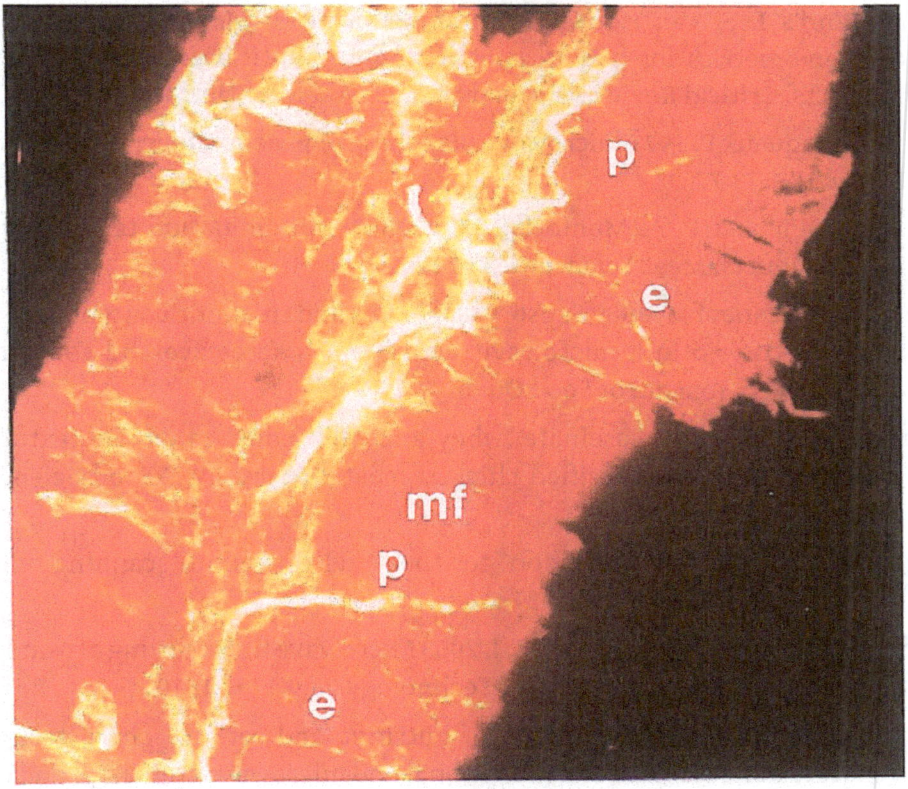

Figure 17. Another microscopic view of industrial microwave heating of goat SM muscle. mf: muscle fiber, c: connective tissue, scale bar: 10 microns.

6. CONCISE SUMMARY

Microstructural analysis was utilized to examine the qualitative characteristics of meat and meat products. Microscopic techniques are employed to characterize the structure of meat. Optical microscopy is applicable for examining the distribution of fat globules and protein gel in sausages. Among microscopic techniques, environmental scanning electron microscopy (ESEM) offers distinct benefits for assessing microstructural alterations in meat, particularly in comparing the effects of different heat treatments. Due to the extensive use of microwave technology in meat processing (including tempering, thawing, and cooking) and the quality control of meat and meat products, further research is required to examine the impacts of microwave processing on these items.

7. REFERENCES

Anderson E. T. Berry B. W. 2001 Effects of inner pea fiber on fat retention and cooking yield in high fat ground beef, Food Research International 34 8 689 694 .

Arvanitoyannis I. S. van Houwelingen-Koukaliaroglou M. 2003 Implementation of chemometrics for quality control and authentication of meat and meat products. Critical Reviews in Food Science and Nutrition 43 173 218 .

Avisse C. Varaquaux P. 1977 Microwave blanching of peaches, Journal of Microwave Power 12 1 73 77 .

Aymerich T. Picouet P. A. Monfort J. M. 2008 Decontamination technologies for meat products. Meat Science 78 114 129 .

Barbeau W. E. Schnepf M. 1989 Sensory attributes and thiamine content of roasting chickens cooked in a microwave, convection microwave and conventional electric oven, Journal of Food Quality 12 3 203 213 .

Barbut S. 2006 Effects of caseinate, whey and milk powders on the texture and microstructure of emulsified chicken meat batters. LWT-Food Science and Technology 39 6 660 664 .

Baumgarten N. 1990 Introduction to the environmental scanning electron microscope. Scanning, 12 36 37 .

Beaudry C. Raghavan G. S. V. Rennie T. J. 2003 Microwave finish drying of osmotically dehydrated cranberries, Drying Technology 21 9 1797 1810 .

Begum S. Brewer M. S. 2001 Chemical, nutritive and sensory characteristics of tomatoes before and after conventional and microwave blanching and during frozen storage, Journal of Food Quality, 24 1 1 15 .

Bhartia P. Stuchly S. S. Hamid M. A. K. 1973 Experimental results for combinatorial microwave and hot air drying, Journal of Microwave Power 8 245 252 .

Bialod D. Jolion M. Legoff R. 1978 Microwave thawing of food products using associated surface cooling. Journal of Microwave Power 13(3): 269.

Boggaard C. Christensen L. B. Jespersen B. L. 2003 Reflection mode microwave spectroscopy for on-line measurement of fat in trimmings. Presented at 49th International Congress of Meat Science and Technology. 31 August-5 September in Campinas, Brazil.

Boles J. A. Parrish F. C. 1990 Sensory and chemical characteristics of precooked microwave-reheatable pork roasts, Journal of Food Science 55 3 618 620 .

Brakenhoff G. J. 1979 Imaging modes in confocal scanning light microscopy. Journal of microscopy 117 233 236 .

Brakenhoff G. J. van der Voort G. T. M. Spronsen E. A. Nanniga N. 1988 Three-dimensional imaging of biological structures by high resolution confocal scanning laser microscopy. Scanning Microscopy 2 33 40 .

Cáceres E. García M. L. Selgas M. D. 2008 Effect of pre-emulsified fish oil-As source of PUFA n-3-on microstructure and sensory properties of mortadella, a Spanish bologna-type sausage. Meat Science 80 2 183 193.

Chattong U. Apichartsrangkoon A. Bell A. E. 2007 Effects of hydrocolloid addition and high pressure processing on the rheological properties and microstructure of a commercial ostrich meat product "Yor" (Thai sausage). Meat Science 76 3 548 554.

Chen S. C. Collins J. L. Mccarty I. E. Johnston M. R. 1971 Blanching of white potatoes by microwave energy followed by boiling water, Journal of Food Science, 36 742-743.

Cipra J. S. Bowers J. A. 1971 Flavour of microwave- and conventionally-reheated turkey, Poultry Science 50 3 703-706 .

Clerjon S. Damez J. L. 2007 Microwave sensing for meat and fish structure evaluation. Meat Science and Technology 18 4 1038 1045 .

Clerjon S. Daudina J. D. Damez J. L. 2003 Water activity and dielectric properties of gels in the frequency range 200 MHz-6 GHz. Food Chemistry 82 1 87 97 .

Crawford C. K. 1979 Charge neutralization using very low energy ions. Scanning Electron Microscopy 2 31 46 .

Danilatos G. D. 1981 The examination of fresh or living plant material in an environmental scanning electron microscope. Journal of Microscopy 121 235 238 .

Danilatos G. D. 1983 A gaseous detector device for an environmental SEM. Micron and Microscopica Acta 14 307 318 .

Danilatos G. D. 1989 Environmental SEM; A new instrument, a new dimension. Proceedings of the Institute of Physics Electron Microscopy and Royal Microscopy Society Conference 98 1 455 458 .

Danilatos G. D. 1990 Theory of the gaseous detector device in the ESEM. Advances in Electronics and Electron Physics 78 1 102 .

Danilatos G. D. 1991 Review and outline of environmental SEM at present. Journal of Microscopy 162 391 402 .

Danilatos G. D. Postle R. 1982 The environmental scanning electron microscope and its application. Scanning Electron Microscopy 1 1 16 .

Decareau R. V. Peterson R. A. 1986 Current state of microwave processing. In Microwave-Processing and Engineering, Ellis Horwood, Chichester, UK, 22 23.

Drew F. Rhee K. S. Carpenter Z. L. 1980 Cooking at variable microwave power levels, Journal of American Dietetic Association 77 4 455 459 .

Dunn N. A. Heath J. L. 1979 Effect of microwave energy on poultry tenderness, Journal of Food Science 44 2 339 342 .

European Chilled Food Federation 1996 Guidelines for Good Hygienic Practice in the Manufacture of Chilled Foods. London, European Chilled Food Federation, 59

Farber J. M. Daoust J. Y. Diotte M. Sewell A. Daley E. 1998 Survival of Listeria spp. on raw whole chickens cooked in microwave ovens. Journal of Food Protection 5 1465 1469 .

Fulton L. Davis C. 1983 Roasting and braising beef roasts in microwave ovens, Journal of American Dietetic Association 83 5 560 563 .

Grynkiewicz G. Poenie M. Tsien R. Y. 1985 A new generation of Ca^{2+} indicators with greatly improved fluorescence properties. Journal of Biologival Chemistry 260 3440 3450 .

Guan D. Cheng M. Wang Y. Tang J. 2004 Dielectric properties of mashed potatoes relevant to microwave and radio-frequency pasteurization and sterilization processes, Journal of Food Science 69 1 30 37 .

Hammernick-Oltrogge M. Prusa K. J. 1987 Sensory analysis and Instron measurements of variable-power microwave-heated baking hen breasts, Poultry Science 66 9 1548 1551 .

Hines R. C. Ramsey C. B. Hoes T. L. 1980 Effects of microwave cooking rate on palatability of pork loin chops, Journal of Animal Science 50 3 446 451 .

Hoda I. Ahmad S. Srivastava P. K. 2002 Effect of microwave oven processing, hot air oven cooking, curing and polyphosphate treatment on physico-chemical, sensory and textural characteristics of buffalo meat products, Journal of Food Science and Technology 39 3 240 245 .

Hong N. Yaylayan V. Raghavan G. S. V. Parea J. Bealanger J. 2001 Microwave-assisted extraction of phenolic compounds from grape seeds, Natural Product Letters 5 3 197 204 .

Howat P. M. Gros J. N. Mcmillin K. W. Saxton A. M. Hoskins F. 1987 Comparison of beef blade roasts cooked by microwave, microwave convection and conventional oven, Journal of Microwave Power 22 2 95 98 .

Hsieh J. H. Baldwin R. E. 1984 Storage and microwave reheating effects on lipid oxidation of roast beef, Journal of Microwave Power 19 3 187 194 .

Hugas M. Garriga M. Monfort J. M. 2002 New mild technologies in meat processing: High pressure as a model technology. Meat Science 62 359 371 .

James C. James S. J. (2010 Freezing Thawing. in F. Toldrá (ed.. Handbook of Meat Processing, A John Wiley & Sons, Inc., Publication, 105 124 .

James C. 2000 Optimising the microwave cooking of bacon, Meat and Poultry 2000 Seminar, CCFRA (Campden & Chorleywood Food Research Association), 17 July 2000.

James S. J. Crow N. 1986 Thawing and tempering: Industrial practice. Meat Thawing/ Tempering and Product Quality, IFR-BL: Subject Day.

James S. J. James C. 2002 Thawing and tempering. In Meat Refrigeration, Woodhead Publishing, Cambridge, 159 190 .

James S. J. 1984 Thawing meat blocks using microwaves under vacuum. Proceedings of the 30th European Meeting of Meat Research Workers, Bristol, Paper 2 61 62 .

James S. J. 1986 Microwave thawing and tempering of frozen meat. BNCE Heating and Processing 1 3000 MHz, Cambridge, 4-7.

Jason A. C. Sanders H. R. 1962 Dielectric thawing of fish. Food Technology 16 101 107 .

Jason A. C. 1974 Microwave thawing-some basic considerations. In Meat Freezing; Why and How? Meat Research Institute Symposium 3 31 43

Jaynes H. O. 1975 Microwave pasteurization of milk, Journal of Milk and Food Technology 38 386 387 .

Kent M. 1990 Hand-held instrument for fat/water determination in whole fish. Food Control 1 47 53 .

Kent M. Knökel R. Daschner F. Berger U. K. 2001 Composition of foods including added water using microwave dielectric spectra. Food Control 12 467 482 .

Kent M. Knökel R. Daschner F. Berger U. K. 2000 Composition of foods using microwave dielectric spectra. European Food Research and Technology 210 359 366 .

Kent M. Lees A. Roger A. 1993 Estimation of the fat content of minced meat using a portable microwave fat meter. Food Control 4 222 227 .

Kent M. Peymann A. Gabriel C. Knight A. 2002 Determination of added water in pork products using microwave dielectric spectroscopy. Food Control 13 143 149.

King A. J. Bosch N. 1990 Effect of NaCl and KCl on rancidity of dark turkey meat heated by microwave, Journal of Food Science 55 6 1549 1551 .

Klose M. J. Webb R. I. Teakle D. S. 1992 Studies on the virus pollen association of tobacco streak virus using ESEM and MDD techniques. Journal of Computer Assisted Microscopy 4 213 220.

Krishnan K. R. Sharma N. 1990 Studies on emulsion-type buffalo meat sausages incorporating skeletal and offal meat with different levels of pork fat. Meat Science 28 1 51 60 .

Kumar S. Aalbersberg B. 2006 Nutrient retention in foods after earth-oven cooking compared to other forms of domestic cooking 2. Vitamins. Journal of Food Composition and Analysis 19 311 320

Levre E. Valentini P. 1998 Inactivation of Salmonella during microwave cooking. Zentralblatt fuÈr Hygiene und Umweltmedizin 6 431 436.

Lin C. C. Li C. F. 1971 Microwave sterilization of oranges in glass pack, Journal of Microwave Power 6 45 48 .

Lyng J. G. Scully M. Mckenna B. M. Hunter A. Molloy G. 2002 The influence of compositional changes in beef burgers on their temperatures and their thermal and dielectric properties during microwave heating, Journal of Muscle Foods 13 2 123 142 .

Maurer R. L. Trembley M. R. Chadwick E. A. 1971 Use of microwave energy in drying alimentary pastes, Proceedings of Microwave Power Symposium IMPI, Monterey, CA, USA.

Meisel N. 1972 Tempering of meat by microwaves, Microwave Energy Application Newsletter 5 3 3 7 .

Mendiratta S. K. Kumar S. Keshri R. C. Sharma B. D. 1998 Comparative efficacy of microwave oven for cooking of chicken meat, Fleischwirtschaft 78 7 827 829 .

Miller P. G. 1971 An elastin stain. Medical Laboratory Technology 28 148 149 .

Moncrieff D. A. Robinson V. N. E. Harris L. B. 1978 Charge neutralization of insulating surfaces in the SEM by gas ionization. Journal of Physics D: Applied Physics 11 17 2315 2325 .

Moody W. G. Bedeau C. Langlois B. E. 1978 Beef thawing and cookery methods: Effect of thawing and cookery methods, time in storage and breed on the microbiology and palatability of beef cuts, Journal of Food Science 43 3 834 838 .

Morin L. A. Temelli F. Mc Mullen L. 2004 Interactions between meat proteins and barley (Hordeum spp.): B-glucan within a reduced-fat breakfast sausage system. Meat Science 68 3 419 430 .

Murano P. 2003 Understanding Food Science and Technology. Belmont, Calif.: Thompson Wadsworth.

Nelson S. O. 1973 Insect control studies with microwaves and other radio frequency energy, Bulletin of the Entomological Society of America 19 3 157 163 .

Nykvist W. E. Decareau R. V. 1976 Microwave meat roasting, Journal of Microwave Power, 11 1 3 24 .

O'Brien G. P. Webb R. K. Uwins P. J. R. Desmarchelier P. M. Imrie B. 1992 Suitability of the environmental scanning electron microscope for studies of bacteria on Mung bean seeds. Journal of Computer Assisted Microscopy 4 225 229 .

Papadima S. N. Arvanitoyannis I. S. Bloukas J. G. 1999 Chemometric model for describing Greek traditional sausages. Meat Science 51 271 277 .

Parker K. Vollmer M. 2004 Bad food and good physics: the development of domestic microwave cookery, Physics Education 39 1 82 90 .

Parsons D. F. Matriccardi V. R. Moretz R. C. Turner J. N. 1974 Electron microscopy and diffraction of wet unstained and unfixed biological objects. Advances in Biological and Medical Physics 15 161 271.

Payton J. Baldwin R. E. 1985 Comparison of top round steaks cooked by microwave-convection, forced-air convection and conventional ovens, Journal of Microwave Power 20 4 255 259 .

Peters K. R. 1990 Introduction to the technique of environmental scanning electron microscopy. Scanning 90 abstracts (71 72). FACMS Inc.

Ramesh M. N. Wolf W. Tevini D. Bognar A. 2002 Microwave blanching of vegetables, Journal of Food Science 67 1 390 398 .

Riffero L. M. Holmes Z. A. 1983 Characteristics of pre-rigor pressurized versus conventionally processed beef cooked by microwaves and by broiling, Journal of Food Science 48 2 346 350 .

Riihonen L. Linko P. 1990 Effect of thawing on the quality of frozen mechanically deboned meat. Journal of Agricultural Science in Finland 62 5 407 415 .

Rosenberg U. Boe G. L. W. 1987 Microwave thawing, drying and baking in the food industry, Food Technology 41(6): 85.

Sanders H. R. 1966 Dielectric thawing of meat and meat products. Journal of Food Technology 1: 183 EOF 192 EOF .

Satchell F. E. Doty D. M. 1951 High-frequency dielectric heating for defrosting frozen pork bellies. Bulletin of the American Meat Institute Foundation 12: 95.

Schiffman R. F. 1986 Food product development for microwave processing, Food Technology 40(6): 94.

Schiffman R. F. 1997 Microwave technology, a half-century of progress, Food Product Design 1 15 .

Severini C. Baiano A. De Pilli T. Romaniello R. Derossi A. 2004 Microwave blanching of sliced potatoes dipped in saline solutions to prevent enzymatic browning, Journal of Food Biochemistry 28 1 75 88 .

Sheppard C. J. R. Gu M. 1993 Modeling of three-dimensional fluorescence images of muscle fibers: an application of the three-dimensional optical transfer function. Journal of microscopy 168 339 345 .

Sheppard C. J. R. Choudhury A. 1977 Image formation in the scanning microscope. Optica Acta 24 1051 1073 .

Sheppard C. J. R. Wilson T. 1978 Depth of field in the scanning microscope. Optics Letter 3 115 117 .

Shivhare U. S. Raghavan G. S. V. Bosisio R. G. 1994 Modeling the drying kinetics of maize in a microwave environment, Journal of Agricultural Engineering Research 57 199 205 .

Sipahioglu O. Barringer S. B. T. A. U. B. T. Yang A. P. P. 2003 Characterization and modeling of dielectric properties of turkey meat, Journal of Food Science 68 2 521 527 .

Sobiech W. 1980 Microwave-vacuum drying of sliced parsley root, Journal of Microwave Power 15 3 143 154 .

Steiner J. B. Johnson R. M. Tobin B. W. 1985 Effect of microwave and conventional reheating on flavor development in chicken breasts, Microwave World 6 2 10 12 .

Sumnu G. 2001 A review on microwave baking of foods, International Journal of Food Science and Technology 36 117 127 .

Swain M. James S. 2005 Thawing and tempering using microwave processing in: Schubert, H. & Regier, M. (ed) The microwave processing of foods, Woodhead Publishing Limited, 189 206 .

Taher B. J. Farid M. M. 2001 Cyclic microwave thawing of frozen meat: experimental and theoretical investigation. Chemical Engineering and Processing 40 4 379 389 .

Taki G. H. 1991 Functional ingredient blend produces low-fat meat products to meet consumer expectations, Food Technology 71 74 .

Tejada M. De las Heras. C. Kent M. 2007 Changes in the quality indices during ice storage of farmed Senegaleses ole (Solea senegalensis). European Food Research Technology 225 225 232 .

Trespalacios P. Pla R. 2007 Simultaneous application of transglutaminase and high pressure to improve functional properties of chicken meat gels. Food Chemistry 100 1 264 272 .

Tulasidas T. N. Raghavan G. S. V. Mujumdar A. S. 1995 Microwave drying of grapes in a single mode cavity at 2450 MHz I: drying kinetics, Drying Technology, 13 1949 1972 .

Tzouros N. E. Arvanitoyannis I. S. 2001 Agricultural produces: Synopsis of employed quality control methods for the authentication of foods and application of chemometrics for the classification of foods according to their variety or geographical origin. Critical Reviews in Food Science and Nutrition 41 287 319 .

Uwins P. J. R. 1994 Environmental scanning electron microscopy. Materials Forum, 18 51 75 .

Virtanen A. J. Goedeken D. L. Tong C. H. 1997 Microwave assisted thawing of model frozen foods using feed-back temperature control and surface cooling. Journal of Food Science 62 1 150 154 .

Wallace H. M. Uwins P. J. R. Mc Chonchie C. A. 1992 Investigation of stigma interaction in Macadamia and Grevillea using ESEM. Journal of Computer Assisted Microscopy 4 231 234 .

Wang Y. Wig T. Tang J. Hallberg L. M. 2003 Dielectric properties of food relevant to RF and microwave pasteurization and sterilization, Journal of Food Engineering 57 3 257 268 .

Welke R. A. Williams J. C. Miller G. J. Field R. A. 1986 Effect of cooking method on the texture of epimysial tissue and rancidity in beef roasts. Journal of Food Science 51 4 1057 1060 .

Wijnaendts-van-Resandt R. W. Marsman H. J. B. Kaplan R. Davoust J. Stelzer E. H. K. Stricker R. 1985 Optical fluorescence microscopy in three dimensions: microtomoscopy. Journal of microscopy 138 29 34 .

Williams O. J. Raghavan G. S. V. Orsat V. Dai J. 2004 Microwave assisted extraction of capsaicinoids from capsicum fruit, Journal of Food Biochemistry 28 2 113 122 .

Yagi S. Shibata K. 2002 Microwave defrosting under reduced pressure. US Patent Application 20020195447.

Yang H. W. Gunasekaran S. 2001 Temperature profiles in a cylindrical model food during pulsed microwave heating, Journal of Food Science 66 7 998 1004 .

Yarmand M. S. Homayouni A. 2009 Effect of microwave cooking on the microstructure and quality of meat in goat and lamb. Food Chemistry, 112 782 785 .

Yarmand M. S. Homayouni A. 2010 Quality and microstructural changes in goat meat during heat treatment. Meat Science, 86 451 455 .

Yarmand M. S. Sarafis V. 1997 An improved maceration technique with polarized light microscopy in the study of muscle. Egyptian Journal of Food Science, 25 425 431 .

Yilmaz I. Arici M. Gümüs T. 2005 Changes of microbiological quality in meatballs after heat treatment. European Food Research and Technology 221 281 283 .

Zagrodzki S. Niedzielski Z. Kulogawska A. 1977 Meat thawing under controlled conditions. Acta Alimentaria Polonica 3 21 1 11 .

Zhang H. F. U. C. Zheng X. X. I. Y. Jiang W. Wang Y. 2004 Control of post harvest rhizopus rot of peach by microwave treatment and yeast antagonist, European Food Research and Technology 218 6 568 572 .

Zhang L. Lyng J. G. Brunton N. P. 2006 Quality of radio frequency heated pork leg and shoulder ham. Journal of Food Engineering 75 275 287 .

PART 4
TECHNOLOGY

7 Chapter

REITERABILITY AND SCALABILITY OF MICROWAVE-ASSISTED REACTIONS

1. INTRODUCTION: AN OVERVIEW

High-speed microwave synthesis has garnered significant attention during the past two decades.

Since 2000, the volume of papers pertaining to MAOS has surged significantly. The heightened interest in microwave heating can be attributed to the emergence, at the beginning of the 21st century, of specialized monomode and multimode devices equipped with suitable temperature and pressure controls, facilitating result repeatability. However, when the microwave technology was established two decades ago, the majority of responses were conducted in household ovens lacking adequate temperature regulation.

Recent reports have revealed the reproducibility of results between monomode and multimode microwave instruments in solution chemistry, their application in parallel and combinatorial chemistry, and comparisons between homogeneous and heterogeneous systems and reactions conducted in closed or open vessels. Similarly, it has been demonstrated that solvent-free reactions can be modernized and utilized in microwave reactors, with an emphasis on the impact of polarity on the reproducibility of these processes. Moreover, computer analyses can aid in elucidating and/or forecasting the reproducibility of a process.

Potential methods to enhance microwave-assisted reactions encompass continuous-flow reactors, small-scale batch stop-flow procedures, or large-scale single-batch reactors. The scalability of microwave-induced processes poses a significant challenge due to various factors, primarily increased heat loss, alterations in absorption, limited penetration depth of radiation into the reaction medium (only a few centimeters at 2.45 GHz), and additional microwave reflection. These inherent challenges have restricted the application of microwave reactors to capacities beyond a few liters, hence impeding the utilization of this method for the synthesis of higher material quantities.

Conversely, certain researchers and microwave manufacturers have investigated the possibilities of continuous flow microwave systems. This technique has numerous benefits for processing adaptability, safety, reaction monitoring, and optimization. This processing method notably circumvents the constraints linked to the construction of scaled microwave cavities, including the related expenses.

This chapter aims to emphasize the pivotal reports addressing the issues of reproducibility and scalability in microwave-assisted processes, ranging from initial reactions conducted in unmodified domestic ovens to contemporary syntheses employing specialized equipment or appropriate flow instrumentation.

2. REPRODUCIBILITY UNDER MICROWAVES

The microwave applicator is the element in a processing system where energy is imparted to the product for heating purposes. In a household oven, the radiation output is typically regulated by the on-off cycles of the magnetron. Prolonged switching intervals are unfavorable in chemistry due to the lack of irradiation between switching phases. In this scenario, certain regions may obtain substantial energy, while others may acquire less energy. Generally, reliably monitoring the reaction temperature is not feasible. Heating organic solvents in open containers may result in dramatic explosions caused by electric arcs within the cavity or sparking due to the magnetron's switching. Conversely, operating sealed vessels without pressure monitoring may result in significant mishaps. Consequently, the utilization of such apparatus in chemistry cannot be endorsed. Dedicated microwave reactors for synthesis incorporate integrated magnetic stirrers, direct temperature measurement of the reaction mixture via fibre-optic probes or IR sensors, and software that facilitates real-time temperature and pressure control through modulation of microwave power (Kappe & Stadler, 2005).

Multimode instruments facilitate the utilization of larger reaction vessels or enhance reaction throughput through the implementation of multivessel rotors for parallel synthesis or scale-up. A primary drawback of multimode instruments is their suboptimal performance in small-scale studies (less than 3 mL). Despite the high generated microwave power (1000–1400 W), the power density of the field is often poor.

Monomode instruments provide a singular, extremely uniform energy field characterized by elevated power density. These devices effectively couple with small samples, with the maximum power output being constrained to 300 W (Ondruschka et al., 2006).

The issues of repeatability and scale-up will invariably pertain to the reaction circumstances. Moreover, the reaction medium (phase) significantly influences this type of power input in contrast to classical reactions. In addition to molecular mass, the polarity of the reaction mixture is a crucial component for microwave power absorption.

2.1. Reproducibility in scaled-up microwave-assisted reactions

One of the limitations of microwave scale-up technology is the restricted penetration depth of microwave irradiation into absorbing materials. This means that solvent or reagents in the centre of large reaction vessel are heated by convection and not by direct 'in core' microwave dielectric heating. For this reason, many researchers have studied the scale-up and reproducibility in batch of the reactions performed on a small scale.

In 1998 Hamelin studied the large-scale synthesis of dioxolanes, dithiolanes and oxathiolanes from 2, 2-dimethoxypropane and 3, 3-dimethoxypentane in the absence of solvent. These reactions had previously been performed in a Prolabo Synthewave 402 apparatus on a 10 mmol scale and were reproduced in a Synthewave 1000 reactor on a 2 mol scale (Perio et al., 1998) (Scheme 1). The authors found that the 2-mol scale was easier than the 10-mmol scale owing the possibility of continuous distillation of methanol under irradiation in the Synthewave 1000 with a packed column, a set-up that is not possible in the smaller Synthewave 402.

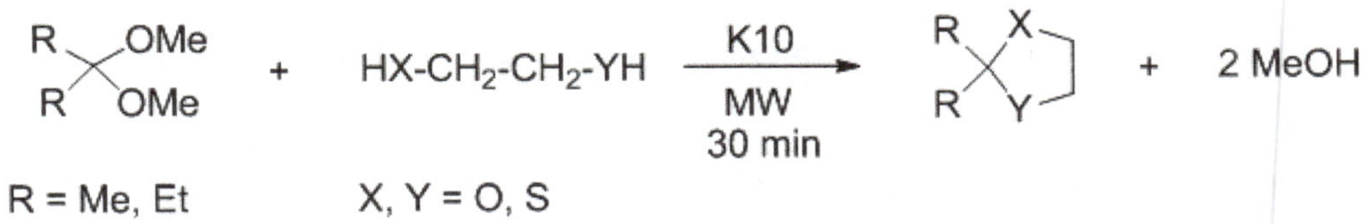

Scheme 1. Synthesis and scale-up of dioxolanes, dithiolanes and oxathiolanes

Similar results were reported by Loupy for the reproducibility and scale-up of several reactions (potassium acetate alkylation, regioselective phenacylation of 1,2,4-triazole, deethylation of 2-ethoxyanisole, and typical examples in carbohydrate chemistry) to several hundred grams in a larger batch reactor (Synthewave 1000), with yields equivalent to those obtained under similar conditions (temperature, reaction time) in laboratory-scale experiments (Synthewave 402) (Cléophax et al., 2000).

Luthman studied the appropriate reaction conditions in terms of choice of solvent, reaction temperature and reaction time to allow the fast and reproducible production of Mannich bases from small (2 mmol) to large (40 mmol) scale reactions in moderate to high yields (18–60%) and high purity (F. Lehmann et al., 2003) (Scheme 2).

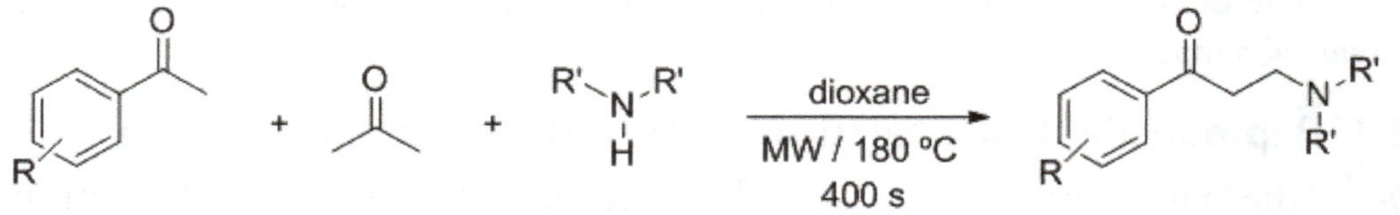

Scheme 2. Microwave-assisted Mannich reactions using substituted acetophenones

Two different microwave systems designed for large-scale operation, such as the Anton Paar 3000 multimode batch reactor (8 × 100 mL PTFE vessels in a ceramic vessel jacket, filling volume of 60–70 mL for each vessel) and the CEM Voyager SF stop-flow reactor (80 mL glass vessel, filling volume 50 mL), were evaluated by Lehmann (H. Lehmann & LaVecchia, 2005) for special use in a kilolab. As a model reaction, the aromatic substitution of aryl halides with nucleophiles like phenolates or amines was chosen (Scheme 3).

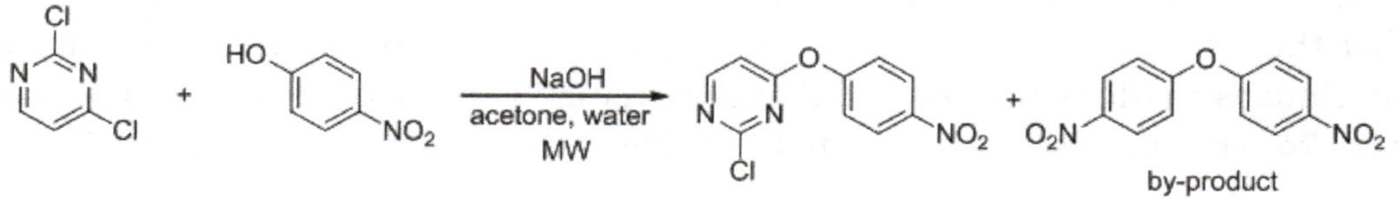

Scheme 3. Aromatic substitution of 2,4-dichloropyrimidine with p-nitrophenol

These reactions were previously performed in a small-scale Emrys Optimizer microwave reactor (20 mL glass vials). The study was focused on temperature/pressure limits, handling of suspensions, ability for rapid heating and cooling, robustness, and overall processing time.

A green approach to *N*-heterocyclization reactions ranging in scale from 20 mmol (CEM Discover) to 1 mol (CEM MARS) performed under microwave irradiation in open vessels was investigated by Barnard (Barnard et al., 2006) (Scheme 4). On using water as the solvent and in the absence of transition metal catalysts, *N*-heterocycles were formed in a fraction of the time needed on using classical heating. These reactions had previously been performed on a small scale by Varma in sealed vessels (Ju & Varma, 2006). Analogously, representative Suzuki couplings in water using low catalyst concentrations in conjunction with microwave heating have been transferred from sealed-vessel to open-vessel reaction conditions, and scale-up to the mole scale using the dedicated multimode apparatus CEM MARS (Arvela et al., 2005; Leadbeater et al., 2006) (Scheme 5). In both processes, single-mode and multimode cavities were used for open-vessel synthesis without changing the reaction times and these produced similar yields.

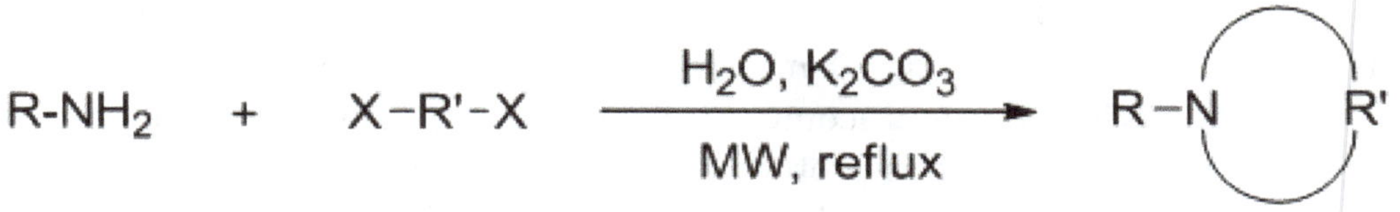

Scheme 4. N-Heterocyclization reactions in water

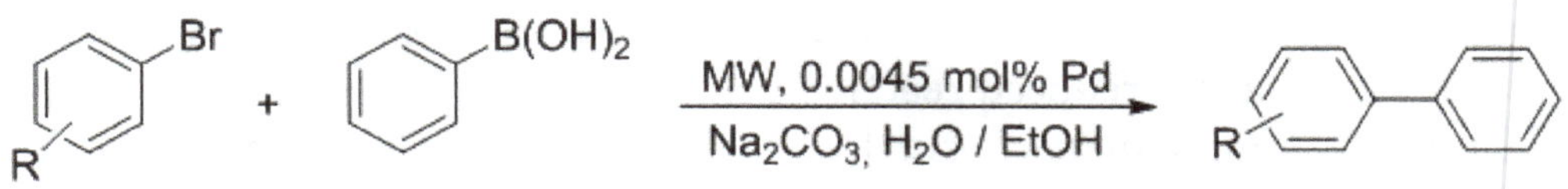

Scheme 5. Open-vessel microwave-promoted Suzuki reactions in water

2.2. Reproducibility in parallel microwave-assisted reactions

In general, two approaches have been used for microwave-assisted parallel and combinatorial chemistry: On the one hand, series' of compounds can be prepared sequentially in an automated single-mode instrument, which allow for control of temperature and pressure in each reaction independently. On the other hand, compounds can be prepared in parallel arrays using a multimode instrument. The use of a single-mode instrument offers the advantage of full control of each reaction. However, in this case, all reactions must be processed sequentially and this could lead to a bottleneck in productivity, especially for large series' of compounds. Multimode instruments offer the possibility of performing multiple reactions in one irradiation experiment, but reactions are usually performed without appropriate control of the temperature, which in turn limits the reproducibility of the experiments – especially when unmodified domestic ovens are used. In an effort to overcome these problems, several dedicated microwave reactors have been developed since 2003.

Nüchter demonstrated that a multiPREP reactor system (36 pressure reactors or 15 reflux reactors), combiCHEM (plates with 24, 48 and 96 reactions vessels sealed with Teflon-laminated silicon mats), and an HPR system (6–10 small autoclaves for pressurized reactions at up to 50 bar and 250 ºC) allow parallel reactions to be carried out with up to 96 reactor chambers in a microwave field with total reproducibility in yield and/or selectivity (Nüchter & Ondruschka, 2003a). The authors studied different processes such as nucleophilic substitutions, condensation reactions, oxidations, and multi-centre reactions.

Also in 2003, Kappe scaled different organic reactions from the 1 mmol to 100 mmol scale using a prototype multimode microwave reactor from Anton Paar that allowed parallel processing in either quartz or PTFE-TFM vessels with maximum operating limits of 300 ºC and 80 bar of pressure (eight-vessel rotor employing either 100 mL PTFE-TFM or 80 mL quartz vessels) (Stadler et al., 2003). The transformations included multicomponent chemistries, transition metal-catalyzed carbon-carbon cross-coupling protocols, solid-phase organic synthesis, and Diels–Alder cycloaddition reactions using gaseous reagents in prepressurized reaction vessels.

Alcázar investigated the possibility of scaling up the synthesis of several compounds simultaneously in one irradiation experiment (Alcázar et al., 2004). This parallel approach would allow reduced reaction times in comparison with the sequential irradiation of single samples in single-mode instruments. The authors studied an *N*-alkylation reaction and used optimized reaction conditions to probe the total reproducibility of five parallel reactions performed in a multimode system (3 mmol scale) and the same reaction carried out in a single-mode reactor on one-tenth of the scale (0.3 mmol) (Scheme 6, Figure 1).

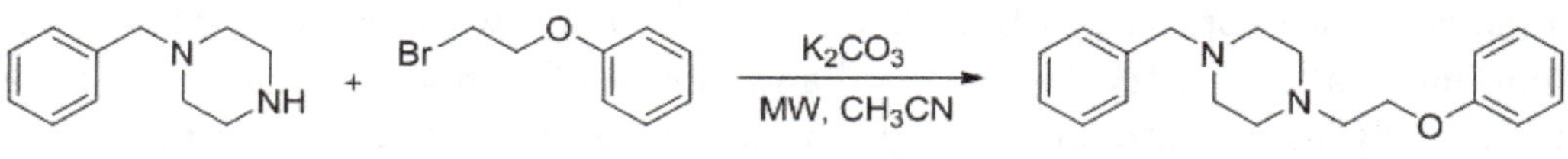

Scheme 6. N-Alkylation of N-benzylpiperazine

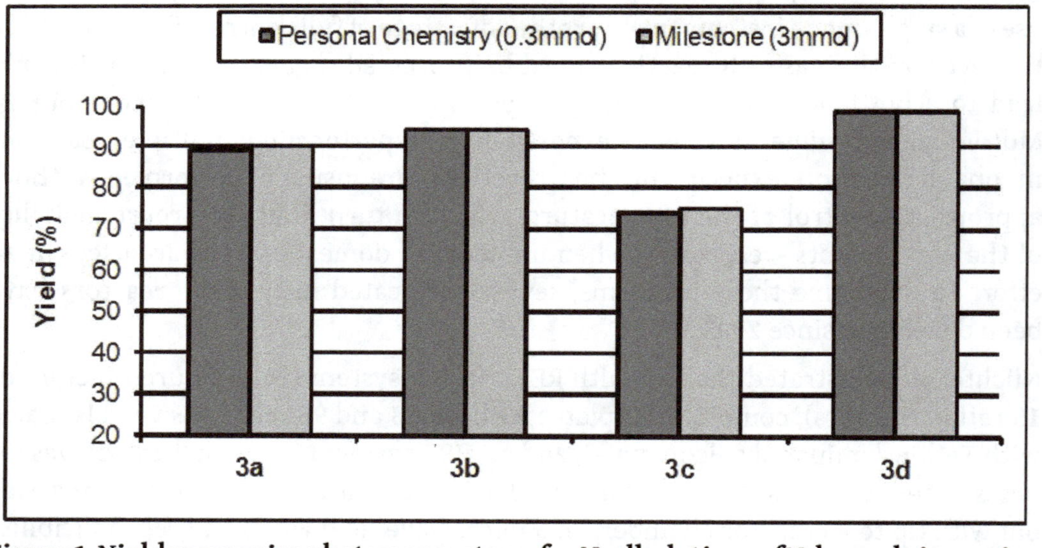

Figure 1. Yield comparison between systems for N-alkylations of N-benzylpiperazine

On the basis of these results, a multimode combiCHEM reactor was used to assess the heating homogeneity of a multiwell plate with 24 equal reactions (Alcázar, 2005). Interestingly, reaction yields were slightly higher in the multimode instrument than in a single-mode set-up, and the author concluded that the former system offered enough reproducibility to develop further parallel chemistry. Once the performance of the multiwell plate had been evaluated, a set of 24 different compounds was prepared in parallel using the fully optimized conditions by combining four amines with six alkylating agents. In order to make an appropriate correlation across instruments, the same reactions were performed in a single-mode reactor. The results showed that good reproducibility was achieved between the two instruments: on average, the difference in yield was only 2.3%. However, to complete the whole sequence of reactions the single-mode instrument required 2.5 h, whereas the multimode instrument required only 40 min to achieve comparable results. This system combined the advantages of the parallel approach and microwave heating.

As mentioned above, for the scale-up from gram to kilogram scales two different approaches have been used historically – batch and continuous-flow. The major limitation of the continuous microwave reactor is that these systems are unsuitable for heterogeneous mixtures and viscous liquids. For these reasons, Maes performed a comparative study between a stop-flow system (CEM Voyager system based on the Discover single-mode platform equipped with an 80 mL glass vessel, a peristaltic pump and two valves) and two multi-mode platforms that allow parallel processing of several vessels per batch: Milestone MicroSYNTH equipped with a rotor with 10 vessel positions (vessel volume 100 mL) and CEM MARS equipped with a fourteen rotor positions (80 mL) (Loones et al., 2005). The reaction model used was the Buchwald–Hartwig amination (Scheme 7). They found that rapid Pd-catalyzed amination of aryl chlorides under microwave irradiation can be easily scaled-up on both single-mode and multi-mode platforms with similar yields if BTF (trifluoromethylbenzene) is used as the solvent. However, the stop-flow Voyager system was preferred since it is a completely automated unit that allows the continuous production of reaction product without the need to manually load and unload reaction vessels. Moreover, the Voyager system allows pumping of heterogeneous mixtures, which is problematic in continuous-flow units.

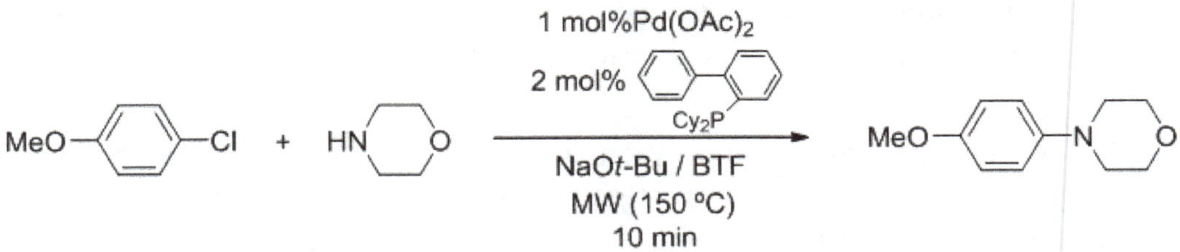

Scheme 7. Scale-up of the Pd-catalyzed amination of 4-chloroanisole with morpholine

2.3. Reproducibility across instruments

Reworking some of the reactions described in the literature that were originally performed in household microwave devices showed that it is absolutely necessary to analyze any result obtained in these specific devices. Due to the incomplete description of many reactions reported in microwave chemistry, reproducibility is difficult to assess in most cases. Only in rare cases are identical microwave devices available. Additionally, devices of the same series can have different field homogeneity. Only in a few cases are all reaction conditions known precisely and reproducibly. Even the use of different amounts of starting materials can produce different results, a factor that can usually be neglected for conventionally heated reactions. Moreover, the applied settings of the microwave devices are often insufficiently reported or the importance of these parameters is completely ignored (Nüchter et al., 2003b).

In 1994 Login reported that commercial domestic ovens have limitations that can result in non reproducible fixation results in pathology and he developed a calibration protocol to identify the best locations for fixation within household microwave ovens (Login & Dvorak, 1994).

Microwave-assisted extraction (MAE) is a process in which microwave energy is used to heat solvents in contact with a sample in order to partition analytes from the sample matrix into the solvent. In most cases, recoveries of analytes and reproducibility are improved compared to conventional techniques. The basic principles and main studies, including reproducibility, on the use of microwave energy for extraction have been reviewed (Eskilsson & Björklund, 2000).

Mastragostino reported the microwave-assisted synthesis of $Ag_2V_4O_{11}$ starting from V_2O_5 and Ag_2CO_3 (Arbizzani et al., 2007). Although acceptable reproducibility may be attained in the exploratory phase by using low-cost domestic ovens modified to read the temperature of the reaction mixture (the irradiation was manually stopped when the temperature of the sample reached a set value), properly controlled synthesis conditions can be achieved only with scientific microwave systems with control of irradiation power and reaction temperature (such as the CEM Discover single-mode reactor).

The influence of microwave energy on the rate of cleavage of –C–S– and –S–S– bonds by azide ions has been investigated using a domestic oven. It was found that microwaves do not accelerate the cleavage of the –C–S– bond whereas –S–S– bond cleavage is faster. However, the reproducibility of this reaction is very poor and the microwave technique is not recommended for quantitative determination (Kurzawa & Stachowiak, 2001).

When the microwave methodology was introduced twenty-five years ago, most reactions were performed in domestic ovens without appropriate temperature and pressure controls. As a consequence, there is a plethora of interesting organic transformations that have been reported to take place in domestic ovens. However, reactions performed in this kind of instrument, i.e. without appropriate temperature and pressure controls, are generally considered as not reproducible, thus limiting the application of such reactions.

The polarity of the solvent is the most important parameter to consider when microwave reactions are performed in solution. Polar solvents absorb the microwave radiation directly and the polarity of the substrate is relatively unimportant. In the case of non-polar solvents, however, the radiation is absorbed by the substrates, but the differences in absorption of the substrates are moderated by the solvent, especially in dilute solution. In both cases, the reaction temperature is limited by the solvent boiling point and microwave-assisted reactions in solution are easily controlled. In neat reactions, radiation is again absorbed by the substrates but there is no solvent to stabilize the temperature. In this situation, the nature of the substituents and the polarity of the substrates influence the absorption of microwave energy and, hence, the yield. Moreover, in the absence of solvent the temperature is not limited by the boiling point of the solvent. A process involving highly polar reagents, intermediates or transition states can hardly be controlled under microwave irradiation. Hence, it does not follow that a solvent-free reaction previously performed in a domestic oven will be controllable and reproducible in dedicated microwave reactors in a similar way to reactions in solution, as the field density is very different from one instrument to another (Nüchter et al., 2004).

In 2007 Díaz-Ortiz & Alcázar showed for the first time that solvent-free reactions performed in domestic ovens could be reproduced, scaled and parallelized in controlled microwave monomode and multimode reactors (Díaz-Ortiz et al., 2007). The 1,3-dipolar cycloaddition of nitrile *N*-oxides with nitriles (Scheme 8) was selected as a model reaction, and this was previously performed by the same authors using a domestic oven (Díaz-Ortiz et al., 1996). The study was structured in four steps. The first step involved the translation of the 1,3-dipolar cycloaddition reaction conditions from the domestic oven to the single-mode CEM Discover microwave reactor by increasing the temperature gradually in successive experiments to prove that a linear dependence between the yield and temperature is observed for this solvent-free reaction (Figure 2). The second step was to study the reproducibility of the reaction under non-optimized conditions because under such conditions reactions are more sensitive to temperature differences. In this case, four identical reactions were performed in parallel in the MicroSYNTH multimode reactor at 100 ºC on a scale five times larger than the one used in the monomode instrument. It can be concluded that there is good reproducibility between the two systems (Figure 2).

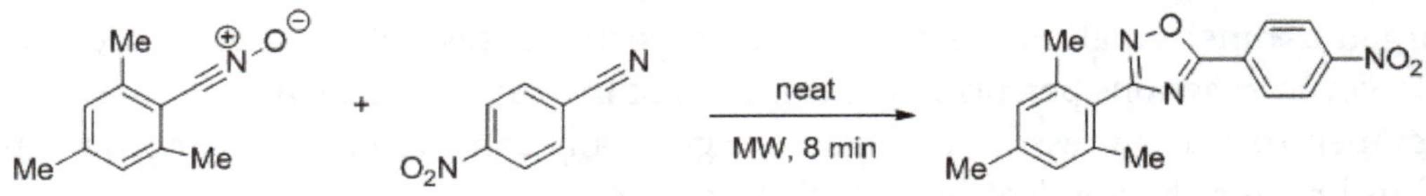

Scheme 8. 1,3-Dipolar cycloaddition of nitrile with nitrile N-oxide

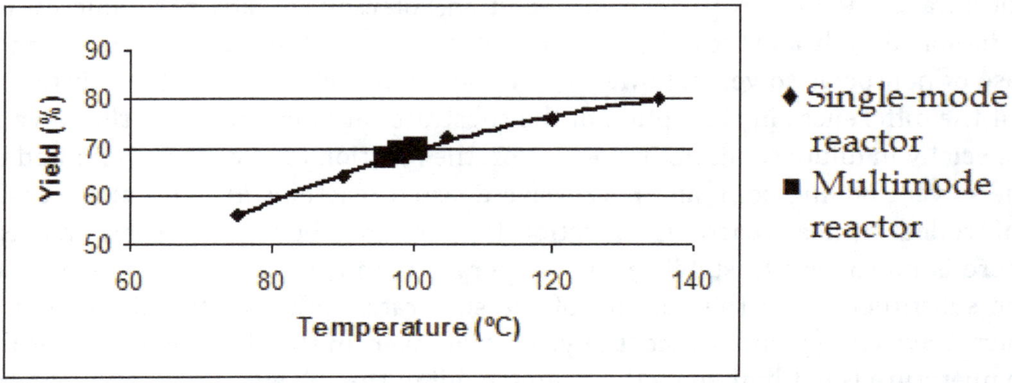

Figure 2. Plot of yield versus temperature for 1,3-dipolar cycloaddition reactions

The third step in the study concerned the scalability of the reaction. For this purpose, eight different nitriles were reacted with a nitrile *N*-oxide using optimized reaction conditions. Reactions were performed in both single-mode and multimode microwave ovens. On average, the difference in yield between the two instruments was 2.2%. The fourth step was to study the preparation of 24 different compounds in a multiwell plate under microwave irradiation to provide an effective combination of productivity and speed. To overcome the problem of different absorption levels in each reaction, due to the different polarity induced by the substituents, the authors took advantage of a Weflon well plate (Nüchter, 2003a). Finally, the results obtained for the 8 substrates used in all experiments (single-mode, multi-mode, Weflon plate and isolated yield) were compared and good reproducibility was found for each compound under each set of conditions: the average yield difference was 3.75% (Figure 3). Similar conclusions were reported by Leadbeater in a study into the rapid optimization of reaction conditions (Leadbeater & Schmink, 2007).

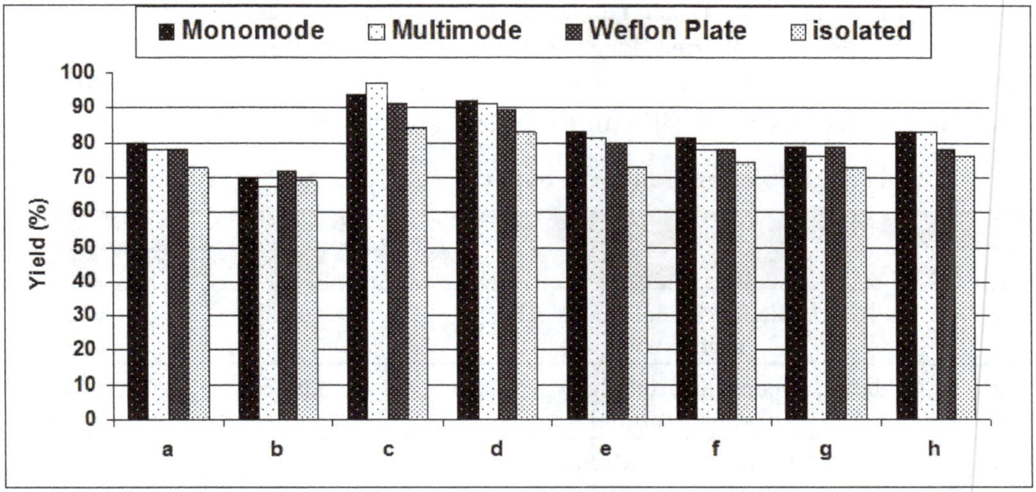

Figure 3. Comparative results for eight nitriles in different applications and isolated yields obtained

In order to extend the preliminary results to see whether any solvent-free reaction previously performed in a domestic oven can be reproduced in dedicated apparatus, the same authors studied four new reactions that cover a wide range of chemical transformations (Díaz-Ortiz et al., 2011). It was found that *N*-alkylation of (1*H*)-benzotriazoles, condensation of anilines with urea (or thiourea) and Beckmann rearrangement of ketoximes are reproducible in both single-mode and multimode microwave instruments using temperature-controlled conditions. However, all attempts to reproduce and control the oxidation of benzylic bromides with pyridine *N*-oxide in single-mode or multimode reactors failed (Scheme 9).

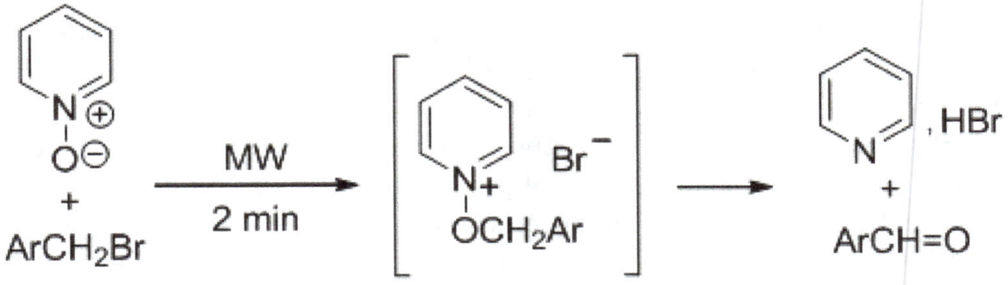

Scheme 9. Oxidation of benzylic bromides

With the aim of explaining why the oxidations of benzylic bromides are not reproducible processes under microwave irradiation, a computational study on this reaction and the aforementioned 1,3-dipolar cycloaddition – which is easily reproduced - was performed (Table 1). The results show that reactions involving a moderate or medium increase in polarity in the pathway from reactants to products (Table 1, Entry 1) are relatively easy temperature-controlled processes under microwaves. In contrast, large increases in polarity during the reaction path (Table 1, Entries 2 and 3) give rise to extreme absorptions of microwave energy and make these processes more difficult to control and reproduce.

Table 1. Dipole moments (Debyes) of selected stationary points

Entry	Process	Dipole Moments (Debyes)	
		Reactants	Intermediate
1	1,3-Dipolar cycloaddition.solution (acetonitrile)	5.51	11.27
2	Oxidation of benzylic bromides S_N2 solution (bromobenzene)	5.05	25.02
3	Oxidation of benzylic bromides S_N1 solution (bromobenzene)	5.55	52.04

A combined (multimode and single-mode) microwave device (MLS ETHOS 1200 Combi) was used by Ondruschka to perform a comparative study of both single-mode and multimode microwave irradiation methods (Nüchter et al., 2003b). The authors studied a lipase-catalyzed transesterification process, and Biginelli and Hantzsch reactions. The results were similar for both single-mode and multimode microwave devices in all reactions. Differences in the yields of the pure products were due to different workup procedures and small systematic errors. The authors completed their work with a useful study of the heating behaviour of polar systems in the multimode microwave field: energy input in pure substances, and mixtures of polar and non-polar substances.

3. MICROWAVE-ASSISTED LARGE-SCALE PRODUCTION

The majority of documented examples of microwave-assisted chemistry have been conducted on a scale below 1 g, with a standard reaction volume ranging from 1 to 5 mL. The primary applications have been to expedite and enhance recognized and proven synthetic methods. Microwave-assisted synthesis requires the development of systems capable of producing goods on a multikilogram scale for this method to gain full acceptance as an industrial technology. Consequently, the advancement of large reactors is necessary, at a minimum on the pilot plant scale, to facilitate multikilogram production of lead compounds. The subject of scale-up was initially examined in the mid-1990s (Raner et al., 1995; Cablewski et al., 1994; Roberts & Strauss, 2005). Since then, numerous conceptual and commercial microwave scale-up reactors have been documented for both batch and continuous operations, primarily by chemical engineering teams.The scale-up of microwave chemistry clearly has several benefits to offer over conventional heating. However, there are some problems that make the scale-up of microwave chemistry difficult to achieve. The big challenge for process scale-up involving microwave technology is to establish a reliable and safe process setup where issues like physical properties (i.e., penetration depth), temperature control, and reactor design have to be carefully considered. On using batch reactors, the user is confronted with problems in heating large volumes due to the limited penetration depth of microwave irradiation (Nüchter et al., 2004).

The scale-up of microwave-mediated transformations can be defined in different ranges, leading to the use of different concepts and varying instrumentation. Depending on the user, the term "scale-up" will have different meanings. In the case of method development, scale-up starts with processing a 50 mL reaction mixture, corresponding to a 10- to 100-fold scale-up of reactions performed in standard single-mode microwave vials with a processing volume of 0.5–5.0 mL. A possibility for further scale-up would be the use of the "numbering up" approach, i.e. running the same reaction several times in sequence. This approach is aided by existing robotic equipment, which allows unattended vessel transfer in and out of the microwave cavity. Alternatively, such reactions can also be performed by parallel synthesis in corresponding multivessel rotor systems on switching to multimode instruments. From an industrial viewpoint, scale-up means process development and the highest possible throughput, a situation that virtually excludes the use of batch reactors. In fact, it is the productivity and not the size of the vessel that is important, which clearly indicates that flow systems, regardless of whether applied in stop-flow (SF) or continuous flow (CF) mode, have distinct benefits over batch process reactors.

The approaches can be categorized as: scale-up in parallel, scale-up in sequence, and scale-up by continuous flow (Figure 4). Hybrid approaches have also been devised.

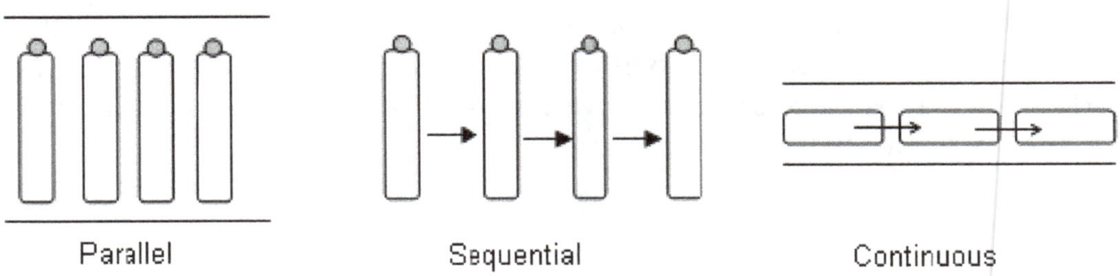

Figure 4. Strategies for microwave scale-up

3.1. Scale-up in batch

The advantages of microwave heating in chemistry results in the need for scale-up possibilities. Moreover, the direct homogenous heating of reaction mixtures under microwave irradiation is believed to facilitate direct scaling without heat and mass transport issues (Kappe & Stadler, 2005). In fact, direct scaling of microwave-assisted synthesis in batch mode, without the need for process optimization, has already been demonstrated in organic synthesis on using open or closed reaction vessels (Leadbeater, 2010).

3.1.1. Scale-up in a single large open vessel

When considering batch scale-up, one approach is to use a single, larger, reaction vessel. This avoids the problem of tedious charging and discharging of multiple reaction vessels. However, penetration depth and/or power density could become issues as the vessel size increases.

If the reaction can be carried out in the absence of solvents, or superheated solvents are not required, or if the process can be performed successfully below the boiling point of the solvent, the large reaction can be run in an open-vessel model. At present there are several commercially available microwave ovens to carry out this scale-up mode, including CEM MARS, MILESTONE Micro-SYNTH and MINILABOTRON SAIREN.

This methodology has some disadvantages. For example, the penetration depth or power density may be an issue and of course reactions that need elevated temperatures and pressures cannot be carried out. However, considerable scale-up can be – and has been – achieved in such reactors.

Open-vessel conditions prove advantageous for driving equilibrium reactions to completion when the product or by-product is a more volatile component than the starting materials. A Milestone MicroSYNTH reactor has been used to perform a fractional distillation in the synthesis of a tricyclic heterocycle (Razzaq & Kappe, 2007). *N*-Methylaniline was reacted with two moles of diethyl malonate in a cyclocondensation reaction with elimination of four moles of ethanol to give the product pyranoquinoline on a 0.2 molar scale. On this scale, the distillation took only 82 min as compared to 3-5 h conventionally. When the reaction was performed in a sealed-vessel microwave system only <5% product formation was achieved (Scheme 10).

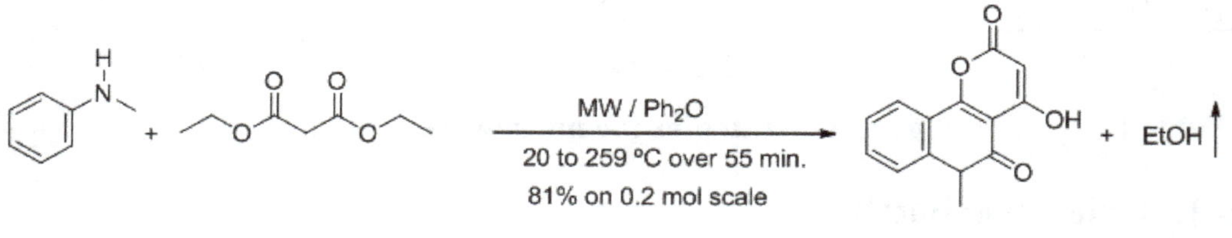

Scheme 10. Reaction between N-methylaniline and diethyl malonate

The Hantzsch 1,4-dihydropyridine synthesis (Bowman et al., 2008a) was also scaled up successfully to the 0.5–1 mol level in the MARS unit and this involved heating up to 1 L of reaction mixture. One batch provided 250 g of dihydropyridine product in less than 1 h.

Leadbeater et al. evaluated a new batch reactor, designed by AccelBeam Synthesis, that allows reactions to be performed on scales from 2–12 L (Schmink et al., 2010). A number of reactions have been investigated, including palladium-mediated transformations, condensation reactions, nucleophilic aromatic substitution reactions, and alkylations. One important aspect of this work was that a linear scaling approach was taken and changes were not made to the protocol on moving from the small, developmental scale to larger scales. In some cases reactions were scaled over 18,000-fold on moving from small (0.1–1 mmol) to large (1–18 mol) runs. It is noteworthy that in order to simulate a situation where multiple sequential microwave steps were employed to reach a desired target compound, these authors developed a sequence of reactions as a medicinal chemist might do at the 10 mmol scale in order to synthesize a drug-like molecule (Scheme 11). Overall, the sequence employed three microwave steps and afforded 473 g of the target molecule in 38% yield, almost identical to the 39% obtained in the small scale process.

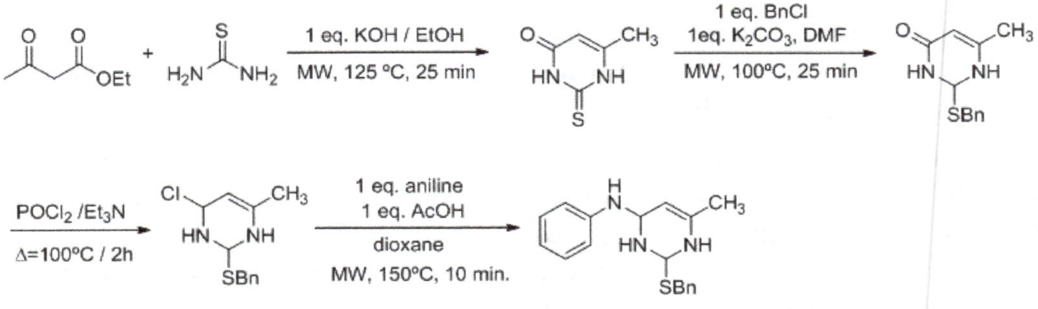

Scheme 11. Preparation of 2-(benzylthio)-6-methyl-4-(phenylamino)pyrimidine

Horikoshi et al. employed a batch reactor to synthesize silver nanoparticles in aqueous media by reduction of a diaminosilver(I) complex with carboxymethylcellulose (Horikoshi et al., 2010). The microwave process yielded smaller silver particles with a narrower distribution than the conventional heating process.

3.1.2. Scale-up in a single large sealed vessel

Another approach to perform the scale-up involves the use of a single large sealed vessel. In this case, it is possible to carry out reactions in heterogeneous conditions, in the presence of solvents, and at pressure. Despite this versatility, the limiting factor is the vapour pressure generated by a superheated solvent due to safety concerns. While moderately high pressures (~20 bar) are easily contained in small scale microwave reactors, the use of large vessel sizes requires more safety features and greater engineering expertise in the system design. This results in more complex reactors that are less easy to use and more expensive. In this sense, to contain the pressures likely to be generated, strong materials must be used to construct the reaction vessels, but they must also be microwave transparent. This rules out the use of metals on this scale and leaves thick-walled (quartz) glass, ceramics, and polytetrafluoroethene (PTFE or Teflon), or combinations thereof. However, these materials are all thermal insulators. In either case, cooling is further compromised by the necessarily thick-walled vessels such that, for larger microwave reactors, the cooling time is often notably longer than the combined heat-up and reaction periods. This situation increases the cycle time per batch and reduces the overall output, thus lessening the benefit of rapid microwave heating. Some of these problems have been solved in certain systems. For example, flash evaporation using the mechanical pressure of superheated solvent in the vessel can discharge the reaction mixture while it is still hot into a collection vessel, thus allowing the cooling to occur off-line from the reactor.

Commercially available microwave reactors in which sealed-vessel scale-up can be performed are the BIOTAGE ADVANCER 350, Batch SYNTH, Ultraclave, MILESTONE ETHOS 1600, Anton Paar monowave 300, and Anton Paar Masterwave BTR.

Scale-up in a single sealed vessel represents perhaps the least utilized scale-up option. This may be because the reactors are large and expensive and that more expertise is required for their use.

One particular example of pharmaceutical interest was the Grandberg synthesis of 2-methyltryptamine (Scheme 12) (Bowman et al., 2008b). A conventional approach to this reaction had already been scaled up to 20 kg at Novartis (Slade et al., 2007). However, the Advancer reactor was used to give results on a 0.2 mol scale that are comparable to those achieved by the Novartis group.

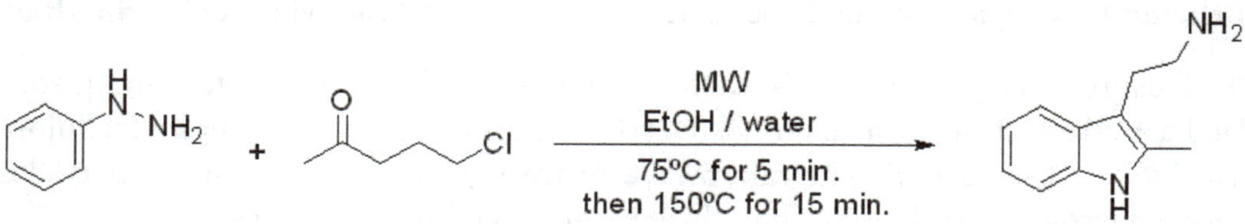

Scheme 12. Grandberg synthesis of 2-methyltryptamine

Schubert reported ring-opening polymerizations in batch mode of 2-ethyl-2-oxazoline on scales up to 100 g without the need for process optimization (Paulus et al., 2006). Nevertheless, it should be noted that this pressurized process cannot be scaled indefinitely in batch mode for safety reasons.

Deetlefs reported the use of a CEM MARS system for the synthesis of a wide range of ionic liquids based on nitrogen-containing heterocycles (Deetlefs & Seddon, 2003). The process can be performed on a range of reaction scales (50 mmol to 2 mol) in either sealed or open vessels.

3.2. Scale-up in parallel synthesis

Laboratory-scale batch microwave reactors generally offer a maximum batch size of 1 L reaction volume and in most cases this is divided into several smaller reaction vessels (multivessel rotor systems). This approach does not exactly match the "one vessel" philosophy, but the parallel setup allows for the processing of several batches of either the same reaction mixture or of closely related mixtures. In fact, this combination of batch synthesis and parallel processing can furnish either compound libraries on the gram scale or a larger amount of one single compound in a short time. Although the Milestone UltraCLAVE system provides a 3.5 L single vessel cavity, it has proven to be more effective to heat smaller volumes in parallel rather than one big batch, given that identical microwave output power is applied (Loones et al., 2005; Cléophax et al., 2000).

The demands made on industry, especially the pharmaceutical industry, are changing at an unprecedented pace, making speed critical in modern chemistry. For this reason, high-speed microwave synthesis and combinatorial chemistry have attracted a considerable amount of attention in recent years. These approaches offer significant advantages to the synthetic chemist: reduced reaction times and improved yields as a result of microwave heating and increased productivity due to the implementation of combinatorial chemistry. However, the evolution of microwave instrumentation is delivering new systems that allow reactions to be carried out in specialized multiwell plates under temperature and pressure controlled conditions (Nüchter & Ondruschka, 2003; Alcázar, 2005; Kremsner et al., 2007).

Caliendo et al. disclosed the synthesis of series of piperazine derivatives as 5-HT1A agonists using microwave irradiation in all the reactions (Caliendo et al., 2001; Caliendo et al., 2002) (Scheme 13). They compared the results obtained with microwave heating with those obtained using traditional heating. In all cases, compounds were obtained in higher overall yields and reaction times were reduced from hours to minutes on using microwave irradiation. Reactions were performed in a multimode instrument using parallel racks that allow the preparation of sets of compounds in a single irradiation experiment.

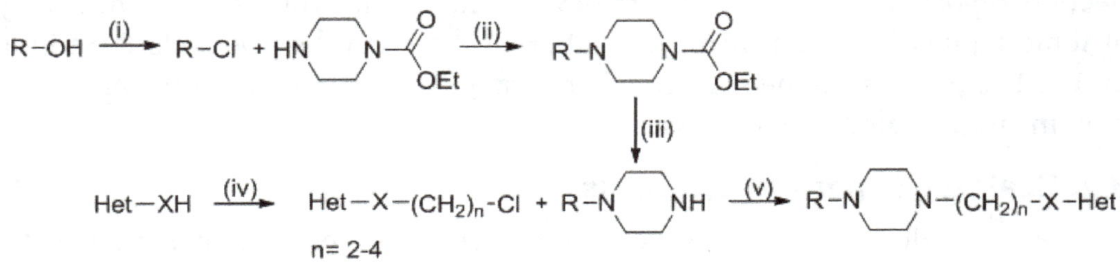

Scheme 13. Synthesis of N-substituted piperazine derivatives. (i) $SOCl_2$, toluene, 55 °C, 20 min, MW; (ii) K_2CO_3, NaI, DMF, reflux, 1 h, MW; (iii) KOH, water, methanol, reflux, 1 h, MW; (iv) $Br(CH_2)_nCl$, K_2CO_3, DMF, MW; (v) K_2CO_3, DMF, 70 min, MW

Researchers at Johnson & Johnson Pharmaceuticals have described tropanylidene derivatives as combined mu/delta opioid receptor agonists (Costas et al., 2004). The preparation of these analogues was achieved by developing a parallel solution phase strategy under microwave irradiation, which allows multiple sequential reactions to occur in a single reaction vessel (Scheme 14).

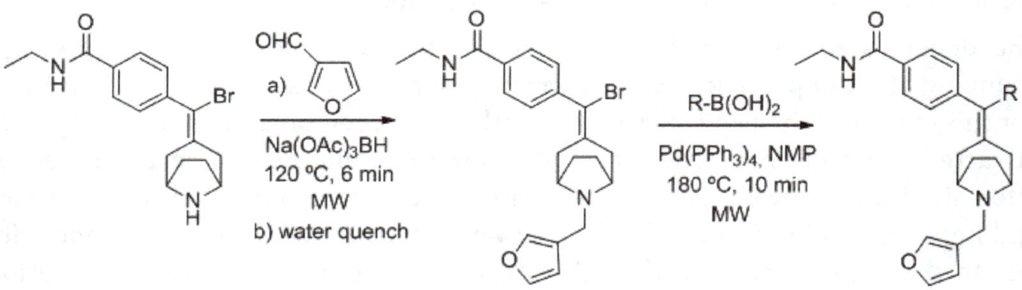

Scheme 14. Preparation of tropanylidene derivatives

This approach allowed the synthesis of 192 compounds in a quicker and more efficient manner than using the corresponding solid phase approach. Additionally, the parallel solution phase approach allowed rapid scale-up of selected compounds for further *in vivo* **studies.**

Caliendo et al. described an efficient, facile, and practical parallel combinatorial synthesis of substituted-benzoxazines under microwave irradiation (Caliendo et al., 2004). The procedure involved the use of a specially designed microwave oven for organic synthesis that was suitable for the parallel synthesis of solution libraries. As a demonstration, a library of 19 substituted *N,N*-dimethyl- and *N*-methyl-benzoxazineamide derivatives, structurally related to the potassium channel opener cromakalim, was generated by both conventional and microwave procedures, with a reduction in the library generation time for the microwave approach from 7 h to 30–36 min (Figure 5).

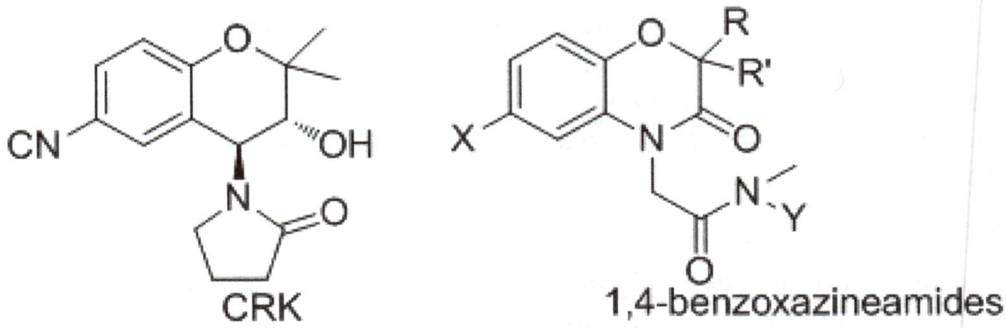

Figure 5. Structure of cromakalin (CRK) and 1,4-benzoxazineamides

To facilitate the preparation of β-peptide libraries in parallel, Murray and Gellman adapted the reaction conditions for the solid-phase synthesis of 14-helical β-peptides for use in a multimode microwave reactor (Murray & Gellam, 2006). The low temperature and pressure requirements of microwave-assisted β-peptide synthesis were found to be compatible with multiwall filter plates composed of polypropylene. Microwave heating of the 96-well plate was sufficiently homogeneous to allow the rapid preparation of a β-peptide library in acceptable purity.

An environmentally friendly method based on microwave radiation has been developed for the synthesis of a library of *N*-acylhydrazones (Andrade & Barros, 2010). With this protocol, the use of solvents and catalysts can be avoided, and the products are obtained in short reaction times and in very good yields. A variety of *N*-acylhydrazones were synthesized under microwave irradiation within 2.5–10 min, starting from benzo, salicyloyl, and isonicotinic hydrazides. The results are reproducible on a 500 mg to 5 g scale and this approach could be applied to a large number of ketones and aldehydes. The reported method could be a useful synthetic path to obtain *N*-acylhydrazones on an industrial scale (Scheme 15).

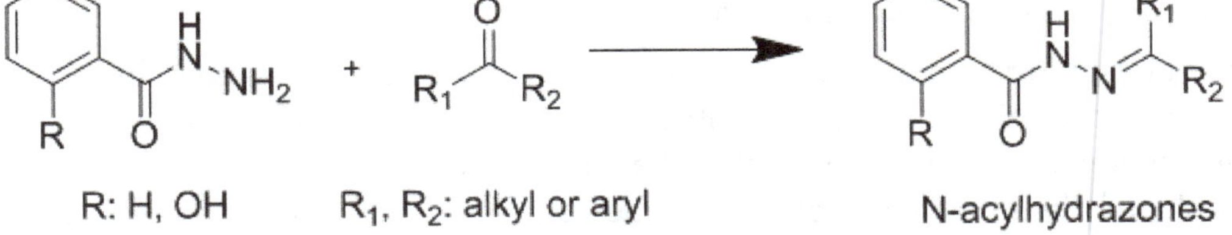

Scheme 15. N-acylhydrazones synthesized under microwave irradiation

The integration of parallel synthesis with microwave chemistry has significantly enhanced the efficiency of derivative preparation. The sequential synthesis of compound libraries using fully integrated single-mode instruments on automated platforms is well known. The incorporation of reaction performance in multiwell plates subjected to microwave irradiation, along with automated platforms for sample preparation and processing, is anticipated to enhance library synthesis further..

3.3. Scale-up in continuous flow systems

Although dedicated modern microwave instruments for MAOS are very successful in small-scale operations, microwave scale-up in batch mode beyond a reaction volume of approx. 1 L is not feasible (Glasnov & Kappe, 2007). The penetration depth of only a few centimetres and the limited dimensions of the standing wave cavity are the main reasons for the development of continuous or stop-flow microwave reactors. In these systems the reaction mixture is passed through a relatively small microwave-heated flow cell, which avoids the aforementioned drawbacks.

It was recognised early on in the development of microwave reactors that flow-based applications offered tremendous advantages in terms of processing capabilities. Two of the early pioneers of microwave chemistry extolled the virtues of continuous microwave reactors as early as the 1990s (Chen et al., 1990; Cablewski et al., 1994). However, such pioneering developments received little recognition or further development due to a combination of inferior technology, poor quality manufacture and the fact that the adoption of continuous flow protocols was in direct contradiction to the emerging, at that time, Combinatorial Chemistry (small scale multiparallel batch processes).

The key requirement of a continuous flow microwave reactor is the ability to continuously monitor and adjust the reaction parameters whilst in operation. This facilitates easy reaction optimization, introduction of automated safety controls and ensures a reliable and prolonged processing capability. Additionally, the combination of such processing techniques with the enabling technology of immobilized reagents, scavengers and catalysts to effect multi-step synthetic transformations has allowed the generation of novel pharmaceutical architectures and more complex natural products.

One of the first successful examples of scale-up was developed by Strauss (Cablewski et al., 1994). These authors prepared hundred gram quantities of a cyclopentenone intermediate using the continuous microwave reactor that they developed themselves (Scheme 16). The standard purification protocol for the product (distillation to remove small residual quantities of the starting material) was ineffective for large scale syntheses and, as a result, a scavenging procedure using a bisulfate-functionalized ion exchange resin to remove the reaction base (NaOH) was employed. The enhanced processing capability provided by this solid phase purification approach demonstrates the superior throughput that can be achieved by an innovative combination of technologies.

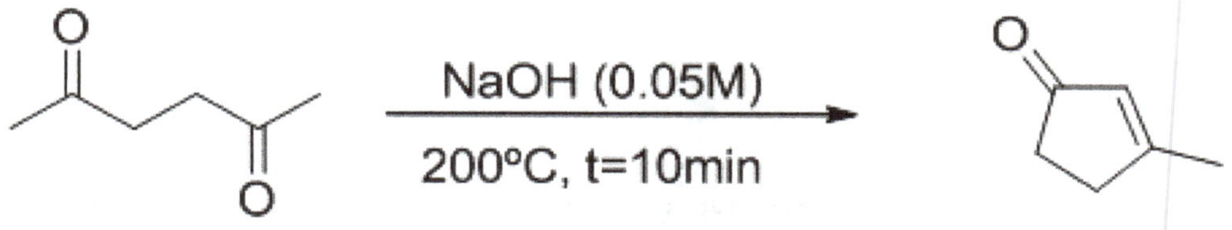

Scheme 16. Preparation of 3-methylcyclopent-2-enone using a flow microwave reactor

Recently published examples of continuous flow organic microwave synthesis involve, for example, the synthesis of 5-amino-4-cyanopyrazoles (Smith et al., 2007). The design consists of fluorinated polymer tubing wound around a Teflon core that is fitted with a dummy pressure cap. This flow device has the basic shape of a 20 mL vial suitable for the Biotage EXP single-mode instrument. The input tube is connected to an HPLC pump and a 7 bar back-pressure regulator is placed at the exit. Both connections are physically located at the bottom of the microwave unit. In addition, with this system, purification can be facilitated by passing the exiting flow stream through columns packed with polymer-supported reagents or scavengers. The synthesis of various 5-amino-4-cyanopyrazoles by reaction of a set of hydrazines with ethoxymethylene malononitrile in methanol was performed at temperatures between 100–120 ºC with a residence time of 0.8–4 min (Scheme 17). Subsequent passage of the reaction mixture through a column with supported benzylamine to scavenge any unreacted malononitrile, followed by a column packed with activated carbon to remove coloured impurities, furnished the desired product in high purities, good to excellent yields and in quantities up to 250 g. A benefit of this flow device is its versatility, since different tubing lengths can be wrapped in the system to provide reactors with different internal volumes. The application of multiple bundle tubings allows different reactions to be performed in parallel in one single reactor.

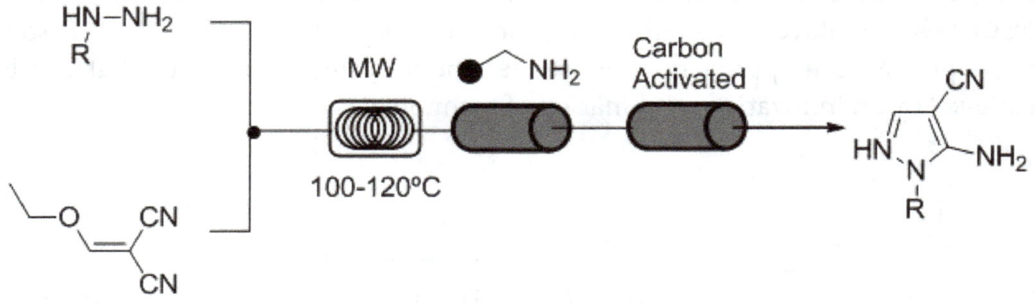

Scheme 17. Flow microwave synthesis of 5-amino-4-cyanopyrazoles

Microwave heating seems to be particularly competitive with transition metal-catalyzed processes, not only bringing long reaction times down to minutes but also minimizing the levels of undesired side products and preventing collapse of the catalytic system (Nikbin et al., 2007).

The combination of reactor design with immobilization techniques is very important for the flow process as it facilitates maximal interaction between reagents and catalysts without causing clogging problems. Moreover, the immobilized catalysts could be used over several cycles without a significant drop in activity. Kappe described a Heck cross-coupling reaction using a continuous-flow reactor based on a megaporous glass carrier material with suitable polymer functionalization introduced into the pore volume of this support (Kunz et al., 2005). For this purpose, a basic ion-exchange resin-loaded monolith was used in order to create the close neighbourhood of ionic sites and Pd(0) sites (Scheme 18). Pure products were collected without the need for extensive purification steps. The combination of composite-based flow-through reactors with microwave irradiation may lead to new and effective methods to scale up organic reactions.

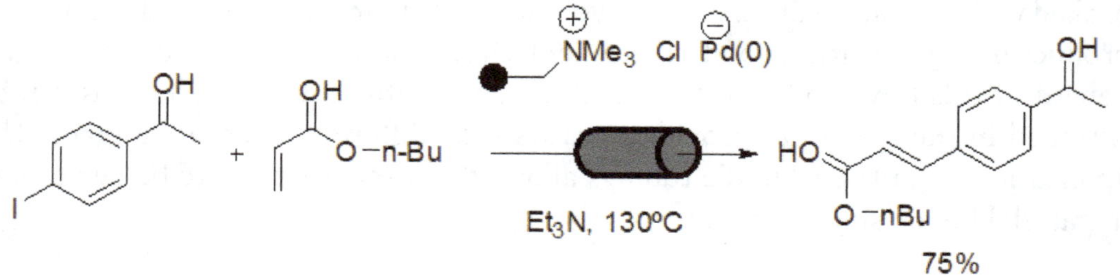

Scheme 18. Heck transformation applying a continuous-flow reactor

Recently, Organ has prepared highly porous Pd films composed of nanometer-sized particles (Comer & Organ, 2005). These Pd films served as an excellent catalyst for Heck reactions under continuous flow microwave conditions. The authors showed that 10 mg of product can be obtained from this reactor system within 1 min. Gram quantities of the products were obtained by flowing reaction mixtures through a single capillary for about 90 min.

Another interesting study was described by Wilson (Wilson et al., 2004), who developed a continuous microwave reactor that eliminates the potential reaction parameter re-optimization (time and temperature) typically required when methods are transferred from small-volume, single-mode systems to larger (but limited)-volume multimode systems (Figure 6). The representative chemistries explored include S_NAr, esterification, and the Suzuki cross-coupling reaction, all of which were successfully and safely scaled up to multigram quantities using a home-made continuous flow microwave cell (Scheme 19).

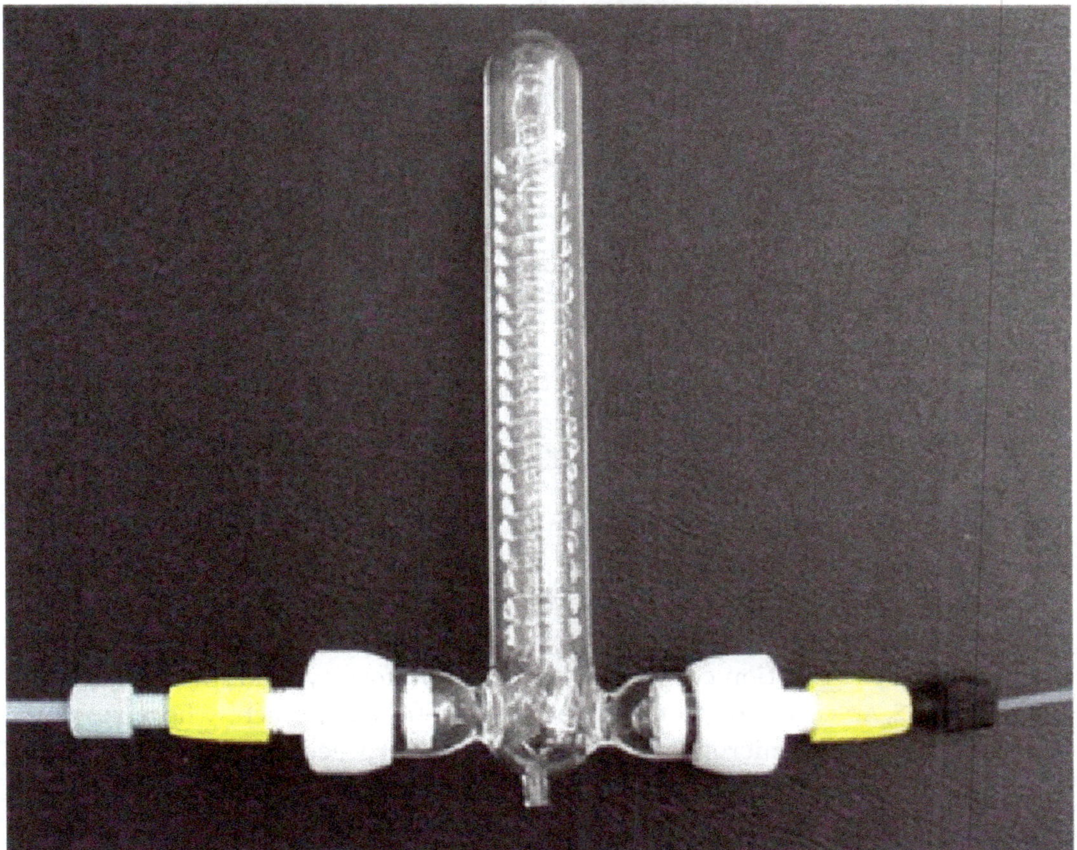

Figure 6. Glass coiled flow cell

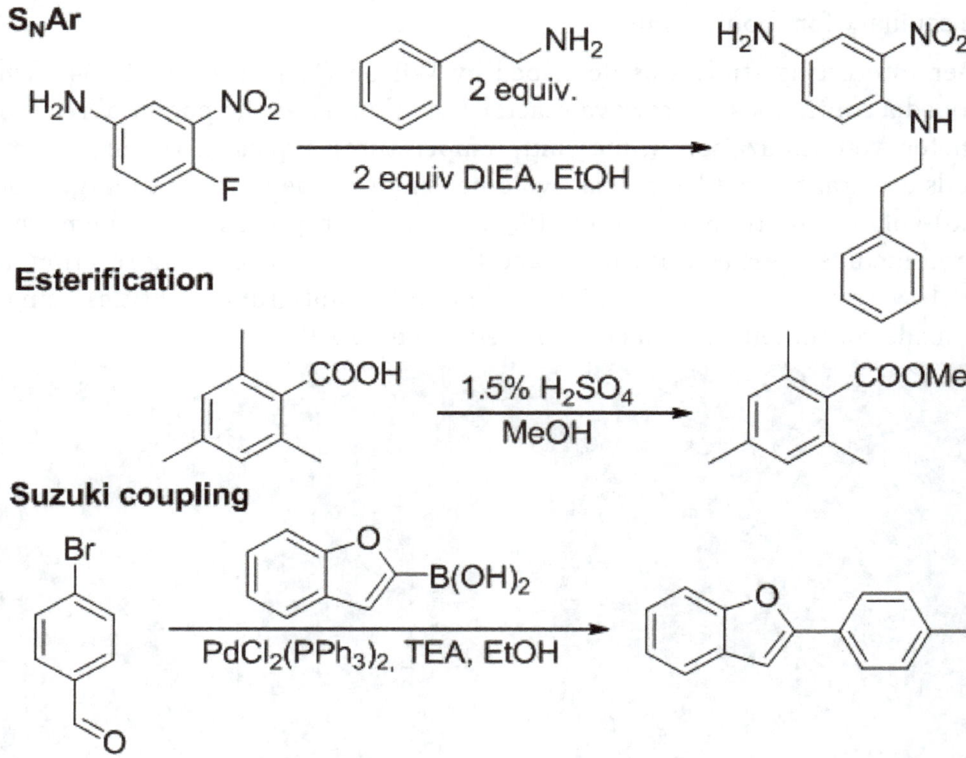

Scheme 19. S$_N$Ar, esterification, and a Suzuki cross-coupling reaction in a microwave-flow reactor

When reactions need to be performed on several hundred gram or kilogram scales, typical for pilot or manufacturing plants, microwave synthesis is not easily scalable (Singh et al., 2008). As a result, the development of stop-flow or continuous flow microwave reactors, capable of working with slurries, are key for the successful translation of the reaction conditions. In this way Moseley (Marafield & Moseley, 2010) used S$_N$Ar reactions to evaluate the stop-flow approach, where reagents were loaded into the microwave vial, the reaction proceeded and finally the vial was emptied in an automatic process without manual intervention. This approach gave production rates of >0.5 Kg per day, which would meet the manufacturing requirements of early clinical pharmaceutical development.

The same transformation shown in Scheme 20 was used in a recent article to validate the Milestone FlowSYNTH, a commercially available continuous flow microwave reactor designed for working at manufacturing scale (Bergamelli et al., 2010). This instrument gave productivities up to 0.65 mol/h (~170 g/h). Even higher productivities were achieved for the ortho Claisen rearrangement, in which nearly 30 kg of material per day could be produced in one unit. The rearrangement was included in a three-step microwave synthesis of naphthofuran, which was successfully scaled up, although the first reaction, naphthol *O*-alkylation, was performed in a 2L batch reactor as pumps became blocked by the heavy slurry (Scheme 21).

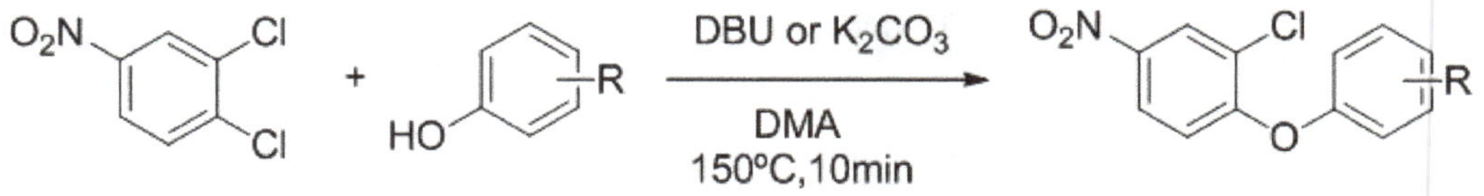

Scheme 20. S_NAr evaluated in a stop-flow reactor

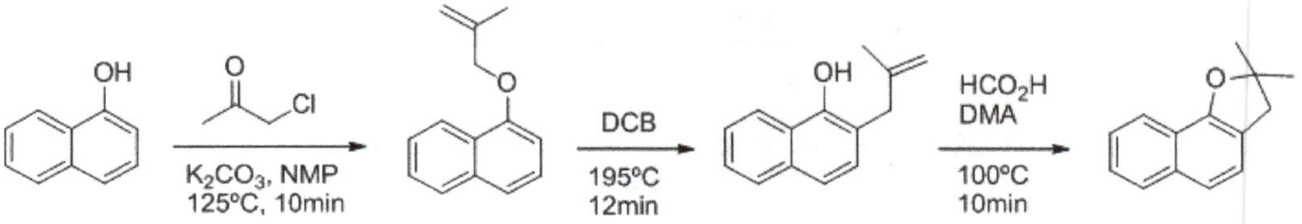

Scheme 21. Synthesis of naphthofuran

In order to improve the performance of the system, Dressen modified five key points to reduce the potential blockage by solids (Dressen et al., 2010). The modified system was assessed in the enantiomerically selective esterification of (*R*,*S*)-1-phenylethanol with vinyl acetate using Novozym 435 and the esterification of (*S*)-pyroglutamic acid with *n*-decanol (Scheme 22). Despite these improvements, plugging problems still occurred as a result of solid reactant /product deposition. Despite these issues the authors concluded that scaling up heterogeneous reactions by microwave irradiation is still feasible as the results obtained are comparable to those from the batch procedure.

To sum up this section, the scale up of processes using microwave irradiation in continuous flow systems has proved to be a valuable tool when homogeneous reactions are involved. In the case of heterogeneous reactions, solid reagents in cartridges have proven efficacy on the multigram scale. In the case of kilogram scale heterogeneous reactions, improvements in the handling of heavy slurries have been achieved, even though further development of this technology is still needed to overcome clogging and plugging problems. Other factors that must be taken into account, as introduced by different authors (Bergamelli et al., 2010; Dressen et al., 2010), concern the concept of energy efficiency. Energy cost is a key factor when deciding between different technologies.

Enantiomerically selective esterification of (R,S)-1-phenylethanol.

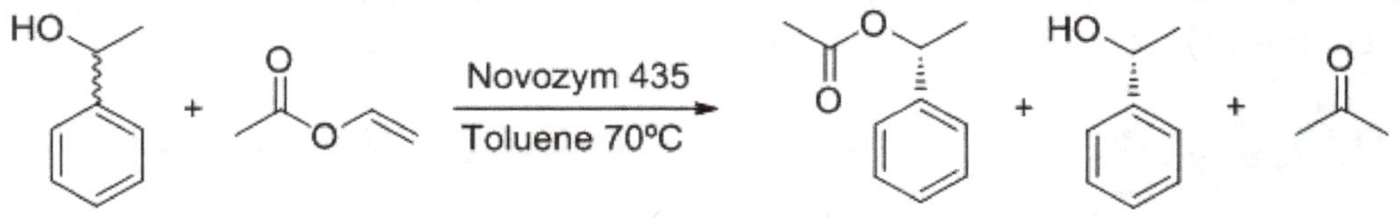

Esterification of (S)-pyroglutamic acid with n-decanol.

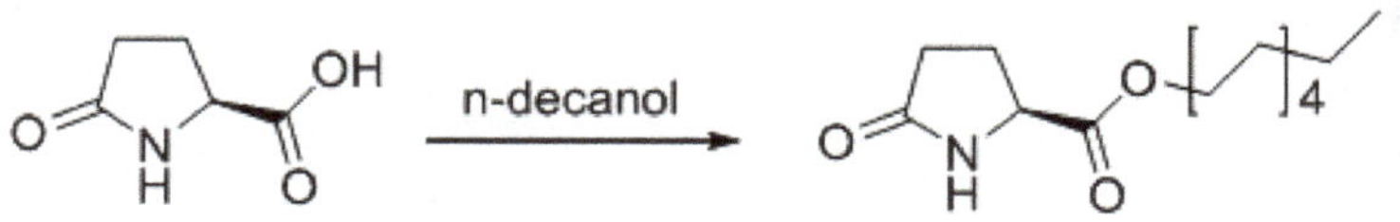

Scheme 22. Esterification reactions of (R,S)-1-phenylethanol and (S)-pyroglutamic acid

4. CONCISE SUMMARY

The application of Microwave Assisted Organic Synthesis at the laboratory scale has proven highly effective, particularly with the advent of specialized microwave instruments. Nonetheless, the proliferation of this technology as the preferred method for conducting chemical synthesis has been impeded by the following constraints:

- The use of various instruments from different companies and with different characteristics (power density, volume, waveguides etc.).
- The indistinct use of monomode and multimode instruments with and without control of the incident power, respectively.
- The general use, until 2000, of domestic kitchen-type microwave instruments, the results of which are considered to be non-reproducible.
- The low penetration depth of microwave irradiation, which hinders the scale up of reactions assisted by microwaves.

This introductory segment of the assessment delineates the measures implemented to resolve repeatability concerns.Reproducibility in microwave equipment can be achieved through meticulous manipulation of reaction conditions, especially when temperature is the variable under control.This reproducibility is applicable to kitchen-type microwave devices, monomode and multimode reactors, and during scaled-up reactions.Reproducibility can be attained regardless of solvent polarity and in solvent-free environments, when substrate polarities markedly influence microwave irradiation absorption. The only exceptions are the reactions in which polarity markedly increases along the reaction coordinate. The subsequent section of the review discusses methodologies for enhancing microwave processes. One approach involves employing large microwave cavities (volumes up to 12 L) in batch processes, and, in some cases, utilizing specially designed multimode microwave devices. A secondary strategy involves parallel reactions in multimode systems, with the most effective option being the use of continuous flow systems. These systems combine the advantages of microwave irradiation, volumetric heating with quick heating rates, and flow reactors, especially in terms of efficient heat transfer and the possibility of uniform reaction conditions from laboratory size to pilot plant.

5. REFERENCES

Alcázar J. Diels G. Schoentjes B. 2004 Reproducibility Across Microwave Instruments: First Example of Genuine Parallel Scale-Up of Compounds under Microwave Irradiation. QSAR & Combinatorial Chemistry, 23 10 (December 2004), 906 910 , 0161-1020X.

Alcázar J. 2005 Reproducibility Across Microwave Instruments: Preparation of a Set of 24 Compounds on a Multiwell Plate under Temperature-Controlled Conditions. Journal of Combinatorial Chemistry, 7 3 (May/June 2005), 353 355 , 1520-4766

Andrade M. M. Barros M. T. 2010 Fast Synthesis of N-Acylhydrazones Employing a Microwave Assisted Neat Protocol. Journal of Combinatorial Chemistry, 12 2 (March/April 2010), 245 247 . 1520-4766

Arbizzani C. Beninati S. Damen L. Mastragostino M. 2007 Power and Temperature Controlled Microwave Synthesis of SVO. Solid State Ionics, 178 5-6 , (March 2007), 393 398 , 0167-2738

Arvela R. K. Leadbeater N. E. Collins M. J. Jr 2005 Automated Batch Scale-Up of Microwave-Promoted Suzuki and Heck Coupling Reactions in Water Using Ultra-Low Metal Catalyst Concentrations. Tetrahedron, 61 36 (September 2005), 9349 9355 , 0040-4020

Barnard T. M. . Vanier G. S. Collins M. J. Jr 2006 Scale-Up of the Green Synthesis of Azacycloalkanes and Isoindolines under Microwave Irradiation. Organic Process Research & Development, 10 6 (November 2006), 1233 1237 , 1083-6160

Bergamelli F. Iannelli M. Marafie J. A. Moseley J. D. 2010 A Commercial Continuous Flow Microwave Reactor Evaluated for Scale-up. Organic Process Research & Development, 14 4 (July-August 2010) 926 930 , 1083-6160

Bowman M. D. . Holcomb J. L. . Kormos C. M. . Leadbeater N. E. Williams V. A. 2008 Approaches for Scale-Up of Microwave-Promoted Reactions. Organic Process Research & Development, 12 1 (January/February 2008), 41 57 , 1083-6160

Bowman M. D. . Scmink J. R. . Mc Gowan C. M. . Kormos C. M. Leadbeater N. E. 2008 Scale-Up of Microwave-Promoted Reactions to the Multigram Level Using a Sealed-Vessel Microwave Apparatus. Organic Process Research & Development, 12 6 (November-December 2008) 1078 1088 . 1083-6160

Cablewski T. Faux A. F. Strauss C. R. 1994 Development and Application of a Continuous Microwave Reactor for Organic Synthesis. Journal of the Organic Chemistry, 59 12 (June 1994), 3408 3412 , 0022-3263

Caliendo G. Fiorino F. Perissutti E. Severino B. Gessi S. Cattabriga E. Borea P. A. Santagada V. 2001 Synthesis by Microwave Irradiation and Binding Properties of Novel 5-HT1A Receptor Ligands. European Journal of Medicinal Chemistry, 36 11-12 , (November/December 2001), 873 866 , 0223-5234

Caliendo G. Fiorino F. Perissutti E. Severino B. Scolaro D. Gessi S. Cattabriga E. Borea P. A. Santagada V. 2002 A Convenient Synthesis by Microwave Heating and Pharmacological Evaluation of Novel Benzoyltriazole and Saccharine Derivatives as 5-HT1A Receptor Ligands. European Journal of Pharmaceutical Science, 16 1-2 , (July 2002), 15 28 , 0928-0987

Caliendo G. Perissutti E. Santagada V. Fiorino F. Severino B. Cirillo D. D'Emmanuele de Villa. Bianca R. Loppolis L. Pinto & Sorrentino, R. 2004 Synthesis by Microwave Irradiation of a Substituted Benzoxazine Parallel Library with Preferential Relaxant Activity for Guinea Pig Trachealis. European Journal of Medicinal Chemistry, 39 10 (October 2004), 815 826 , 0223-5234

Chen S. T. Chiou S. H. Wang K. T. 1990 Preparative Scale Organic Synthesis Using a Kitchen Microwave Oven. Journal of the Chemical Society, Chemical Communications, 11 (June 1900), 807 809 , 0022-4936

Cléophax J. Liagre M. Loupy A. Petit A. 2000 Application of Focused Microwaves to the Scale-Up of Solvent-Free Organic Reactions. Organic Process Research & Development, 4 6 (November/December 2000), 498 504 , 1083-6160

Comer E. Organ M. G. 2005 A Microreactor for Microwave-Assisted Capillary (Continuous Flow) Organic Synthesis. Journal of the American Chemical Society, 127 22 (June 2005), 8160 8167 , 0002-7863

Cotas S. J. Schulz M. J. Carson J. R. Codd E. E. Hlasta D. J. Pitis P. M. Stone D. J. Zhang S. P. Colburn R. W. Dax S. L. 2004 Parallel Methods for the Preparation and SAR Exploration of N-ethyl-4-[(8-alkyl-8-aza-bicyclo[3.2.1]oct-3-ylidene)-aryl-methyl]-benzamides, Powerful mu and delta Opioid Agonists. Bioorganic & Medicinal Chemistry Letters, 14 22 (November 2004), 5493 5498 , 0096-0894X.

Deetlefs M. Seddon K. 2003 Improved Preparations of Ionic Liquids Using Microwave Irradiation. Green Chemistry, 5 2 (March 2003), 181 186 , 1463-9262

Díaz-Ortiz A. Díez-Barra E. de la Hoz A. Moreno A. Gómez-Escalonilla M. J. Loupy A. 1996 1,3-Dipolar Cycloaddition of Nitriles under Microwave Irradiation in Solvent-Free Conditions. Heterocycles, 43 5 (May 1996), 1021 1030 , 0385-5414

Díaz-Ortiz A. de la Hoz A. Alcázar J. Carrillo J. R. Herrero M. A. Fontana A. Muñoz J. de M. 2007 Reproducibility and Scalability of Solvent-Free Microwave-Assisted Reactions: From Domestic Ovens to Controllable Parallel Applications. Combinatorial Chemistry & High Throughput Screening, 10 3 (March 2007), 163 169 , 1386-2073

Díaz-Ortiz A. de la Hoz A. Alcázar J. Carrillo J. R. Herrero M. A. Fontana A. Muñoz J. de M. Prieto P. de Cózar A. 2011 Influence of Polarity on the Scalability and Reproducibility of Solvent-Free Microwave-Assisted Reactions. Combinatorial Chemistry & High Throughput Screening, 14 2 (February 2011), 109 116 , 1386-2073

Dressen M. H. C. L. Kruijs B. H. P. van de Meuldijk J. Vekemans J. A. J. M. Hulshof L. A. 2010 Flow Processing of Microwave-Assisted (Heterogeneous) Organic Reactions. Organic Process Research & Development, 14 3 351 361 , 1083-6160

Eskilsson C. S. Björklund E. 2000 Analytical-Scale Microwave-Assisted Extraction. Journal of Chromatography A, 902 1 (December 2000) 227 250 , 0021-9673

Glasnov T. N. Kappe C. O. 2007 Microwave-Assisted Synthesis under Continuous-Flow Conditions. Macromolecular Rapid Communications, 28 4 (February 2007), 395 410 , 1022-1336

Horikoshi S. . Abe H. . Torigoe K. . Abe M. Serpone N. 2010 Access to Small Size Distributions of Nanoparticles by Microwave-Assisted Synthesis. Formation of Ag Nanoparticles in Aqueous Carboxymethylcellulose Solutions in Batch and Continous-Flow Reactors. Nanoscale, 2 8 (May 2010), 1441 1447 . 2040-3364

Ju Y. Varma R. S. 2006 Aqueous N-Heterocyclization of Primary Amines and Hydrazines with Dihalides: Microwave-Assisted Syntheses of N-Azacycloalkanes, Isoindole, Pyrazole, Pyrazolidine, and Phthalazine Derivatives. Journal of the Organic Chemistry, 71 1 (January 2006), 135 141 , 0022-3263

Kappe C. O. Stadler A. 2005 Microwaves in Organic and Medicinal Chemistry, Wiley-VCH, 3-52731-210-2

Kremsner J. M. Stadler A. Kappe C. O. 2007 High-Throughput Microwave-Assisted Organic Synthesis: Moving from Automated Sequential to Parallel Library-Generation Formats in Silicon Carbide Microtiter Plates. Journal of Combinatorial Chemistry, 8 2 298 291 , 1520-4766

Kunz U. Kirschning A. Wen H. L. Solodenko W. Cecilia R. Kappe O. Turek T. 2005 Monolithic Polymer/Carrier Materials: Versatile Composites for Fine Chemical Synthesis. Catalysis Today, 105 3-4 , (August 2005), 318 324 , 0920-5861

Kurzawa J. Stachowiak S. 2001 Investigation of Influence of Microwave Energy on Cleavage of S-S and C-S Bonds by Azide Ions. Chemia Analityczna, 46 6 (March 2001), 907 910 , 0009-2223

Leadbeater N. E. Williams V. A. Barnard T. M. Collins M. J. Jr 2006 Open-Vessel Microwave-Promoted Suzuki Reactions Using Low Levels of Palladium Catalyst: Optimization and Scale-Up. Organic Process Research & Development, 10 4 (July 2006), 833 837 , 1083-6160

Leadbeater N. E. Schmink J. R. 2007 Use of a Scientific Microwave Apparatus for Rapid Optimization of Reaction Conditions in a Monomode Function and then Substrate Screening in a Multimode Function. Tetrahedron, 63 29 (July 2007), 6764 6773 , 0040-4020

Leadbeater N. E. . C. R. C. Press 2010 Microwave Heating as a Tool for Process Chemistry: From Microwave Heating as a Tool for Sustainable Chemistry, Taylor & Francis Group, 978-1-43981-269-3 Boca Raton.

Lehmann F. Pilotti A. Luthman K. 2003 Efficient Large Scale Microwave Assisted Mannich Reactions Using Substituted Acetophenones. Molecular Diversity, 7 2-4 , (June 2003), 145 152 , 1381-1991

Lehmann H. La Vecchia L. 2005 Evaluation of Microwave Reactors for Prep-Scale Synthesis in a Kilolab. Journal of the Association for Laboratory Automation, 10 6 (December 2005), 412 417 , 1540-2452

Login G. R. Dvorak A. M. 1994 Application of Microwave Fixation Techniques in Pathology. Journal of Neuroscience Methods, 55 2 (December 1994), 173 182 , 0165-0270

Loones K. T. J. Maes B. U. W. Rombouts G. Hostyn S. Diels G. 2005 Microwave Assisted Organic Synthesis: Scale-Up of Palladium-Catalyzed Aminations Using Single-Mode and Multi-Mode Microwave Equipment. Tetrahedron, 61 43 (October 2005), 10338 10348 , 0040-4020

Marafield J. A. Moseley J. D. 2010 The Application of Stop-Flow Microwave Technology to Scaling-Out SNAr Reactions Using a Soluble Organic Base. Organic & Biomolecular Chemistry, 8 9 (February 2010), 2219 2227 , 1477-0520

Murray J. K. Gellam S. H. 2006 Microwave-Assisted Parallel Synthesis of a 14-Helical β-Peptide Library. Journal of Combinatorial Chemistry, 8 1 (January-February 2006), 58 65 , 1520-4766

Nikbin N. Ladlow M. Ley S. V. 2007 Continuous Flow Ligand-Free Heck Reactions Using Monolithic Pd [0] Nanoparticles. Organic Process Research & Development, 11 3 (May-June 2007), 458 462 , 1083-6160

Nüchter M. Ondruschka B. 2003 Tools for Microwave-Assisted Parallel Syntheses and Combinatorial Chemistry. Molecular Diversity, 7 2-4 , (June 2003), 253 264 , 1381-1991

Nüchter M. Müller U. Ondruschka B. Tied A. Lautenschläger W. 2003 Microwave-Assisted Chemical Reactions. Chemical & Engeneering Technology, 26 12 (December 2003), 1207 1216 , 0930-7516

Nüchter M. Ondruschka B. Bonrath W. Gum A. 2004 Microwave-Assisted Synthesis-A Critical Technology Overview. Green Chemistry, 6 3 (February 2004), 128 141 , 1463-9262

Ondruschka B. Bonrath W. Stuerga D. 2006 Chapter 2, In: Microwaves in Organic Synthesis, 2nd. Edition, Loupy, A., Ed., 62 107 , Wiley-VCH, 3-52731-452-0

Paulus R. M. . Erdmenger T. Hoogenboom R. . Pilotti A. Schubert U. S. 2006 Scale-Up of Microwave-Assisted Polymerizations in Batch Mode: The Cationic Ring-Opening Polymerization of 2-Ethyl-2-oxazoline. Macromolecular Rapid Communications, 27 18 (September 2006), 1556 1560 , 1022-1336

Perio B. Dozias M. J. Hamelin J. 1998 Ecofriendly Fast Batch Synthesis of Dioxolanes, Dithiolanes, and Oxathiolanes without Solvent under Microwave Irradiation. Organic Process Research & Development, 2 6 (November-December 1998), 428 430 , 1083-6160

Raner K. D. Strauss C. R. Trainor R. W. Thorn J. S. 1995 A New Microwave Reactor for Batchwise Organic Synthesis. Journal of the Organic Chemistry, 60 8 (April 2005), 2456 2460 , 0022-3263

Razzaq T. Kappe C. O. 2007 Rapid Preparation of Pyranoquinolines Using Microwave Dielectric Heating in Combination with Fractional Product Distillation. Tetrahedron Lett., 48 14 (April 2007), 2513 2517 , 0040-4039

Roberts B. A. Strauss C. R. 2005 Toward Rapid, "Green", Predictable Microwave-Assisted Synthesis. Accounts of Chemical Research, 38 8 (August 2005), 653 661 , 0001-4842

Schmink J. R. . Kornos C. M. . Devine W. G. Leadbeater N. E. 2010 Exploring the Scope for Scale-Up of Organic Chemistry Using a Large Batch Microwave Reactor. Organic Process Research & Development, 14 1 (January-February 2010), 205 214 , 1083-6160

Singh B. K. Kaval N. Tomar S. Van der Eycken E. Parmar V. S. (2008 (2008). Transition Metal-Catalyzed Carbon-Carbon Bond Formation Suzuki, Heck, and Sonogashira Reactions Using Microwave and Microtechnology. Organic Process Research & Development, 12 12 3 (May-June 2008), 468 474 , 1083-6160

Slade J. . Parker D. . Girgis M. . Wu R. . Joseph S. Repic O. 2007 Optimization and Scale-Up of the Grandberg Synthesis of 2-Methyltryptamine. Organic Process Research & Development, 11 4 (July-August 2007), 721 725 , 1083-6160

Smith C. J. Iglesias-Sigüenza F. J. Baxendale I. R. Ley S. V. 2007 Flow and Batch Mode Focused Microwave Synthesis of 5-Amino-4-cyanopyrazoles and Their Further Conversion to 4-Aminopyrazolopyrimidines. Organic & Biomolecular Chemistry, 5 17 (July 2007), 2758 2761 , 1477-0520

Stadler A. Yousefi B. H. Dallinger D. Walla P. Van der Eycken E. Kaval N. Kappe C. O. 2003 Scalability of Microwave Assisted Organic Synthesis. From Single-Mode to Multimode Parallel Batch Reactors. Organic Process Research & Development, 7 5 (September-October 2003), 707 716 , 1083-6160

Wilson N. S. Sarko C. R. Roth G. P. 2004 Development and Applications of a Practical Continuous Flow Microwave Cell. Organic Process Research & Development, 8 3 (May-June 2004), 535 538 , 1083-6160

9 Chapter

APPLICATION OF MICROWAVES IN COAL RESEARCH AND MATERIALS SYNTHESIS

1. INTRODUCTION: AN OVERVIEW

This chapter reviews the application of microwave heating in resolving issues in coal research and in the synthesis of inorganic compounds. The review examines the fundamental principles of microwave heating and its application in particular instances. This review comprises three sections: Section 2 addresses the underlying theory of microwave heating, while Sections 3 and 4 focus on its application in coal research and material synthesis, respectively. References to specific works are also provided. This list of references is not comprehensive; readers are encouraged to visit these sources for prior studies in particular domains.

2. MICROWAVE HEATING

When a time varying electric field $E = E_0 exp(i\omega t)$ is applied to a liquid or solid, polarization develops which in general can come from three sources viz. dipolar, ionic and electronic (Kittel, 1996). The dipolar contribution is present in polar materials viz. materials with permanent dipoles such as water as these dipoles re-orient in the direction of the applied field E. The relaxation time τ associated with this reorientation is in picoseconds (ps) and therefore this phenomena is observable when frequency $f = (\omega/2\pi)$ is in the microwave range (f = 0.1 GHz to 100 GHz). The ionic contribution to the polarization results from the various normal modes of vibration of the atoms relative to each other. These frequencies in the IR range are from 300 GHz to 100 THz. Finally, the electronic contribution resulting from the relative displacements of the nucleus with positive charge and electron shell of the atoms occurs in the ultraviolet (UV) region of the frequencies. Thus microwave heating as shown below is most significant in materials having permanent dipoles such as water.

A number of experimental studies (Kaatze, 1989; Merabat & Bose, 1988) have shown that dielectric relaxation in water is adequately described by the Debye relation (Debye, 1929)

$$\varepsilon(\omega) = \varepsilon(\infty) + [\varepsilon(0) - \varepsilon(\infty)]/(1 + i\omega\tau) \quad (1)$$

where $\varepsilon(0)$ and $\varepsilon(\infty)$ are the dielectric constant in the limit $\omega \to 0$ and $\omega \to \infty$ respectively. For water, $\varepsilon(\infty)$ includes the polarization effects at the infrared and optical-uv frequencies. Separating the in phase and the out of phase components of $\varepsilon = \varepsilon' - i\varepsilon''$ in Eq.(1) leads to

$$\varepsilon' = \varepsilon(\infty) + [\varepsilon(0) - \varepsilon(\infty)]/(1 + \omega^2\tau^2) \quad (2)$$

and

$$\varepsilon'' = \omega\tau[\varepsilon(0) - \varepsilon(\infty)]/(1 + \omega^2\tau^2) \quad (3)$$

Eliminating $\varepsilon(\infty)$ through substitutions in Eq. (2) and **(**3) leads to the following relation between $\varepsilon'$and ε'':

$$\varepsilon' = \varepsilon(0) - \varepsilon''\cdot\omega\cdot\tau \quad (4)$$

Once experimental data for ε' and ε'' are available for different ω, a plot of ε' vs. $\varepsilon''\cdot\omega$ allows one to determine the relaxation time τ. Such an analysis for water in the frequency range of 1 GHz to 60 GHz and in the temperature range of 0°C and 60°C have shown (Kaatze, 1989) that at T = 25°C, τ = 8.27 ps, $\varepsilon(0)$ = 78.36 and $\varepsilon(\infty)$ = 5.2 and that with increase of temperature, $\varepsilon(0)$, $\varepsilon(\infty)$ and τ decrease rapidly so that at 60°C, $\varepsilon(0)$ = 66.7, $\varepsilon(\infty)$ = 4.2 and τ = 4.0 ps.

Microwave heating involves absorption of electrical power by water from the microwaves. The power absorbed P can be shown to be given by (Agmon, 1996)

$$P = 2\pi|E_o|^2\varepsilon_o f\cdot\varepsilon'', \quad w/m^3 \quad (5)$$

where $\varepsilon_0 = 8.854 \times 10^{-12}$ A s/V m, E_0 is the amplitude of the electric field and $\omega = 2\pi f$. The important point to note is that P is proportional to $f\cdot\varepsilon''$so that P depends upon the frequency f of the microwaves and ε''**of** the medium (water). From Eq. (3) it can be easily shown that ε'' peaks at $\omega\tau = 1$.

In order to provide visualization of the frequency variation of the quantities ε', ε'' and $f\cdot\varepsilon''$, the latter representing the power absorbed P, we have used the data for water given in a tabular form (Kaatze, 1989). The simulated plots of the frequency dependence of ε', ε'' and $f\cdot\varepsilon''$ for the data valid at 25°C are shown as lines in Fig. 1 where the circles are experimental values **(**Kaatze, 1989). The data up to 60 GHz fit quite well with the Debye relaxation with a single τ. Later studies have suggested that two relaxation times are necessary to explain the data at frequencies higher than 100 GHz (Barthel et al., 1990; Agmon, 1996).

An interesting feature of Fig. 1 is the frequency dependence of $f\cdot\varepsilon''$ representing power absorbed. This plot, usually not avaliable in literature, shows that power absorbed saturates only near 100 GHz. At f = 2.45 GHz, the conventional frequency used in microwave ovens, domestic and industrial, the absorbed power is only about

1.6% of the power absorbed at 100 GHz. However, the wavelengths of microwaves λ = 0.3 cm at 100 GHz compared to λ = 12.2 cm at 2.45 GHz, thus reducing the size of the microwave ovens to impractical size of only about 0.3 cm at the higher frequency. At 17 GHz where ε'' has the maximum value of 36, λ = 1.76 cm and power absorbed is nearly 50% of the maximum value at 100 GHz.

As water heats up, τ, $\varepsilon(0)$ and $\varepsilon(\infty)$ decrease (Kaatze, 1989) affecting the peak positions for $\varepsilon''$and $f{\cdot}\varepsilon''$. In Fig. 2 simulated frequency dependence of ε', ε'' and $f{\cdot}\varepsilon''$ is compared at two temperatures viz. 25°C and 60°C using the magnitudes of $\varepsilon(0)$, $\varepsilon(\infty)$ and τ (Kaatze, 1989). The important point to note is that the power absorbed at a fixed frequency below about 30 GHz decreases as the temperature of water increases.

The frequency derivative of power absorbed by water given by $\partial(f{\cdot}\varepsilon'')/\partial f$, is plotted in Fig. 3. The maximum slope occurs at 11 GHz for 25°C and the position of this maximum changes to 23 GHz at 60°C. Analytically, using relation (3), it can be shown that $\partial(f{\cdot}\varepsilon'')/\partial f$ is maximum at $f = 1/\pi\tau\sqrt{12}$.

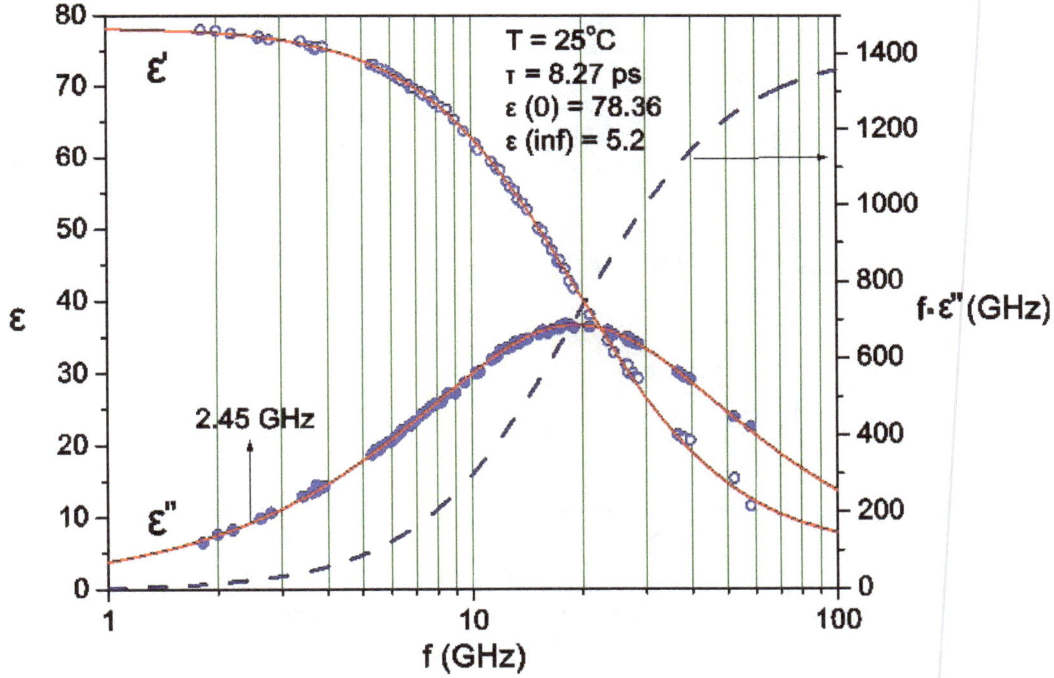

Figure 1. The simulated plots of the frequency dependence of ε', ε'' and $f{\cdot}\varepsilon''$ for water using the data (circles) at 25°C from Kaatze (1989).

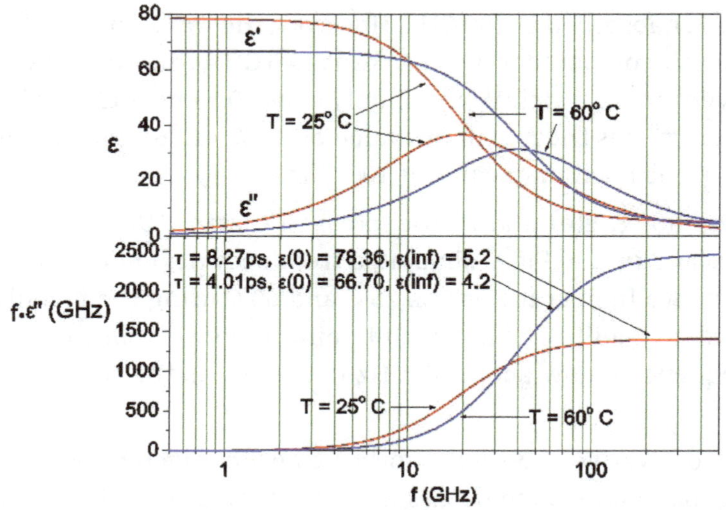

Figure 2. Simulated frequency dependence of ε′, ε″ and f•ε″ is compared for water at two temperatures viz. 25°C and 60°C using the magnitudes of ε(0), ε(∞) and τ

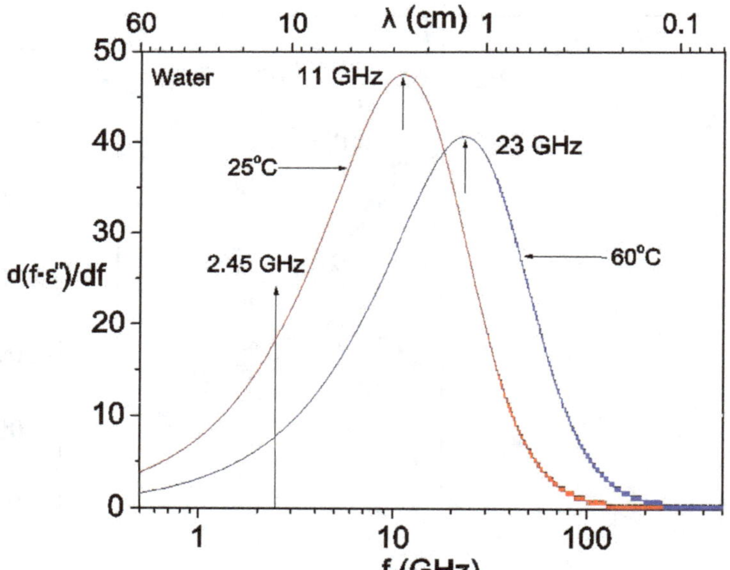

Figure 3. Derivatives of power absorbed from Fig. 2 vs. frequency and wavelength at 25°C and 60°C.

3. MICROWAVE HEATING IN COAL RESEARCH

3.1. Microwave properties of coals

In the recent paper, (Marland et al., 2001) have reported the measurements of ε′and ε″ at 60°C for over a dozen coals and for the ash components normally found in most coals. In these studies, data were reported at f = 0.615 GHz, 1.413 GHz and

2.216 GHz for the coals and the minerals pyrite, quartz, dolomite, kaolin, mica and calcite. Although there is some variations in the magnitudes of ε' and ε'' from coal to coal, the range of values were between 1.6 to 3.5 for ε' and 0.07 to 0.31 for ε''at f = 2.216 GHz. The frequency dependence among the three frequencies was found to be relatively weak. Among the minerals the highest magnitude was found to be for pyrite with ε'= 7.07 and ε''= 1.06 at 2.216 GHz and at 60°C. For the other minerals, $\varepsilon' \sim 2$ to 3 and $\varepsilon'' \sim 0.05$-0.06. Moisture levels in these coals varied from a low of 1% to a high of 13%. From Fig. 1 $\varepsilon'' \simeq 10$ for water at 2.45 GHz. Since power P absorbed is proportional to $f \cdot \varepsilon''$ (Eq. 1), the highest power absorbed will be by moisture, followed by pyrite. Thus the largest effect of microwave heating of coal is the removal of moisture and heating of pyrite FeS_2.

Pyrite FeS_2 when heated above 250°C is known to convert to pyrrhotites $Fe_{1-x}S$ (x = 0 - 0.125) whose mangetization at room temperature is ferrimagnetic like, with the magnitude depending on the x value (Jagadeesh & Seehra, 1981). Whereas FeS_2 is only paramagnetic with feeble magnetization (Burgardt & Seehra, 1977) pyrrhotites $Fe_{1-x}S$ are strongly magnetic (Jagadeesh & Seehra, 1981). Therefore conversion of pyrites to pyrrhotites after microwave heating make them amenable to separation by magnetic methods. Thus microwave heating of coals at an appropriate frequency followed by magnetic separation of the impurities can lead to both dewatering and desulphurization, since a major source of sulphur in coals is pyrites. A combination of these techniques is likely being used in the recently developed and marketed technology by CoalTek (http://www.coaltek.com/).

3.2. Measuring moisture content by microwaves

Another area in which microwaves have been used in coal research is the on-line measurements of moisture in coals employing the frequencies of 2.45 GHz (Cutmore et al., 2000) and 1 GHz (Ponte et al., 1996). In these techniques, microwave horns are used to transmit and receive microwaves through the coal sample. The use of the lower microwave frequencies and hence corresponding larger wavelengths of λ = 12.2 cm (30 cm) for 2.45 GHz (1 GHz) allows the use of larger samples encountered at coal power plants. As microwaves pass through a coal sample, large ε' of water produces a frequency and phase shift of the microwaves and ε'' results in the attenuation of the microwave signal. Both phase shift and attenuation are measured by microwave network analyzer. However, experiments show that phase shift provides a better linear correlation with the moisture content (Ponte et al., 1996; Cutmore et al., 2000). Since attenuation is due to absorption of microwaves primarily by moisture present, this technique can also result in sufficient dewatering of the coals. This topic of dewatering or drying of coals by using microwave is discussed next.

3.3. Microwave dewatering of coals

Dewatering or drying of coals is an important consideration in coal industry since removal of moisture reduces the transportation costs and improves the BTU value of the coals. The use of microwaves for dewatering of coals has been discussed in a number of earlier studies (Lindroth, 1985; Chatterji & Mishra, 1991; Perkin, 1980; Standish el al., 1988; Marland el al., 2000; Lester & Kingman, 2004) and in a more recent study from author's laboratory (Seehra et al., 2007). In microwave dewatering of coals, the basic mechanism for power absorption by water discussed in Section 2 is operative.

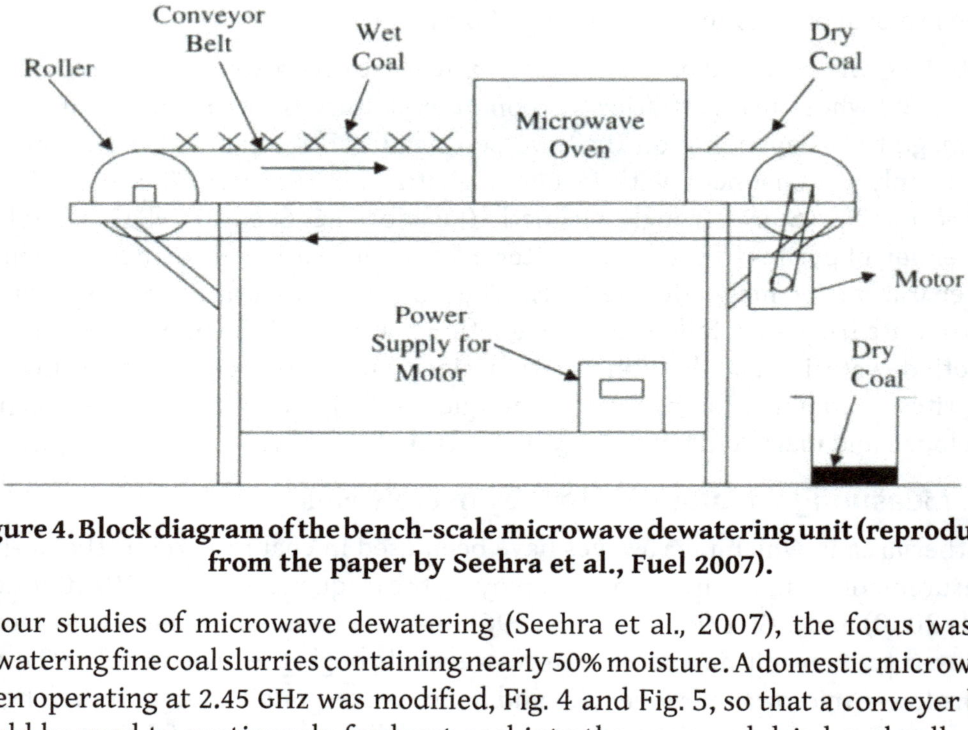

Figure 4. Block diagram of the bench-scale microwave dewatering unit (reproduced from the paper by Seehra et al., Fuel 2007).

In our studies of microwave dewatering (Seehra et al., 2007), the focus was on dewatering fine coal slurries containing nearly 50% moisture. A domestic microwave oven operating at 2.45 GHz was modified, Fig. 4 and Fig. 5, so that a conveyer belt could be used to continously feed wet coal into the oven and dried coal collected at the other end. The whole apparatus was shielded with a microwave shield. This bench scale unit was designed to simulate a practical situation at a coal plant. The website of CoalTek shows such a conveyer belt approach and perhaps the only way to achieve large scale continous operation.

In our experiment using a 0.8 kW domestic oven, the amount of moisture removed depended strongly on the belt speed; The slower belt speed removed more moisture but consumed more energy. This data on the relation between belt speed, electrical energy consumed and more moisture lost is shown in Fig. 6 (Seehra et al., 2007). Belt speed of around 13 cm/min was found to be most energy efficient (~ 83%) for removal of water, the remaining power perhaps lost to surroundings and small heating of the minerals in coal and coal itself. In the experiments of Lindroth using higher power magnetrons (9.7 kW) operating at 2.45 GHz, the reported efficiency varied from 52% to 97% for drying of fine coals. A cost estimate of $ 3/ton to remove 10% moisture from coal was made based only on the amount of electrical power consumed (Seehra et al., 2007).

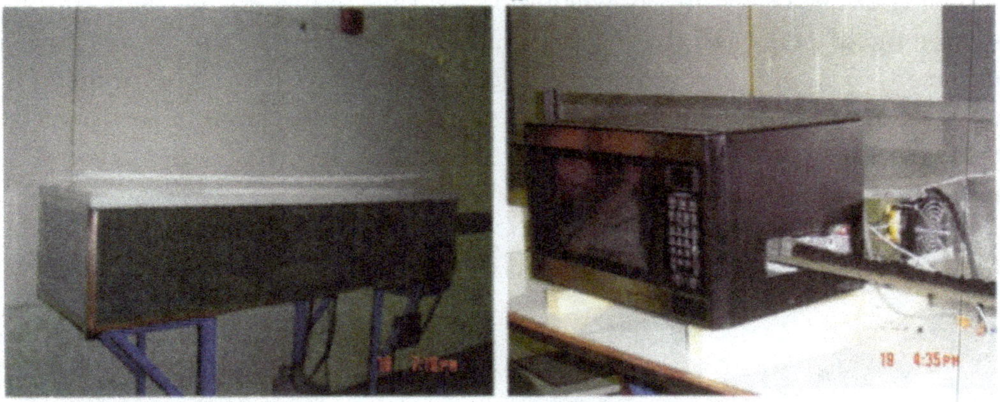

Figure 5. (Left) Picture of the bench-scale microwave dewatering units with microwave shield in place. (Right) Close-up of the unit without the microwave shield showing the conveyor belt for feeding wet coal into the microwave oven (reproduced from the paper by Seehra et al., Fuel 2007).

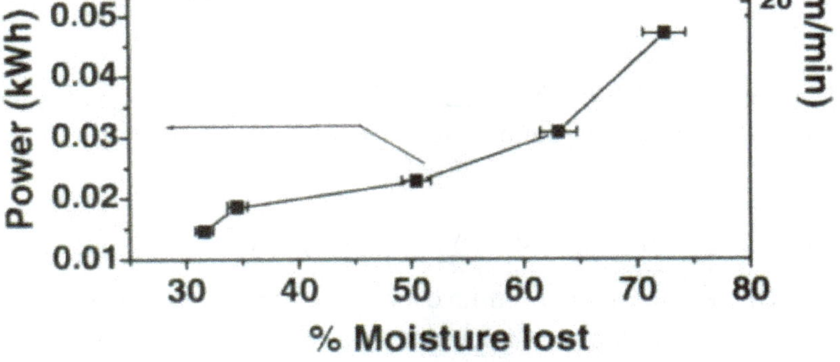

Figure 6. Power consumed and belt speed (cm/min) vs. % moisture lost. The region of the least slope (13 cm/min) yields the optimum speed (reproduced from the paper by Seehra et al., Fuel 2007).

Dewatering efficiencies of microwave heating vis-à-vis thermal heating for the coal slurry samples are compared in Fig. 7 (Seehra et al, 2007). In this comparison, 20 gms of coal slurry sample was heated in a thermal oven set at 110°C followed by taking the sample out after every 30 s for quick weighing and reinserting the sample in the oven in less than 5 s. Similar experiments were done using a domestic microwave oven. The results of weight loss vs. time in Fig. 7 **show that complete dewatering of the coal slurry takes about 5 min with microwave heating compared to about an hour with thermal heating. Thus this experiment shows an order of magnitude reduction in time in microwave heating vis-à-vis thermal heating.**

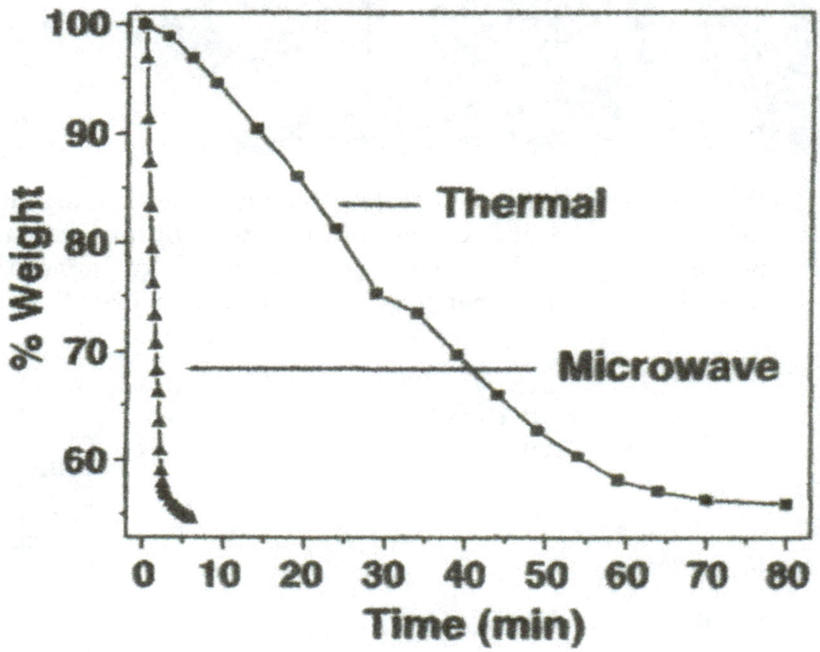

Figure 7. Comparison of the efficiencies of thermal heating vis-a`-vis microwave heating using 20 g sample in each case (reproduced from the paper by Seehra et al., Fuel, 2007).

4. USE OF MICROWAVES IN MATERIAL SYNTHESIS

In section 2, the mechanism of microwave heating was described in terms of dielectric constants ε' **and** ε'' with specific examples given for microwave heating of water. Same argument can be used for all polar materials with known ε' **and** ε''. As an example, Barthel et al. (1990) have discussed the microwave heating using lower alcohols (methanol, ethanol, 1-propanol and 2-propanol).

In the chemical synthesis of a variety of materials, use of a conventional heating which has been a norm for a long time, is being replaced by microwave heating. In conventional heating, energy is first delivered to the surface of a material, followed by heat transfer to the material through conduction, convection and radiation. This results in non-uniform heating producing poor crystallinity and often wider distribution of particle sizes. By contrast, in microwave heating, energy is transferred throughout the material via molecular interaction thus producing uniform heating without thermal gradients and stress. This often results in materials with finer microstructures and particles with narrow size distribution. The reader is referred to several recent papers (Barison et al., 2010; Wu et al., 2008; Prasad et al. 2002; Utchariyajit et al. 2010; Chang et al. 2009; Gopinath et al. 2002; Vincente et al 2010) where examples of these advantages are described in detail. Thus, compared to conventional thermal heating, microwave heating has the advantages of speed, energy efficiency, finer microstructures and more uniform particle size distribution in the synthesis of materials in line with the data shown in Fig. 7.

Microwave heating in combination with hydrothermal processing has shown positive effects on crystallinity of samples. Samples prepared by conventional hydrothermal treatment showed poor crystallinity in an amorphous phase, and they required higher temperatures and longer times than that of microwave hydrothermal treatment (Kim et al. 2001). Crystalline samples have been obtained without increase in particle size (Addamo et al. 2008). Other observations have shown the particles to have higher crystallinity indicating larger particle size (Vincente et al. 2009), and it avoids the post treatment such as annealing required after sol-gel synthesis (Hu et al. 2007). The treatment time under microwave heating has shown increase in the crystallite size due to Oswald mechanism of crystallite growth. Thus for applications which involve surface and phase characteristics, and the presence of porosity used in catalytic applications, microwave synthesis has been very successful (Wilson et al. 2006).

From a review done by Yanshou et al. in 2008, synthesis using microwave heating has been found to have the Si/Al ratio increased from 1.4 to 1.8 in zeolite membranes. It is clearly pointed out that the difference between microwave heating and conventional heating is due to different heating mechanisms (Li &Yang, 2008). Shape effects have also been noticed by Chen et al. 2009, where the concentration of solvents of diols have significantly affected the length of the Si-MFI crystals compared to the widths and thickness having a lesser effect. This effect could not be reproduced using a conventional synthesis method (Chen et al., 2009). In general, microwave hydrothermal treatment has proved to be more efficient than conventional hydrothermal treatment because of energy saving and time. For preparing microwave-assisted organic materials, readers are referred to the review by Lidström et al. (2001).

All items placed within a microwave oven during operation must be "microwave safe," meaning they should not be effective electrical conductors, as is widely recognized in domestic microwave cooking. The concentrated electric field generated by the microwaves traverses the good conductor, leading to electrical sparks and potential fires. Ceramic and polymer containers, recognized for their insulating properties and labeled as "microwave safe," are now accessible for material synthesis applications.

Historically, the initial application of microwave heating in material synthesis utilized domestic microwave ovens that were suitably adapted by each researcher. Recently, microwave ovens specifically engineered for material synthesis have become available, allowing for in situ monitoring of temperatures and pressures within an autoclave. The use of autoclaves within microwave ovens for the synthesis of materials under hydrothermal conditions has significantly broadened the application of microwave heating. (for example, refer to Chang et al. 2009 and Kim et al. 2001). Readers may readily identify suppliers of such commercial units in each country via online searches.

This chapter on microwave heating in material synthesis presents a tabulated organization of the known examples. The subsequent table enumerates the synthesized elements, the circumstances employed in their synthesis, and the accompanying references, arranged chronologically from the most recent (2010) to the earliest (1998). The list indicates that microwave heating in material synthesis is more prevalent and is expected to grow further in the future.

Table 1. Summary of selected materials prepared using microwave heating with conditions and apporiate references listed.

SYSTEM	CONDITIONS	REFERENCES
$BaCe_{0.65}Zr_{0.20}Y_{0.15}O_{3-\delta}$	P = 350, 500 W	Barison S., Fabrizio M., Fasolin S., Montagner F. &Mortalò C., (2010). A microwave-assisted sol-gel Pechini method for the synthesis of $BaCe_{0.65}Zr_{0.20}Y_{0.15}O_{3-\delta}$powders. *Mater. Res. Bull.*, Vol. 45, pp. 1171-1176.
$Zn_4O(BDC_3)$	t = 30 min, P = 1 kW	Lu C., Liu J., Xiao K.& Harris A., (2010). Microwave enhanced synthesis of MOF-5 and its CO_2 **capture ability at moderate temperatures across multiple capture and release cycles.** *Chem. Eng. J.*, Vol. 156, pp. 465-470.

SYSTEM	CONDITIONS	REFERENCES
Capillary MFI-type Zeolite-Ceramic Membrane	T = 70 - 180 °C, P = 250 or 400 W, t = 10 - 150 min	Sebastian V., Mallada R., Coronasa J., Julbe A., Terpstra R.&Dirrix R., (2010). Microwave-assisted hydrothermal rapid synthesis of capillary MFI-type zeolite-ceramic membranes for pervaporation application.*J. Membrane Sci.*, Vol. 355, pp. 28-35.
Gold Nanoparticles	t = 30 sec, P = 800 W t = 9 min, P = 1 kW	Vargas-Hernandez C., Mariscal M., Esparza R. &Yacaman M., (2010). A synthesis route of gold particles with using a reducing agent.*Appl. Phys. Lett.* Vol. 96, pp. 213115-(1-3).
Silicalite (Si-MFI) Crystals	T = 180 °C, P = 400 W,t = 10 min	Chen X., Yan W., Cao X. &Xu R., (2010). Quantitative correlation between morphology of silicalite-1 crystals and dielectric constants of solvents. *Microporous andMesoporous Materials*, Vol. 131, pp. 45-50.
Saponite.M_x [Mg_6Al_x $Si_{8-x}O_{20}(OH)_4$] (M = Na, Li, NH_4)	T = 453 K,t = 6 hrs	Vicente I., Salagre P., Cesteros Y., Medina F.&Sueiras J., (2010). Microwave-assisted synthesis of saponite. *Appl. Clay Sci.*, Vol. 48, pp. 26-31.
$K_8Ln_3Si_{12}O_{32}NO_3{\cdot}\mathbf{H}_2O$ (Ln = Eu, Tb, Gd, Sm).	T = 363 K, t = 20 min	Wang X., Li J., Wang G., Han Y., Su T., Li Y., Yu J.&Xu R., (2010). Synthesis, characterization and properties of microporous lanthanide silicates:$K_8Ln_3Si_{12}O_{32}NO_3{\cdot}\mathbf{H}_2O$ (Ln = Eu, Tb, Gd, Sm). *Solid State Sciences*, Vol. 12, pp. 422-427.
Calcia Stabilized Zirconia Nanoparticles	T = 200 °C, t = 90 min	Rizzuti A., Corradi A., Leonelli C., Rosa R., Pielaszek R. &Lojkowski W., (2010). Microwave technique applied to the hydrothermal synthesis and sintering of calcia stabilized zirconia nanoparticles. *J. of Nanopart. Res.*, Vol. 12, pp. 327-335.
Double-walled Carbon Nanotubes	T = 200 °C, t = 20 min	Dunens O. M., MacKenzie K. J. & Harris A. T., (2010). Large-Scale Synthesis of Double-Walled Carbon Nanotubes in Fluidized Beds. *Ind. Eng. Chem.Res.*, Vol. 49, No. 9, pp. 4031-4035).
One-Dimensional Nano- Structured Silver Titanates	T = 200 °C, P = 300 W, t = 1 hrs	Li Q., Kako T. & Ye J., (2010). Strong adsorption and effective photocatalytic activities of one-dimensional nanostructured silver titanates. *Appl. Catalysis A: General*, Vol. 375, pp. 85-91.

SYSTEM	CONDITIONS	REFERENCES
Mesoporous Zeotype SAPO-5 (Silicoaluminophosphate)	T = 180 - 200 °C, t = 0.5 - 2 hrs	Utchariyajit K. &Wongkasemji S., (2010) Effect of synthesis parameters on mesoporous SAPO-5 with AFI-type formation via microwave radiation using alumatrane and silatrane precursors. *Microporous and Mesoporous Materials*, Vol. 135, pp. 116-123.
$Ru/La_{0.75}Sr_{0.25}Cr_{0.5}Mn_{0.5}O_{3-\delta}$	P = 750 W, t = 1 min	Barison S., Fabrizio M., Mortalò C., Antonucci P., Modafferi V.&Gerbasi R., (2010). Novel Ru/ $La_{0.75}Sr_{0.25}Cr_{0.5}Mn_{0.5}O_{3-\delta}$ **catalysts for propane reforming in IT-SOFCs.** *SolidState Ionics*, Vol. 181, pp. 285-291.
Tungsten Oxide Nanorods	T = 180 °C, t = 20 min	Li Y., Su X., Jian J. & Wang J., (2010). Ethanol sensing properties of tungsten oxide nanorods prepared by microwave hydrothermal method. *Ceramics International*, Vol. 36, pp. 1917-1920.
Saponites with Mg^{2+}	T = 180 °C, P = 600 W, t = 1, 2, 3, 4, 8 or 16 hrs	Trujillano R., Rico E., Vicente M., Herrero M. & Rives V., (2010). Microwave radiation and mechanical grinding as new ways for preparation of saponite-like materials. *Appl. Clay Sci.*, Vol. 48, pp. 32-38.
Titanate Nanotubes	T = 403 K, P = 70, 400 or 700 W, t = 3 hrs	Chen Y., Lo S. &Kuo J., (2010). Pb(II) adsorption capacity and behavior of titanate nanotubes made by microwave hydrothermal method. *Colloids and Surfaces A:Physicochemical and Engineering Aspects*, Vol. 361, pp. 126-131.
Mg, Al-Layered Double Hydroxides	T = 125 °C, t = 10 - 180 min	Benito P., Herrero M., Labajos F. & Rives V., (2010). Effect of post-synthesis microwave hydrothermal treatment on the properties of layered double hydroxides and related materials. *Appl. Clay Sci.*, Vol. 48, pp. 218-227.
Y_2O_3 **and Yb-Y_2O_3**	T = 140, 170 °C, P = 150 - 300 W, t = 5 - 22 min, 2 hrs	Serantoni M., Mercadelli E., Costa A., Blosi M., Esposito L.&Sanson A., (2010). Microwave-assisted polyol synthesis of sub-micrometer Y_2O_3 **and Yb-Y_2O_3 particles for laser source application.** *Ceramics International*, Vol. 36, pp. 103-106.
Co-Zn-Al catalysts	T = 125 °C, t = 10 - 300 min	Benito P., Herrero M., Labajos F. M., Rives V., Royo C., Latorre N. &Monzon A., (2009). Production of carbon nanotubes from methane: Use of Co-Zn-Al catalystsprepared by microwave-assisted synthesis. *Chemical Engineering Journal*, Vol. 149, pp. 455-462.

SYSTEM	CONDITIONS	REFERENCES
Hectorite $M_x [Li_x Mg_{6-x} Si_8 O_{20} (OH)_4]$ (M = Na, Li, NH_4)	T = 393 K, t = 16 hrs	Vicente I., Salagre P., Cesteros Y., Guirado F., Medina F.&Sueiras J., (2009). Fast microwave synthesis of hectorite. *Appl. Clay Sci.*, Vol. 43, pp. 103-107.
Titania Nanowires	T = 210 °C, P = 350 W, t = 5 hrs	Chang M., Lin C., Deka J., Chang F. & Chung C., (2009). Nanomechanical characterization of microwave hydrothermally synthesized titania nanowires. *J. Phys. D:Appl. Phys.*, Vol. 42, pp. 145105.
Graphitic Carbon Particle Chains	T = 160 - 220 °C, t = 40 min	Harris A., Deshpande S. &Kefeng X., (2009). Synthesis of graphitic carbon particle chains at low temperatures undermicrowave irradiation. *Materials Lett.*,Vol. 63, pp. 1390-1392.
Silicalite (Si-MFI)	T = 80 °C, P = 250 W, t = 90 min T = 180 °C, P = 400 W, t = 60 min	Chen X., Yan W., Cao X., Yu J. &Xu R., (2009). Fabrication of silicalite-1 crystals with tunable aspect ratios by microwave assisted solvothermal synthesis. *Microporous and Mesoporous Materials*, Vol. 119, pp. 217-222.
$ZnWO_4$ **Nanoparticles**	T = 160 °C, t = 1, 3 hrs	Bi J., Wu L., Li Z., Ding Z., Wang X. & Fu X., (2009). A facile microwave solvothermal process to synthesize ZnWO4 nanoparticles. *J. Alloys and Compd.*, Vol. 480, pp. 684-688.
Multiferroic $BiFeO_3$	t = 30 min, P = 800 W	Prado-Gonjal J., Villafuerte-Castrejón M., Fuentes L. &Morán E., (2009). Microwave-hydrothermal synthesis of the multiferroic $BiFeO_3$. *Mater. Res. Bull.*, Vol. 44,pp. 1734-1737.
$NaYF_4:Yb^{3+},Tm^{3+}$	T = 180 °C, t = 4 hrs	Chen X., Wang W., Chen X., Bi J., Wu L., Li Z.&Fu X., (2009). Microwave hydrothermal synthesis and upconversion properties of $NaYF_4:Yb^{3+},Tm^{3+}$ **with microtube morphology.** *Mater. Lett.*,Vol. 63, pp. 1023-1026.
Carboxylate-Intercalated Layered Double Hydroxides	T = 85, 100, 125 °C, t = 10, 30, 180 min	Benito P., Labajos F., Mafra L., Rocha J. &Rives V., (2009). Carboxylate-intercalated layered double hydroxides aged under microwave-hydrothermal treatment. *J. Solid State Chem.*, Vol. 182, pp. 18-26.
Rapid Synthesis of Titania Nanowires	T = 210 °C, P = 350 W, t = 2 hrs	Chung C., Chung T. & Yang T., (2008). Rapid Synthesis of Titania Nanowires by Microwave-Assisted Hydrothermal Treatments. *Ind. Eng. Chem. Res.*, Vol. 47, No. 7, pp. 2301-2307.

SYSTEM	CONDITIONS	REFERENCES
TiO_2 Photocatalysts	T = 393, 423 K, t = 15, 30, 60 min	Addamo M., Bellardita M., Carriazo C., Paola A., Milioto S., Palmisano L. & Rives V., (2008). Inorganic gels as precursors of TiO_2photocatalysts prepared by low temperature microwave or thermal treatment. *Appl. Catalysis B: Environmental*, Vol. 84,pp. 742-748.
Titanate Nanowires	T = 210 °C, P = 350 W, t = 5 hrs	Chang M., Chung C., Deka J., Lin C. & Chung T., (2008). Mechanical properties of microwave hydrothermally synthesized titanate nanowires. *Nanotechnology*, Vol. 19, pp. 025710.
Mesoporous$AlPO_4$-5 (AFI) Zeotype	T = 180 - 200 °C, t = 0.5 - 2 hrs	Utchariyajit K. &Wongkasemjit S., (2008). Structural aspects of mesoporous $AlPO_4$-5 (AFI) zeotype using microwave radiation and alumatrane precursor. *Microporous and Mesoporous Materials*, Vol. 114, pp. 175-184.
Zn,Al-CO_3 **Layered Double Hydroxides**	T = 100, 125 °C, t = 30, 60, 180, 300 min	Benito P., Guinea I., Labajos F., Rocha J. & Rives V., (2008). Microwave-hydrothermally aged Zn,Alhydrotalcite-like compounds: Influence of the composition and the irradiation conditions. *Microporous and Mesoporous Materials*, Vol. 110, pp. 292-302.
Ni-Al, Zn-Al Hydrotalcite	T = 100, 125, 150, 175 °C, t = 5 - 300 min	Benito P., Herrero M., Barriga C., Labajos F. & Rives V., (2008). Microwave-Assisted Homogeneous Precipitation of Hydrotalcites by Urea Hydrolysis. *Inorg. Chem.*, Vol. 47, No. 12, pp. 5453-5463.
Synthesis, Passivation and Stabilization of Nano (particles, rods & wires)	P = 650 W (60%), t = 10 sec to mins	Abdelsayed V., Panda A. B., Glaspell G. P. & El-Shall M. S., (2008). Synthesis, Passivation and Stabilization of Nanoparticles, Nanorods and Nanowires by Microwave Irradiation. *ACS Symposium Series*, Vol. 996, pp. 225-247.
$LaPO_4$:RE (RE=Ce,Tb,Eu) and In_2O_3:Sn (ITO)	T = 300 °C, t = 0 - 30 sec	Buhler G., Stay M. &Feldmann C., (2007). Ionic liquid based approach to nanoscale functional materials. *Green Chemistry*, Vol. 9, pp. 924-926.
$ZnIn_2S_4$ Submicrospheres	T = 200 °C, t = 10 min	Hu X., Yu J., Gong J. &Quan Li., (2007). Rapid Mass Production of Hierarchically Porous $ZnIn_2S_4$Submicrospheres via a Microwave-SolvothermalProcess.;*Cryst. GrowthDes.*,Vol. 7, No. 12, pp. 2444-2448.
Co^{2+} **in Layered Double Hydroxides**	T = 100, 125 °C, t = 10, 30, 60, 180 min	Herrero M., Benito P., Labajos F. &Rives V., (2007).Stabilization of Co^{2+} **in layered double hydroxides (LDHs) by microwave-assisted ageing.** *J. Solid State Chem.*, Vol. 180, pp. 873-884.

SYSTEM	CONDITIONS	REFERENCES
MFI Zeolite Membranes	T = 80, 120 °C, P = 250, 400 W, t = 60, 90 min	Motuzas J., Heng S., Lau P., Yeung K.,Beresnevicius Z. &Julbe A., (2007). Ultra-rapid production of MFI membranes by coupling microwave-assisted synthesis with either ozone or calcination treatment. *Microporous and Mesoporous Materials*, Vol. 99, pp. 197-205.
Titanate Nanotubes	T = 110, 130, 150, 175 °C, P = 70, 200, 400, 700 W,t = 1.5, 3, 6, 12 hrs	Ou H., Lo S. &Liou Y., (2007).Microwave-induced titanate nanotubes and the corresponding behavior after thermal treatment. *Nanotechnology*, Vol. 18, pp. 175702.
α-**Fe**$_2$O$_3$ Nanoarchitectures	T = 120 - 220 °C, t = 10 min	Hu X., Yu J. &Gong J., (2007). Fast Production of Self-Assembled Hierarchical α-Fe_2O_3Nanoarchitectures. *J. Phys. Chem. C*, Vol.111 , No. 30, pp. 11180-11185.
Nanosized Cobalt Oxides	T = 100 °C, t = 10, 30, 60, 180, 300 min	Herrero M., Benito P., Labajos F. & Rives V., (2007). Nanosize cobalt oxide-containing catalysts obtained through microwave-assisted methods. *Catalysis Today*, Vol. 128, pp. 129-137.
Nanorods and Nanoplates of Rare Earth Oxides	P = 650 W (70%)	Panda A. B., Glaspell G. & El-Shall M. S., (2007). Microwave Synthesis and Optical Properties of Uniform Nanorods and Nanoplates of Rare Earth Oxides. *The Journal of Physical Chemistry B Letters*. Vol. 111, pp. 1861-1864.
$AlPO_4$-18 Films	T = 150 - 200 °C, P = 1 kW	Vilaseca M., Yagüe C., Coronas J. &Santamaria J., (2006).Development of QCM sensors modified by $AlPO_4$-18 films. *Sensors and Actuators B: Chemical*, Vol. 117, pp. 143-150.
Modification of TiO_2	T = 145 °C, P = 1.2 kW, t = 5 -360 min	Wilson G., Matijasevich A., Mitchell D., Schulz J. &Will G., (2006). Modification of TiO_2 **for Enhanced Surface Properties: Finite Ostwald Ripening by a Microwave Hydrothermal Process.** *Langmuir*, Vol. 22, No. 5, pp. 2016-2027.
Nano-sized Hydroxyapatite ($Ca_{10}(PO_4)_6(OH)_2$)	T = 85 - 300 °C, P = 250 - 650 W, t = 4 min	Han J., Song H., Saito F. &Lee B., (2006). Synthesis of high purity nano-sized hydroxyapatite powder by microwave hydrothermal method. *Mater. Chem. Phys.*, Vol. 99, No. 2-3, pp. 235-239.
Ba-exchanged zeolite A	T = 150, 200, 230 °C, P = 1kW, t = 40, 60, 120 min	Ferone C., Esposito S. &Pansini M., (2006). Microwave assisted hydrothermal conversion of Ba-exchanged zeolite A into metastable paracelsian. *Microporous andMesoporous Materials*, Vol. 96, pp. 9-13.

SYSTEM	CONDITIONS	REFERENCES
Nanocrystalline Anatase TiO_2	P = 950 W, t = 2 - 6 min	Murugan A., Samuel V. &Ravi V., (2006). Synthesis of nanocrystallineanatase TiO_2 **by microwave hydrothermal method.** *Materials Lett.*,Vol. 60, pp. 479-480.
Semiconductor Rods and Wires	P = 650 W (60%), t = 30 sec to 3 mins	Panda A. B., Glaspell G. & El-Shall M. S., (2006). Microwave Synthesis of Highly Aligned Ultra Narrow Semiconductor Rods and Wires. *J. Am. Chem. Soc.*, Vol. 128, No. 9, pp. 2790-2791.
TiO_2 **Nanocrystals**	P = 650 W (33%), t = 30 mins	Glaspell G., Panda A. B. & El-Shall M. S., (2006). Reversible paramagnetism to ferromagnetism in transition metal-doped TiO_2 **nanocrystals prepared by microwave irradiation.** *Journal of Applied Physics*, Vol. 100, pp. 124307-12412.
Silicalite-1 Seeds	T = 70 - 180 °C, P = 250, 400 W,t = 10 - 150 min	Motuzas J., Julbe A., Noble R. D., Guizard C., Beresnevicius Z. J. &Cot D., (2005).Rapid synthesis of silicalite-1 seeds by microwave assisted hydrothermal treatment. *Microporous and MesoporousMaterials*,Vol. 80, pp. 73-83.
Coaxial Ag/C Nanocables	T = 200 °C,t = 20 min	Yu J., Hu X., Lib Q. &Zhang L., (2005).Microwave-assisted synthesis and in-situ self-assembly of coaxial Ag/C nanocables. *Chem. Commun.*, pp. 2704-2706.
M-Doped ZnO (M = co, Cr, Fe, Mn, Ni)	P = 650 W (33%),t = 10 min	Glaspell G., Dutta P. & Manivannan A., (2005). A Room-Temperature Synthesis of M-Doped ZnO (M = Co, Cr, Fe, Mn & Ni), *Journal of Cluster Science*, Vol. 16, No. 4, pp. 523-536.
Au and Pd Nanoparticle Catalysts	P = 650 W (33%), t = 10 mins	Glaspell G., Fuoco L. & El-Shall M. S., (2005). Microwave Synthesis of Supported Au and Pd Nanoparticle Catalysts for CO Oxidation. *The Journal of Physical Chemistry B Letters*, Vol. 109, pp. 17350-17355.
$AlPO_4$-5	T = 180 °C, P = 500 W, t = 17 min	Lin C., Dipre J. &Yates M., (2004).Novel Aluminum Phosphate-5 Crystal Morphologies Synthesized by Microwave Heating of a Water-in-Oil Microemulsion. *Langmuir*, Vol. 20,No. 4, pp. 1039-1042.

SYSTEM	CONDITIONS	REFERENCES
Crystallization of Zeolite A	P = 180, 360, 900, 1260 W, t = 10, 30, 40, 50 min P = 360, 900 W, t = 5,10, 20,30, 50 min P = 360 W, t = 15, 30, 40 min	Bonaccorsi L. &Proverbio E., (2003).Microwave assisted crystallization of zeolite A from dense gels. *J. Cryst. Growth*, Vol. 247, No. 3-4, pp. 555-562.
Sodalite/αAl_2O_3 **Membranes**	T = 160 - 200°C, P = 250 W,t = 15 - 180 min	Julbe A., Motuzas J., Cazevielle F., Volle G. &Guizard C., (2003). Synthesis of sodalite/αAl_2O_3 **composite membranes by microwave heating.** *Separation andPurification Technology*, Vol. 32, pp. 139-149.
$BaTiO_3$ **Film**	T = 100 °C, t = 5, 10, 15, 30, 60 min	Lee J., Kumagai N., Watanabe T. &Yoshimura M., (2002). Direct fabrication of oxide films by a microwave-hydrothermal method at low temperature. *Solid State Ionics*, Vol. 151, pp. 41-45.
Titanium Silicates	T = 175 °C, P = 800 W	Prasad M., Kamalakar G., Kulkarni S., Raghavan K., Rao K., Prasad P. &Madhavendra S., (2002). An improved process for the synthesis of titanium-rich titanium silicates (TS-1) under microwave irradiation. *Catalysis Commun.*,Vol. 3, pp. 399-404.
Hydrodechlorination of Chlorobenzene on Pd/ Nb_2O_5 Catalysts	T = 200 °C, P = 650 W, t = 5 min	Gopinath R., Rao K., Prasad P., Madhavendra S., Narayanan S. &Vivekanandan G., (2002). Hydrodechlorination of chlorobenzene on Nb_2O_5-supported Pd catalysts: Influence of microwave irradiation during preparation on the stability of the catalysts. *J. of Mol. Catalysis A: Chemical*, Vol. 181, pp. 215-220.
Co-, Co-Zn and Ni-Zn Ferrite Powders	T = 120, 150, 180, 200 °C, t = 30 min	Kim C. K., Lee J. H., Katoh S., Murakami R. &Yoshimura M., (2001). Synthesis of Co-, Co-Zn and Ni-Zn ferrite powders by the microwave hydrothermal method. *Mater. Res. Bull.*, pp. 2241-2250.
Uniform Pure and Dye-Loaded $AlPO_4$-5 Crystals	T = 145 - 170 °C, P = 1 kW, t = 5 - 60 min	Ganschow M., Schulz-Ekloff G., Wark M., Wendschuh-Josties M. &Wohrle D., (2001). Microwave-assisted preparation of uniform pure and dye-loaded $AlPO_4$-5 crystals with different morphologies for use as microlaser systems. *J.Mater. Chem.*, Vol. 11, pp. 1823-1827.

SYSTEM	CONDITIONS	REFERENCES
Synthesis of $AlPO_4$-5	T = 180 - 190 °C, P = 900 W, t = 8 min	Braun I., Schulz-Ekloff G. &Lautenschläger W., (1998).Synthesis of AlPO4-5 in a microwave-heated, continuous-flow, high-pressure tube reactor. *Microporous andMesoporous Materials*, Vol. 23, pp. 79-81.

5. REFERENCES

Addamo M. Bellardita M. Carriazo C. Paola A. Milioto S. Palmisano L. Rives V. 2008 Inorganic gels as precursors of TiO2 photocatalysts prepared by low temperature microwave or thermal treatment. Appl. Catalysis B: Environmental, 84 742 748 .

Agmon N. 1996 Tetrahedral Displacement: The Molecular Mechanism behind the Debye relaxation in Water. J. Phys. Chem., 100 1072 1080 .

Barison S. Fabrizio M. Fasolin S. Montagner F. Mortalò C. 2010 A microwave-assisted sol-gel Pechini method for the synthesis of BaCe0.65Zr0.20Y0.15O3-δ powders. Mater. Res. Bull., 45 1171 1176 .

Barthel J. Bachhaber K. Buchner R. Hetzenauer H. 1990 Dielectric Spectra of some common solvents in the Microwave Region: Water and Lower Alcohols. Chem. Phys. Lett., 165 369 373 .

Burgardt P. Seehra M. S. 1977 Magnetic susceptibility of iron pyrite (FeS2) between 4.2 and 620 K. Solid State Commun., 22 2 153 156 .

Chang M. Lin C. Deka J. Chang F. Chung C. 2009 Nanomechanical characterization of microwave hydrothermally synthesized titania nanowires. J. Phys. D:Appl. Phys., 42 145105

Chatterjee I. Misra M. 1991 Electromagnetic and thermal modeling of microwave drying of fine coal. Miner Metall Process, 8 110 114 .

Chen X. Yan W. Cao X. Yu J. Xu R. 2009 Fabrication of silicate-1 crystals with tunable aspect ratios by microwave-assisted solvothermal synthesis. Microporous and Mesoporous Materials, 119 217 222 .

Cutmore N. G. Evans T. G. Mc Ewan A. J. Roger C. A. Stoddard S. L. 2000 Low Frequency Microwave Technique for On-Line Measurement of Moisture. Mineral Engineering, 13 1615 1622 .

Debye P. 1929 Polar Molecules, The Chemical Catalog Company, NY.

Gopinath R. Rao K. Prasad P. Madhavendra S. Narayanan S. . Vivekanandan G. 2002 Hydro dechlorination of chlorobenzene on Nb2O5-supported Pd catalysts: Influence of microwave irradiation during preparation on the stability of the catalysts. J. of Mol. Catalysis A: Chemical, 181 215 220 .

Hu X. Yu J. Gong J. 2007 Fast Production of Self-Assembled Hierarchical α-Fe2O3 Nanoarchitectures. J. Phys. Chem. C, 111 30 11180 11185 .

Jagadeesh M. S. Seehra M. S. 1981 Thermomagnetic studies of conversion of pyrite and marcasite in different atmospheres (vacuum, H2, He and CO). J. Phys. D: Appl. Phys., 14 2153 2167 .

Kaatze U. 1989 Complex permittivity of water as a function of frequency and temperature. J. Chem. Eng. Data, 34 371 374 .

Kim C. K. Lee J. H. Katoh S. Murakami R. . Yoshimura M. 2001 Synthesis of Co-, Co-Zn and Ni-Zn ferrite powders by the microwave hydrothermal method. Mater. Res. Bull., 2241 2250 .

Kittel C. 1996 Introduction to Solid State Physics, (Seventh Edition), John Wiley & Sons, Inc.), NY.

Lester E. Kingman S. 2004 The effect of microwave preheating on five different coals. Fuel, 83 1941 1947 .

Li Y. Yang . W. 2008 Microwave synthesis of zeolite membranes: A review. Journal of Membrane Science, 316 3 17 .

Lidström P. Tierney J. Wathey B. Westman J. 2001 Microwave assisted organic synthesis- a review. Tetrahedron, 57 9225 9283 .

Lindroth D. P. 1985 Microwave drying of fine coal. US Bureau of Mines Report # 9005, US Department of Interior.

Marland S. Han B. Merchant A. Rowson N. 2000 The effect of microwave radiation on coal grindability. Fuel, 79 1283 1288 .

Marland S. Merchant A. Rowson N. 2001 Dielectric properties of coal. Fuel, 80 1839 1849 .

Merabet M. Bose T. K. 1988 Dielectric Measurements of Water in the Radio and Microwave Frequencies by Time Domain Reflectometry. J. Phys. Chem., 92 6149 6150 .

Perkin R. M. 1980 The heat and mass transfer characteristics of boiling point drying using radio frequency and microwave electromagnetic fields. Int J Heat Mass Transfer, 23 687 695 .

Ponte D. G. Prieto I. F. Vair P. F. Luengo J. C. G. 1996 Determination of moisture content in power station coal using microwaves. Fuel, 75 2 133 138 .

Prasad M. Kamalakar G. Kulkarni S. Raghavan K. Rao K. Prasad P. . Madhavendra S. 2002 An improved process for the synthesis of titanium-rich titanium silicates (TS-1) under microwave irradiation. Catalysis Commun.,3 399 404 .

Seehra M. S. Kalra A. Manivannan A. 2007 Dewatering of fine coal slurries by selective heating with microwaves. Fuel, 86 829 834 .

Standish N. Worner H. Kaul H. 1988 Microwave drying of brown coal agglomerates. J. Microwave Power EM Energy, 23 171 175 .

Utchariyajit K. . Wongkasemji S. 2010 Effect of synthesis parameters on mesoporous SAPO-5 with AFI-type formation via microwave radiation using alumatrane and silatrane precursors. Microporous and Mesoporous Materials, 135 116 123 .

Vicente I. Salagre P. Cesteros Y. Guirado F. Medina F. Sueiras J. 2009 Fast microwave synthesis of hectorite. Appl. Clay Sci., 43 103 107 .

Vicente I. Salagre P. Cesteros Y. Medina F. Sueiras J. 2010 Microwave-assisted synthesis of saponite. Appl. Clay Sci., 48 26 31 .

Wilson G. Matijasevich A. Mitchell D. Schulz J. . Will G. 2006 Modification of TiO2 for Enhanced Surface Properties: Finite Ostwald Ripening by a Microwave Hydrothermal Process. Langmuir, 22 5 2016 2027 .

Wu L. Bi J. Li Z. Wang X. . Fu X. 2008 Rapid preparation of Bi2WO6 photocatalyst with nanosheet morphology via microwave-assisted solvothermal synthesis. Catalysis Today, 131 1-4 , 15 20 .

Yanshou L. Weishen Y. 2008 Microwave synthesis of zeolite membranes: A review. Journal of Membrane Science, 316 3 17 .

9 Chapter

MICROWAVE IRRADIATION IS EMPLOYED TO SYNTHESIZE AND PROCESS BIODEGRADABLE AND BIO-BASED POLYMERS

1. INTRODUCTION: AN OVERVIEW

The intrinsic environmental benefits of this class of materials are being increasingly pursued through the continuous pursuit of higher energy efficiency in synthesis and processing steps. Biodegradable polymers and polymers produced from renewable feedstocks are also attracting increasing interest as potential substitutes for conventional plastics. Currently, microwave-assisted organic synthesis (MAOS) is a significant area of research in green chemistry, and a significant number of papers have been published that demonstrate its superiority to traditional thermal heating. However, microwave (MW) energy sources are now being employed in numerous polymerization reactions. In fact, the environmental performance may be significantly enhanced by the reduced heating periods and superior homogeneity offered by MW reactors, which could be instrumental in optimizing production processes. The theoretical principles of microwave-matter interaction are briefly reviewed in the introduction of this chapter. Subsequently, a variety of experimental configurations are analyzed, and, ultimately, thermal and non-thermal specific microwave effects are described and commented upon. The following paragraph is dedicated to a comprehensive survey of synthesis examples found in scientific literature and categorized by polymerization technique. Specifically, the paragraph addresses products of increasing commercial importance, such as poly(ε-caprolactone) (PCL) and poly(lactic acid) (PLA). A third section of the chapter pertains to the utilization of microwave heating for the chemical modification and processing of polymers. The final paragraph summarizes the benefits and disadvantages of microwave-assisted polymer chemistry, with a particular emphasis on energy efficiency and the formulation of conclusions.

1.1. Microwave irradiation: Theoretical principles

Microwaves are electromagnetic radiations characterized by frequencies that extend from 0.3 GHz (corresponding to an energy of 1.2×10^{-6} eV) to 30 GHz (1.2×10^{-3} eV), covering the portion of the spectrum between the less energetic Radio waves and the more energetic Infrared waves. Due to an intensive employment by the information and telecommunication technologies, only few frequencies among the whole microwave band are actually available for scientific, industrial and household heating applications: in order to avoid any interference, the vast majority of devices operates at 2.45 GHz (1.0×10^{-5} eV). Comparing the latter value to the energy associated with some common covalent bonds (C-C single bond: 3.61 eV; C=C double bond: 6.35 eV) or even with a weaker interaction like hydrogen bond (0.04 – 0.44 eV) it is clear that microwaves do not posses enough energy to induce a chemical reaction by breaking any of these. Very often, indeed, the impressive shortening of reaction times, reported by the authors cited in the following paragraphs, is ascribed to peculiarities in the heating process like homogeneity, inverted temperature gradient (i.e. vial walls are colder than reactants in the bulk) and extremely fast temperature increase. The interaction between microwaves and the electric dipoles of irradiated reactants or solvent molecules is what underlies "dielectric heating". Since the order of magnitude of molecular rotational frequencies is lower, but quite close to microwave frequencies, the tendency to align with the oscillating electromagnetic field produces an immediate and simultaneous motion that can't indefinitely follow the oscillating field and eventually results in thermalization. Kinetic energy is transformed into heat via phenomena (i.e. phase shifts and dielectric losses deriving from molecular friction) whose mechanism depends on both molecular mass (more correctly on rotational moment of inertia) and dielectric coefficient "ε" of each component of the irradiated reaction mixture. Dielectric coefficient (or complex permittivity) consists of a real component and of an imaginary component ($\varepsilon = \varepsilon' + i\varepsilon''$): widely adopted parameters to describe the efficiency of radiation-to-heat conversion are the "loss factor"(ε") and the "dissipation factor", indicated with "tan(δ)" or "D" and defined as:

$$\tan(\delta) = D = \varepsilon''/\varepsilon' \tag{1}$$

Polar solvents and reactants normally possess high values of tan(δ), indicating strong microwave absorption; on the other hand low values shown by most non-polar substances indicate transparency and inefficient energy transfer (Fig. 1).

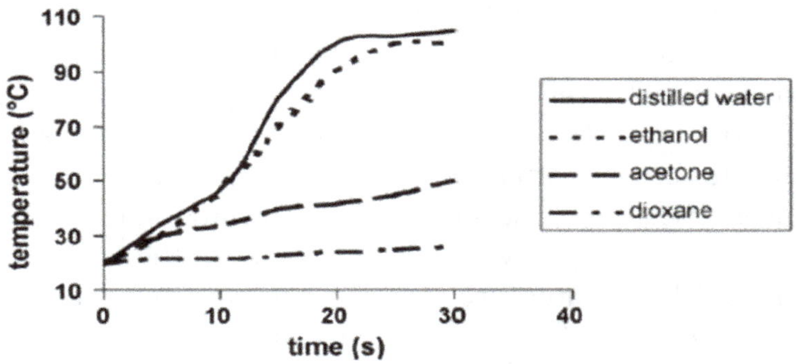

Figure 1. Microwave heating curves of different solvents in a multimodal cavity (power = 150 W) (from C. Leonelli, T.J. Mason, Microwave and Ultrasonic Processing: Now a Realistic Option for Industry, Chemical Engineering and Processing, 2010, 49, 885–90. Copyright Elsevier Ltd. Reproduced with permission).

Even if, commonly, homogeneous heating represents a topic feature of microwave devices it should be noted that highly dielectric materials rapidly diminish the incident radiation intensity, hence recourse to large volumes may cause shielding effects and colder matter at the core. This is the reason why "penetration depth", defined as the point where only 37% (1/e) of the incident MW radiation is still present and calculated as the reciprocal of loss factor, turns out to be another useful parameter in the characterization of an experiment. Besides absorption and transmission, a third phenomenon can take place: radiation can be reflected by the surface of a body (highly conducting metals are important examples). This eventuality should generally be avoided in setting up a reaction since, while only negligible heating is generated in the reflecting material, microwave constructive interferences may arise elsewhere amid the irradiated matter and cause uneven temperature profiles and "hot spots" on account of local super-heating effects.

1.2. Microwave heating setups

Although papers occasionally exhibit inadequate technical detail, the selection of the experimental apparatus and its engineering specifications is of paramount importance in microwave-assisted synthesis, particularly in comparison to conventional chemistry. Reproducibility is significantly influenced by the degree to which the user's configuration corresponds to the actual reaction parameters. Reproducible results that can be productively contrasted with literature are obtained by effectively tuning and monitoring the most critical variables. Unquestionably, significant advancements have been made in this area over the past few decades. As microwave-assisted synthesis has garnered an increasing amount of focus, a number of dedicated reactors have been commercialized. In numerous laboratories, these devices have now supplanted the former domestic MW ovens, setting new standards for safety, efficacy, and parameter regulation.

Reactors are frequently divided into multi- and mono-modal configurations based on the direction of microwaves from the magnetron to the sample. Multimodal devices produce a relatively uniform electromagnetic field that disperses radiations in various directions through a large cavity, where the field intensity is averaged by interference with reflected waves. In contrast, monomodal devices have waveguides that are precisely calibrated, allowing radiation to be reflected in phase to produce standing waves with a maximum at the sample compartment position. Monomodal applicators are exclusively found among dedicated scientific equipment, while domestic MW ovens are examples of multimodal instruments and some high-capacity scientific devices. Some authors (Nuchter et al., 2004) argue that the aforementioned distinction is only valid until the reaction mixture is introduced into the cavity, which disrupts the emitted field and converts monomodal reactors into (partly) multimodal ones. They believe that a more practical classification should be based on radiation intensity or power density. However, it is generally acknowledged that monomodal reactors are the most appropriate for small loads (note that the dimension of the wavelength itself limits the viable volumes for this class). This is due to the fact that they minimize the occurrence of hot spots, a detrimental phenomenon that disrupts reactions in multimodal devices when uneven temperature profiles are not averaged out by a significant amount of absorbing matter. This is the reason why it is frequently advised (Bogdal et al., 2003) to fill multimodal cavities with samples that occupy more than 50% of the available volume. Applicators with radiant geometries may also be advantageously implemented when substantial volumes and consistent field homogeneities are necessary. The capacity to install antennas within the cavity renders the field considerably less susceptible to perturbation by the load (Leonelli & Mason, 2010). It is important to mention that a central technical feature for synthetic purposes, which is typically provided by a professional apparatus, is the ability to emanate continuous (i.e., unpulsed) radiation. This feature ensures that the reaction mixture is heated at a user-assigned power level that remains constant over time. Inexpensive domestic ovens operate in a pulsed profile, irradiating at their utmost power. The on/off time ratio is determined by the selected power level, allowing for the adjustment of the average MW intensity throughout the irradiation period. However, this approach is clearly susceptible to momentary overheating. Temperature control is a critical engineering consideration. While some authors (F. Liu et al., 2010; Sharma & Mishra, 2010) still opt to conduct the measurement immediately following microwave exposure (a conventional thermometer may be employed in this scenario), real-time monitoring is generally preferred in order to operate a more precise control now. In the overwhelming majority of the reported examples, three configurations are implemented: The preferred and most cost-effective sensors appear to be fiber-optic thermometers that are inserted into the reaction mixture using transparent tubes at a laboratory scale, as their cost and fragility limit their industrial application. According to some authors, permanent deterioration is already apparent after a few hours at 250 °C, and the operative range is 0-300 °C (Nuchter et al., 2004). Nevertheless, sapphire-based fiber-optic instruments can be employed to measure temperatures as high as approximately 2000 °C. Optical (IR) pyrometers are a universally viable solution that are widely used in commercial technical devices and can be used in large-scale industrial reactors. They offer a much wider temperature range (-40 °C to 1000 °C), but they must measure the temperature at the vial wall, which means they invariably underestimate the value.. In order to prevent excessive MW absorption by their metallic shields, which would lead to overheating effects and biased measurement, shielded thermocouples are the most cost-effective option. They can withstand temperatures of up to 300 °C. However, they are quite cumbersome. Lossy (polar) solvents are required. The selection of the appropriate solvent for a MW-assisted reaction must consider its dielectric coefficient (ε). It is important to note that dielectric constants (ε') typically decrease significantly with temperature (Fig. 2). For instance, pressurized water at 300 °C exhibits a value of 20, which is very similar to that of acetone at room temperature (Strauss & Trainor, 1995).

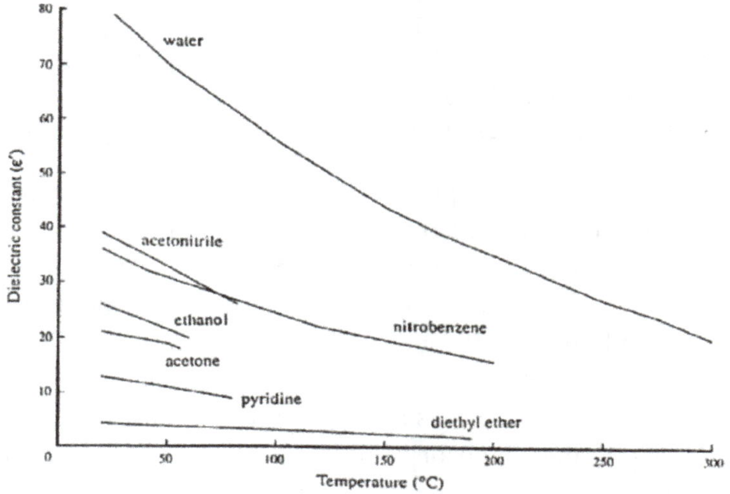

Figure 2. Dielectric constants vs. temperature for different solvents (from Dean, J. A. Ed.; Lange's Handbook of Chemistry, 13th ed.; McGraw-Hill, New York, 1985; p.99, ISBN: 0070161925. Copyright McGrawHill Education. Reproduced with permission).

Non polar solvents may be employed improving their heating efficiency by small additions (few percents in weight) of miscible liquids with high loss factors or even salts, preferably with a sufficient solubility in order to assure homogeneous conditions (Lidström et al., 2001). A similar approach underlies the diffuse exploitation as heating elements of highly MW absorbing solids, like silicon carbide (Kremsner & Kappe, 2006) (Fig. 3), carbon black (F. Liu et al., 2010) or silica particles (P. Liu & Su, 2005). Microwave assisted reactions may also be conducted in environmentally friendly water-miscible ionic liquids which are characterized by high dielectric constants and offer the possibility of isolating hydrophobic products by the addition of water to the homogeneous solution (Guerrero-Sanchez et al., 2007).

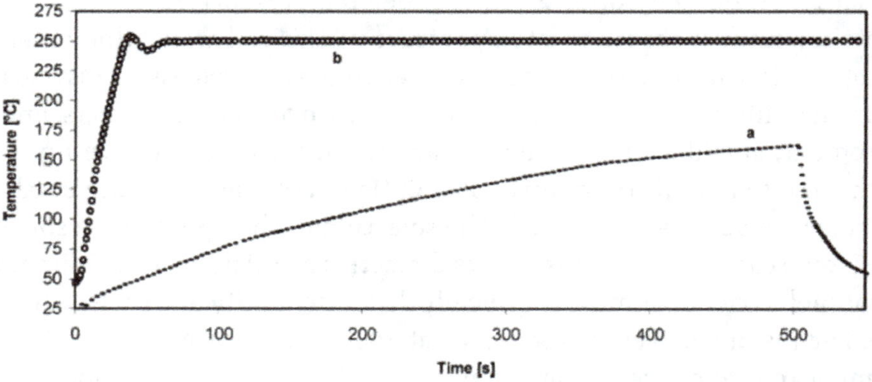

Figure 3. Microwave heating curves for a 0.2 M solution of allyl phenyl ether in toluene: a) no heating element; b) with SiC as heating element (from J. M. Kremsner, C. O. Kappe, Silicon Carbide Passive Heating Elements in Microwave-Assisted Organic Synthesis, J. Org. Chem., 2006, 71, 4651-4658. Copyright American Chemical Society (ACS). Reproduced with permission).

When possible, the solvent-free option should always be favored on the basis of the green chemistry principles: the bulk ring opening polymerization of lactide to yield poly(lactic acid) is a notable case (Frediani et al., 2010). Early experiments were often performed in open vessels, at a temperature below the boiling point of high-boiling solvents (e.g. DMSO or DMF). This simplest configuration may as well comply with solvent-free reactions not requesting modified atmosphere, in which at least one of the reactants is a high-boiling liquid or a low-melting solid capable of efficient microwave absorption. Even if open vessels were tentatively adopted with more volatile solvents like toluene, increasing fire hazards brought by this choice should be seriously considered. Pressurized vials allow to use solvents at a temperature above their boiling point, considerably increasing reaction rates, though it should be noted that gas-tight sealed caps might cause explosions: technical devices usually feature dedicated vials equipped with proper septa capable of withstanding pressures up to a determinate value. Examples of customized reflux systems are frequently found in literature (Waugh & Parkin, 2010; Zhu et al., 2003): this particular setup proved to be safe and easy to scale-up, on the other hand it precludes overheating limiting the advantages of microwave irradiation over conventional heating. It is eventually worth mentioning that industrial-scale MW generators can be engineered to work efficiently under conditions from full vacuum to 1.03 MPa and from 30 °C to 200 °C (Leonelli et al., 2007). Microwave assisted synthesis is gaining increasing appeal for industrial applications, therefore the possibility of scaling-up reactions becomes a pivoting issue. One major limitation that prevents the load of large volumes at one time is the penetration depth of microwaves through lossy solvents or reactants: as already stated, the shielding of radiation by the peripheral matter produces inefficient heating at the core. Besides, heating large quantities of liquids in sealed vessels augments explosion hazard and obligate to employ thick, robust walls that must be MW transparent at the same time. It is eventually worth noting that massive quantities contained in thick vials are very difficult to cool rapidly and evenly with water or pressurized air. This critical points may be circumvented by loading high volumes in smaller, separated vials in parallel (a tedious configuration for large scale) or more efficiently by switching to a continuous flow reactor equipped with a relatively thin reaction tube through which reactants and solvents move in a necessarily homogeneous solution (Moseley & Woodman, 2008; Glasnow & Kappe, 2007). Open vials were successfully chosen for the irradiation of volumes up to 1000 mL (Deetlefs & Seddon, 2003) though it should be pointed out that MW penetration depth and the boiling point of the mixture limit the feasibility of such a layout. When only high-boiling substances are employed, solvent free reactions may also be performed with vials sealed under vacuum. As an example, the bulk ring opening polymerization of Caprolactone was recently performed loading five 1.8 L customized glass tubes with about 800 g of monomer each and irradiating simultaneously into a large multimodal microwave heater (Xu et al., 2010).

1.3. "Thermal" and "non-thermal" specific microwave effects

The remarkable increase in microwave-assisted reaction rates that numerous authors have observed since the early works has raised a number of concerns regarding the rational origins of this phenomenon. A wide range of distinctions have been documented between experiments conducted under conventional or microwave conditions. Typically, these discoveries are elucidated using rational hypotheses; however, the absence of a rigorous and widely accepted terminology occasionally leads to ambiguity when conducting a literature review. A frequent distinction is made between "thermal" and "non-thermal" effects, which are occasionally referred to as "specific microwave" effects. Precisely, also the "thermal" effects are microwave-specific: despite being based on heating phenomena they cannot be obtained by conventional oil baths. Dealing with variations in reaction kinetics, former works (Perreux & Loupy, 2001) sensibly suggested to refer to the Arrhenius equation:

$$k = A \exp(-\Delta G^{\neq}/RT) \quad (2)$$

In the equation three variables may be influenced by the use of microwaves as energy source, hence the rate (k) can be enhanced by three contributions: higher pre-exponential factor (A) due to increased collision efficiency; lower activation energy ($\Delta G^{\neq} = \Delta \mathbf{H}^{\neq} - \mathbf{T}\Delta \mathbf{S}^{\neq}$) due to stabilized polar transition states and eventually a higher actual temperature. The latter can obviously be listed as a "thermal" effect, and comprises the phenomena of local superheating and hotspots, inverted temperature gradient (i.e. wall effect), selective heating and similar. The variations of A and $\Delta G^{\neq}$, on the other hand, may stem from the peculiar mechanism of interaction between matter and electromagnetic field, and therefore could be catalogued as "non thermal" effects. It should be pointed out that this approach holds also for the quite frequent cases of altered or enhanced reaction selectivities: since thermodynamic parameters do not depend upon the energy source, selectivity could be identified, here, as a kinetic competition between two or more reaction pathways, keeping in mind that different molecules may absorb microwaves with a different efficiency. This fairly rigorous paradigm is sometimes not applied in the interpretation of real life experiments since relating the observed phenomena to the terms of the equation is not always straightforward. Certain writers analyze the actual temperatures recorded during conventional and microwave heating methods to see whether the thermal variations are significant enough to account for the inconsistencies in their findings (Jing et al., 2006). The existence of "non-thermal" effects is sometimes disputed, making it a contentious issue, whereas the significance of "thermal" microwave effects is now broadly acknowledged, despite the absence of a comprehensive explanation for their occurrence. Descriptions of these phenomena are occasionally identified and emphasized within the publications examined in the subsequent paragraphs.

2. MICROWAVE ASSISTED SYNTHESIS OF BIODEGRADABLE AND BIO-BASED POLYMERS

This paragraph enumerates and categorizes various works according to the polymerization method, with the subsequent two subparagraphs focusing on step-growth polymerizations and ring-opening polymerizations, respectively. All the examples presented herein pertain to biodegradable or bio-based polymers (i.e., polymers wholly or partially generated from renewable feedstocks) to underscore the ongoing proliferation and advancement of goods with a more environmentally sustainable focus.

2.1. Step growth polymerizations

Step growth polymerizations generally involve bifunctional monomers yielding linear chains with functionalized terminations, capable of reacting with more monomer units. On account of the same principle, multifunctional monomers may be utilized to generate branched structures and thermosetting materials. The most common polymerizations in this class belong to the family of polycondensations, in which every coupling reaction implies the release of a small molecule (i.e. water).

2.1.1. Aliphatic polyesters

Among the conspicuous category of aliphatic polyesters many important examples of biodegradable and bio-based polymers can be found. An eco-friendly, solvent-free synthesis of poly(butylene succinate) (PBS) with a M_w of 1.03×10^4 g/mol and 2.35×10^4 g/mol was achieved by the use of microwave heating at a temperature of, respectively, 200 °C and 220 °C (Scheme 1). An irradiation time of 20 min was needed, while poly(buthylene succinate) with a M_w of 1.02×10^4 g/mol was obtained by conventional heating in 5 h. The polycondensation was performed under atmospheric pressure of N_2, in the presence of 1,3-dichloro-1,1,3,3-tetrabutyldistannoxane as catalyst (Velmathi et al., 2005).

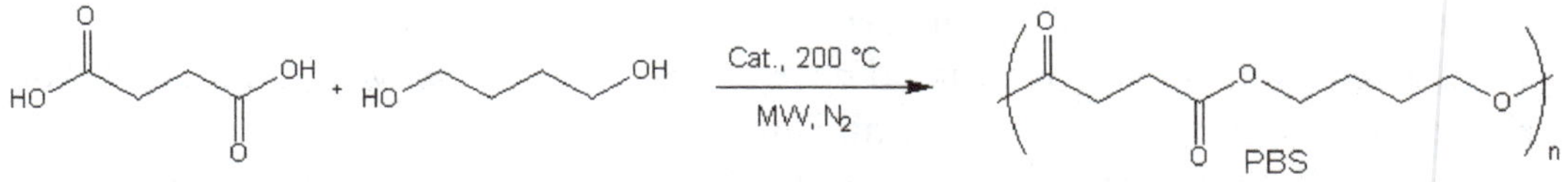

Scheme 1. Synthesis of poly(butylene succinate) (PBS).

Several authors accomplished the direct polycondensation of lactic acid to yield poly(lactic acid) (PLA) under microwave conditions. Ring opening polymerization of Lactide (the cyclic dilactone of lactic acid) is usually favored for the synthesis of relatively high molecular weight PLAs (see paragraph 2.2.3), nonetheless this single-step route is often attempted on account of its simplicity and low cost. In a former work D,L-Lactic acid was polymerized in bulk in a domestic microwave oven for 30 min. at a power level of 650 W, yielding oligomers with molecular masses ranging from 500 to 2000 g/mol. Reaction time was significantly shortened comparing to 24 h of conventional heating at 100°C and molecular masses were found to increase with irradiation time, while yields followed an opposite trend; after 20 min. cyclic oligomers were also formed, affecting the physical properties of the product (Keki et al., 2001). More recently, P(L)LAs with markedly higher molecular masses ($M_w \approx 16,000$ g/mol) were obtained within 30 min. of microwave irradiation employing a single-mode CEM *Discover* reactor set at a temperature of 200 °C. A binary catalyst of $SnCl_2$/p-TsOH was used and the polycondensation was conducted at a reduced pressure (30 mmHg) in order to remove water that progressively formed, avoiding the hydrolysis of the growing chain. Cyclic oligomers were still present as byproducts (Nagahata et al., 2007). Further enhancement was proposed in a later work by the use of solid super-acid (SSA) as a green heterogeneous catalyst: PLAs with molecular masses as high as 30,000 g/mol were synthesized in 2-3 h irradiation times, after a water removal step under vacuum, at a temperature of 260 °C reached gradually. Interestingly, the authors claimed to save about 90% energy with respect to the time consuming (24-48 h) conventional heating procedure (Lei et al., 2009). A phenyl substituted derivative of lactic acid, precisely L-2-hydroxy-3-phenylpropanoic acid, was also polymerized in a domestic microwave oven with a maximum power of 510 W: molecular masses, quite low in any case, were found to increase with irradiation time, ranging from 1,800 g/mol within 30 min. to 5,400 g/mol within 2.5 h (L. Liu et al. 2001).

2.1.2. Other step growth polymerizations

The polycondensation of sebacic acid and ω-amino alcohols was performed in a variable frequency microwave reactor to yield poly(amide-ester)s with impressive mechanical properties yet susceptible to degradability by proteolytic enzymes. Temperature was set at 180 °C, 200 °C and 220 °C depending on the chosen amino alcohol and regulated by an on - off power switching; power level was kept relatively low (50-70 W) due to the strong absorption of the starting materials. Reaction time was reduced from 3h of conventional heating to 1h of microwave irradiation, while product yields, molecular masses, and physico-chemical properties were found to be equal or higher (Borriello et al. 2007). Poly(anhydride)s are studied since long time as drug delivery systems with a tunable surface erosion: the synthesis of several poly(anhydride)s from aliphatic and aromatic diacids were conducted under microwave conditions to reduce reaction times from 3 h to 6-20 min at atmospheric pressure. Molecular weights were similar to the product of conventional heating; working under vacuum, as well as the isolation of an intermediate acetylated prepolymer, was not necessary, drastically simplifying the synthetic procedure (Vogel et al., 2003). With the aim of valorizing renewable feedstocks, interesting poly(ether)s were synthesized from isosorbide (1,4:3,6-dianhydrosorbitol) and 1,8-dibromo- or 1,8-dimesyl-octane using phase transfer catalysis under microwave irradiation. A monomodal *Synthwave 402* reactor, manufactured by Societè Prolabo, allowed to regulate temperature by modulating the emitted field and to monitor the reaction temperature on the surface of the vials by an infrared detector; toluene (1 mL for 5 mg of reagents) was used as a solvent to enhance temperature control and lower the mixture viscosity. Several advantages using microwave heating were reported: similar yields were obtained in 30 min instead of 1 day of reaction in an oil bath; products showed higher molecular weights and better homogeneities; a different termination mechanism generates ethylenic instead of hydroxylated chain ends (Chatti et al., 2002). Few years later the same group described the synthesis of novel poly(ether-ester)s from adipoyl chloride and aliphatic diols derived from isosorbide: polycondensations under microwave conditions proceeded five times faster (95% yield after 5 min at 150 °C) and the polymer showed higher molecular weights (M_w = 4,200 g/mol vs 4,050 g/mol obtained by thermal heating). It is worth noting that, considering the identical temperature profiles generated by the two heating procedures, the better results shown by microwave irradiation are ascribed by the authors to "specific microwave features" such as the extension of polarity in the transition state comparing to the ground state (Chatti et al., 2006). Biodegradable, low molecular weight poly(alkylene hydrogen phosphonate)s were synthesized by a transesterification reaction of dimethyl hydrogen phosphonate with PEG 400 under microwave irradiation. Slightly higher temperatures were reached during the polymerization and markedly reduced reaction times (55 min vs 9h of conventional heating) were observed, which avoided the undesirable thermal degradation of dimethyl hydrogen phosphonate (Bezdushna et al., 2005).

2.2. Ring opening polymerizations

Ring opening polymerizations (ROPs) are nowadays widespread and very popular synthetic routes. The use of cyclic monomers exhibits some unquestionable advantages for the preparation of high molecular weight polymers since no byproduct (e.g. water) is released during the propagation step, avoiding chain scission as well as competition with monomers and consequential early chain termination. In some cases ring strain may also play a role in favoring the polymerization thermodynamically.

2.2.1. Poly(ε-caprolactone)

Poly(ε-caprolactone) (PCL) is not produced from renewable resources, nonetheless, in the continuously developing class of biodegradable polymers, it proved to be one of the most useful and attractive: the number of works in scientific literature concerning this polymer is indeed impressive, and several successful microwave assisted syntheses are reported. An early example of microwave assisted synthesis employed titanium tetrabutylate as catalyst and an impressively refined custom monomodal reactor: conversion, measured by means of viscosity build-up, and number average molecular weights were close to those observed for conventionally heated procedure and did not indicate any advantage in using microwave irradiation (Albert et al., 1996). Poly(caprolactone) with weight-average molar mass (M_w) over 4×10^4 was prepared by metal free, microwave assisted ring opening polymerization initiated by benzoic acid: the authors indicate that the ratio of polymer chain propagation was directly proportional to temperature between 160 °C and 230 °C, while above this interval the synthesized PCL degraded. A multimodal domestic MW oven was employed: power was modulated by power on-off cycles, temperature was monitored by a tin-grounded thermocouple and the reactions were performed in vacuum sealed ampoules. As stated by the authors, the advantage of microwave irradiation was the significantly enhanced propagation ratio of PCL chain while the formation of growing centers was greatly inhibited (Yu & Liu, 2004). Earlier, the same group compared the synthesis of PCL initiated by benzoic acid and chlorinated acetic acids (Fig. 4): in both cases higher M_w values and lower polydispersity indexes were observed with respect to conventional heating for an equal reaction time, nonetheless the molar masses of polymers synthesized employing chlorinated acids were lowered by degradation processes (Yu et al., 2003).

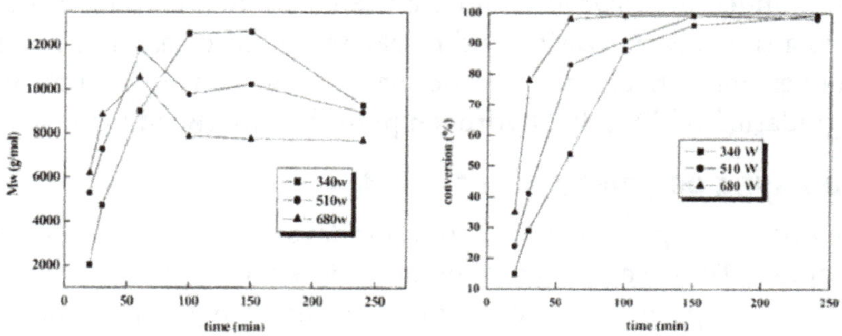

Figure 4. Mw (left) and conversion (right) vs irradiation time for the ROP of ε-caprolactone initiated by chloroacetic acid (M/I = 15) (from Z. J. Yu, L. J. Liu, R. X. Zhuo, Microwave-Improved Polymerization of ε-Caprolactone Initiated by Carboxilic Acids, Journal of Polymer Science: Part A: Polymer Chemistry, 2003, 41, 13–21. Copyright Wiley-VCH Verlag GmbH & Co. KGaA. Reproduced with permission).

A lipase-catalyzed microwave assisted synthesis of PCL was attempted in several solvents at their boiling point and remarkably different behaviors were observed: while the reactions in toluene and benzene proceeded slower compared to oil bath heating, the reaction in diethyl ether showed an increased rate. It is known that the activity of the catalyst Novozyme 435, depends on temperature and on the polarity of the medium: no polymerization was indeed observed in polar solvents like THF or dioxane. Microwave irradiation was performed with a monomodal CEM-*Discover* device, equipped with a CEM fiber optic temperature sensor, for 90 min under reflux conditions (Kerep & Ritter, 2006). A "green" synthesis was described to obtain poly(ε-caprolactone)s with molar masses ranging from 3,000 g/mol to 16,000 g/mol: no solvent was used and non toxic, biologically acceptable lanthanide halides were chosen as initiators. Two different microwave equipments were employed: a monomodal Prolabo *Synthwave S402* equipped with an IR temperature sensor (15 to 45 min of irradiation at 180 - 200 °C) and a Sairem *GMP 12T* provided with a Marconi *F1192-12* water-cooled circulator (3 to 20 min of irradiation at 230 °C). Interestingly, the degradation rate of PCL was also assessed: it was found that the slow hydrolytic process is accelerated by the catalyst, while the faster enzymatic process is inhibited (Barbier-Baudry et al., 2003). The

$Sn(Oct)_2$ catalyzed microwave assisted polymerization of ε-caprolactone (ε-CL) can be initiated by methacrylic acid or acrylic acid yielding radical polymerizable PCL macromonomers. Reactions were carried out in a monomodal CEM *Discover* reactor at 180 °C for 90 min in bulk, under air; it is worth noting that the high temperature reached under microwave irradiation did not imply uncontrolled free radical polymerization processes, though in this case a comparison with classical thermal activation showed no significant acceleration effect (Sinwell et al., 2006). The kinetics of pure and zinc catalyzed ring opening polymerization of ε-caprolactone was also investigated in comparison to conventional heating syntheses and a model based on continuous distribution kinetics with time/temperature-dependent rate coefficients was proposed. From these coefficients the activation energies were calculated: while for pure thermal heating and catalyzed thermal heating the values were respectively 24.3 and 23.4 kcal/mol, the value found for catalyzed microwave heating was 5.7 kcal/mol, reflecting the observed increment in the polymerization rate (Sivalingam et al., 2004). Kinetic investigations were also performed comparing the rate constant for the stannous octoate catalyzed ROP of ε-CL under microwave conditions and conventional heating. Several equilibrium temperatures were reached by means of different power levels: while conventional polymerizations followed Arrhenius' law, the rate of microwave assisted reactions (Fig. 5) showed an abnormal peak value for 180 °C (at 500 W), inducing the authors to suggest the hypothesis of non-thermal microwave effects (Li et al., 2007).

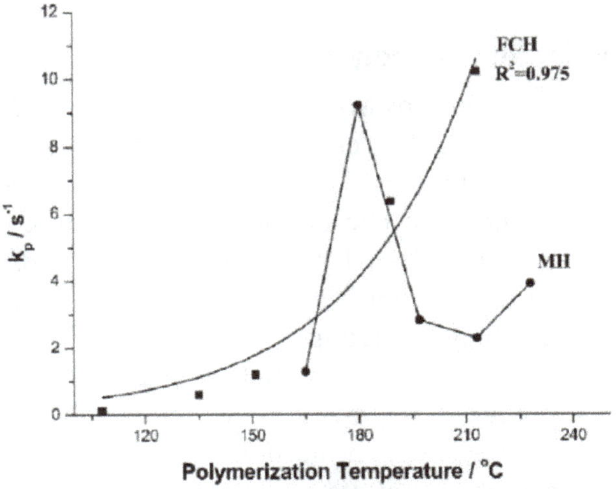

Figure 5. Chain propagation rate constant (k_p) for the ROP of ε-caprolactone by microwave heating (circles) and conventional heating (squares) (from H. Li, L. Liao, L. Liu, Kinetic Investigation into the Non-Thermal Microwave Effect on the Ring Opening Polymerization of ε-Caprolactone, Macromol. Rapid Commun., 2007, 28, 411–416. Copyright Wiley-VCH Verlag GmbH & Co. KGaA. Reproduced with permission).

Combining ring opening polymerization and microwave assisted "click" chemistry, acetylene-functionalized PCL was synthesized and reacted with heptakis-azido-β-cyclodextrin to obtain star shaped polymers. A tetra-arm metallopolymer was also achieved by the "supramolecular click" reaction between di(pyridinyl)piridazine-functionalized macroligands and Copper (I) ions (Hoogenboom et al., 2006). The 1-butyl-3-methylimidazolium tetrafluoroborate ionic liquid has been successfully used as solvent for the ROP of ε-caprolactone: on account of its efficient microwave energy absorption. PCL with M_w of 28,500 g/mol was obtained by means of a multimodal domestic oven at 85 W for 30 min, using Zinc oxide as catalyst (Liao et al., 2006). A large scale synthesis was recently performed using up to 2.45 Kg of monomer to assess the industrialization possibilities of microwave assisted ε-CL polymerization. A self-designed industrial microwave oven, capable of a maximum power of 6,000 W, was employed at different power levels and loaded with different amount of reactants: in any case the stannous octoate catalyzed ring opening polymerizations proceeded smoothly, yielding PCL with M_w ranging from 66,000 g/mol to 122,000 g/mol (Xu et al., 2010). The effects of the presence of diisopropyl hydrogen phosphonate (2.9%) on the microwave assisted synthesis of PCL was examined, concluding that the ROP was initiated by traces of water but could be catalyzed by the oligo(ε-caprolactone)-substituted hydrogen phosphonate resulting from transesterification reactions. A polymer with molecular weight of 8,100 g/mol was obtained in 100 min at 510 W of irradiation power (Tan et al., 2009).

2.2.2. Poly(ε-caprolactone) copolymers

One of the first examples of microwave assisted synthesis of PCL copolymers was performed by means of a variable frequency device operated in the range from 0.4 to 3 GHz: poly(ε-caprolactam-*co*-ε-caprolactone) was obtained by the anionic catalyzed ring opening of the two cyclic monomers (both showed effective radiation absorption). Compared to conventional procedures, higher yields and higher amide content were observed, the latter implying a higher value for the glass transition temperature (T_g) of the copolymer (Fang et al., 2000). More recently, a series of completely biodegradable poly(vinyl alcohol)-*graft*-poly(ε-caprolactone) comb-like copolymers were prepared employing poly(vinyl alcohol) as an initiator for the microwave assisted ROP of ε-CL. The authors report a degree of polymerization (DP) for PCL in the range of 3-24 and a degree of substitution (DS) of PVA from 0.35 to 0.89 and specify that both values were enhanced by microwave heating: irradiation was carried out at three power levels (340, 510 and 680 W), finding that the higher power provided higher DPs and DSs (Yu & Liu, 2007). Amphiphilic di-block PEG-*b*-PCL copolymers were synthesized in a one pot procedure under microwave irradiation using PEG monomethyl ester as oxydrilic initiator and boron trufluoride, Sodium hydride or stannous octoate as alternative catalysts, the latter yielding polymers with the highest purity (Ahmed et al., 2009). By means of a similar procedure, different authors obtained PEG-*b*-PCL copolymers with M_w values ranging from 5,800 g/mol to 14,500 g/mol and very low polydispersity indexes (≤ 1.19). A comparison with oil-bath heating synthesis showed a dramatic reduction of reaction time under microwave conditions: analogous conversions were obtained after 45 min instead of 1 day (Karagöz & Dinçer, 2010). Biodegradable poly(ε-caprolactone)-*co*-poly(*p*-dioxanone) random copolymers with M_w in the 120,000-180,000 g/mol range and polydispersity indexes around 2 were synthesized by ring opening polymerization of the two cyclic co-monomers under microwave irradiation. The reaction was completed within 20 min at 140 °C: after 30 min the conversion of ε-CL remained constant, while the conversion of *p*-DO slightly decreased (Chen et al., 2010).

2.2.3. Poly(lactic acid)

Even though poly(lactic acid) can be synthesized by direct polycondensation of lactic acid (see paragraph 2.1.1), polymers with higher molecular weights and lower polydispersity indexes are commonly obtained by ring opening polymerization of lactide (i.e. the cyclic dilactone of lactic acid) (Scheme 2). LL- and DD-lactide yield semi crystalline polymers named P(L)LA and P(D)LA, while *rac*-lactide is used to synthesize amorphous polymers named P(LD)LA. On account of its attractive physico-chemical properties and its biodegradability, PLA is widely used as bio-based, large scale packaging material and in a number of biomedical applications, hence it represents a subject of extensive research. A stannous octoate catalyzed microwave assisted synthesis of P(LD)LA was reported without the need of an inert atmosphere during irradiation. Using a domestic oven set at a power level of 450 W for 30 min a polymer with viscosity-average molecular weight over 200,000 g/mol was obtained in 85% yield; the reaction was carried out in bulk, using open vessels and adding carborundum (SiC) to the mixture as microwave absorber. It is worth noting that, comparing the temperature in microwave and conventional conditions, the occurrence of both thermal and non-thermal microwave effects is claimed by the authors (Jing et al., 2006). In a different study, by means of an analogous procedure (except in this case the polymerization took place under vacuum) P(LD)LA with weight-average molecular weight ranging from 97,000 g/mol to 185,000 g/mol and polydispersity indexes around 2 were obtained in 90% yield within 10 min.

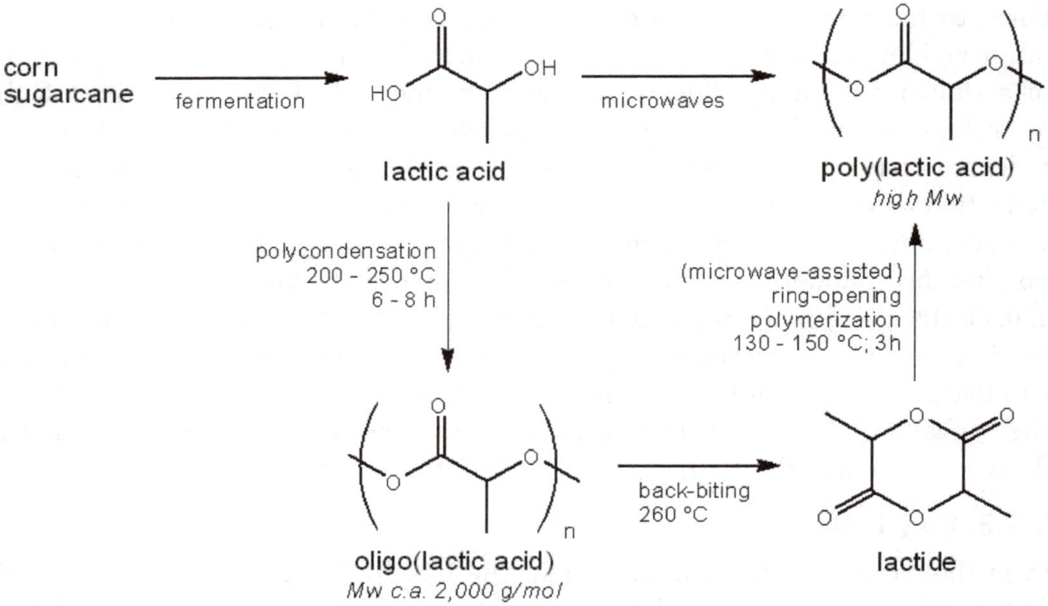

Scheme 2. Synthetic routes to high Mw poly(lactic acid).

Optimal results were achieved setting the power level at 255 W, while reactions carried out at 340 W or more showed a significant degradation of the product (C. Zhang et al., 2004). According to other authors, the reaction rate in the ROP of L-lactide increases linearly with power output (tested up to 1,300 W), though molecular weights (and consequently degradation phenomena) were not monitored for P(L)LA (Koroskenyi & McCarthy, 2002). The synthesis of relatively high molecular weight P(LD)LA (ranging from 26,700 g/mol to 112,500 g/mol) was recently carried out in a CEM *Discover* monomodal MW reactor set at 100 °C for 30 min. Stannous octoate in low quantities with respect to monomer (0.1 mol%; 0.02 mol% and 0.01 mol%) was used as catalyst; the reaction was carried out in bulk, under reduced pressure (Nikolic et al., 2010). P(LD)LA with a high viscosity-average molecular weight (159,200 g/mol) could also be obtained, in similar conditions, by means of a modified domestic oven capable of a continuous microwave irradiation: the authors report a good yield (81.1%) after only 10 min at a low power level of 90 W (Y. Zhang et al., 2009). In a recent work an innovative calix[4]arene-Ti(IV) catalyst (Fig. 6) was employed in the bulk ring opening polymerization of L-lactide under microwave irradiation. A CEM *Discover* S-class monomodal reactor was operated at a maximum power level of 200 W for variable reaction times: a conversion of 85 % was registered after 40 min of irradiation reaching 95 % after 80 min; longer times led to decomposition phenomena. Compared to conventional synthesis, microwave heating markedly reduced reaction times but slighltly lowered the control over molecular weights and polydispersity indexes (Frediani et al., 2010).

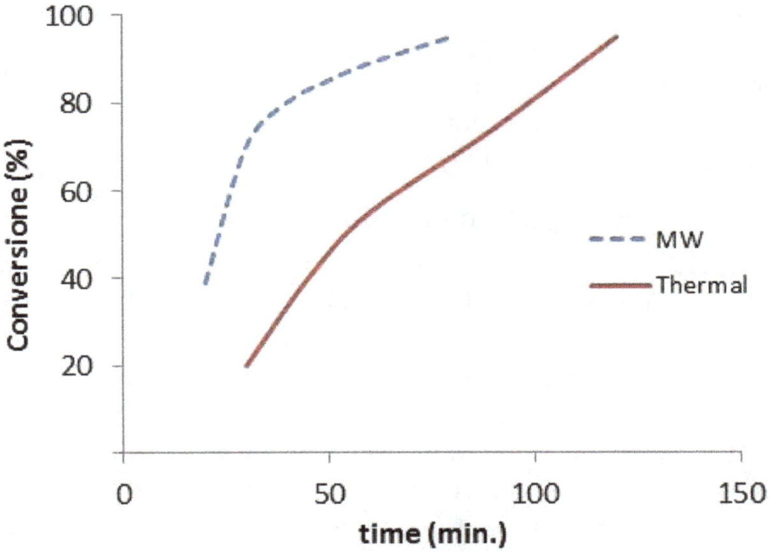

Figure 6. Conversion vs reaction time for the ROP of L-lactide (left) and the employed calix[4]arene-Ti catalyst (see Frediani et al., 2010).

2.2.4. Poly(lactic acid) copolymers

Fully biocompatible and biodegradable poly(glycolic acid-*co*-lactic acid) copolymers with potential applications in the biomedical field were synthesized by ring opening polymerization: stannous octoate was used as catalyst (0.03 mol%) and octadecanol as the oxydrylic initiator (0.01 mol%). Microwave irradiation was performed in a *MAS-II* scientific reactor at a temperature of 120 °C for a variable time (3 min to 9 min), under nitrogen atmosphere: different glycolic acid (GA) : lactic acid (LA) feed ratios were assessed and yields ranged from 69% to 86%. It is worth noting how the products showed higher GA content compared to the feed ratio, and that copolymers with higher GA content were characterized by decreased glass transition temperature and increased melting temperature (G. Li et al., 2010). Amphiphilic, biodegradable PLLA-*b*-PEG-*b*-PLLA triblock copolymers were synthesized by microwave assisted ring opening polymerization of L-lactide employing poly(ethylene glycol) (PEG2000) as a difunctional oxydrylic initiator (Scheme 3).

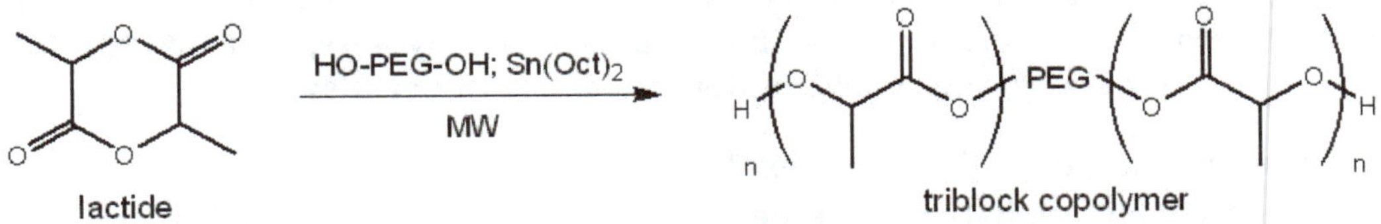

Scheme 3. Synthesis of PLA-b-PEG-b-PLA triblock copolymer.

A surprisingly short irradiation time of 3 min at 100 °C in a CEM *Discover* reactor was enough to achieve 92% lactide conversion, yielding a copolymer with a number average molecular weight of 28,230 g/mol. It was observed that longer PLLA segments (i.e. higher PLLA content) induced higher glass transition and melting temperatures (C. Zhang et al., 2007).

2.2.5. Other ring opening polymerizations

Stannous octoate catalyzed ring opening polymerization of *p*-dioxanone was conducted under microwave irradiation to obtain (63% yield) biodegradable PDDO with a viscosity average molecular weight of 156,000 g/mol. A domestic oven was employed for this study and several power level tested: 270 W for 25 min proved to be the most favorable choice, allowing a dramatic increase in the reaction rate compared to the 14 h of heating required by conventional procedures in analogous conditions (Y. Li et al., 2006). Poly(trimethylen carbonate) (PTMC) is a promising, biocompatible poly(alkyl carbonate) with good mechanical properties: a PTMC-*b*-PEG-*b*-PTMC triblock copolymers with tunable degradation properties were synthesized by catalyst-free microwave assisted ring opening polymerization of TMC using PEG600, PEG1000 and PEG2000 as a difunctional oxydrylic initiators. It was observed that the molar mass of the copolymer and the TMC conversion reached a maximum after 30 min at 120 °C (Fig. 7): further irradiation or higher temperatures induced product degradation phenomena (Liao et al., 2008).

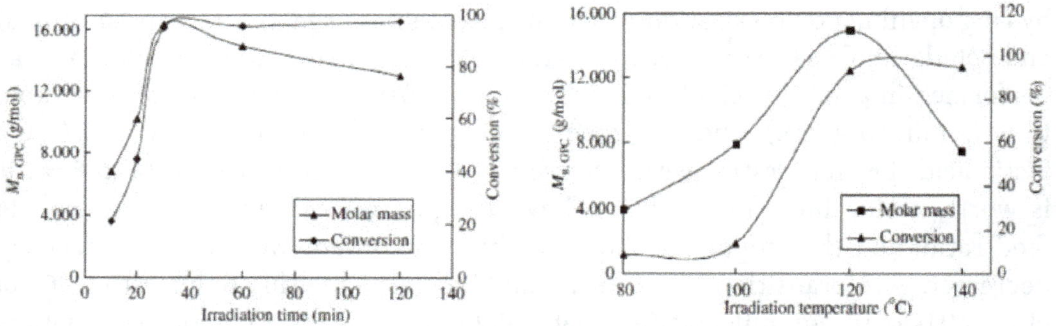

Figure 7. Synthesis of a PTMC-b-PEG-b-PTMC copolymer: effect of irradiation time (left) and temperature (right) on Mn and conversion (from Liqiong Liao, Chao Zhang, Shaoqin Gong, Preparation of Poly(trimethylene carbonate)-block-poly(ethylene glycol)-block-poly(trimethylene carbonate) Triblock Copolymers under Microwave Irradiation, Reactive & Functional Polymers, 2008, 68, 751-758. Copyright Elsevier Ltd. Reproduced with permission).

The ring opening of cyclohexene-oxide in the presence of CO_2 for the synthesis of biodegradable aliphatic poly(ether-carbonate)s was investigated in order to employ a troublesome waste product as feedstock for the production of useful materials. Double metal cyanide complex catalysts based on $Zn_3[Co(CN)_6]$ were profitably utilized and microwave irradiation was chosen as heating method, with strong beneficial effects on polymerization times and CO_2 incorporation grade (Dharman et al., 2008). Poly(oxazoline)s are relatively non toxic polymers with attractive properties for biomedical applications, nonetheless they cannot be classified as biodegradable or bio-based polymers: comprehensive reviews about microwave assisted synthesis in this class of products can be found elsewhere (Adams & Schubert, 2007; Makino & Kobayashi, 2010; Wiesbrock et al., 2004).

3. MICROWAVE ASSISTED MODIFICATION AND PROCESSING

This paragraph reviews several publications demonstrating the application of microwave irradiation for the chemical modification and processing of both naturally abundant and manufactured polymers. Certain research address non-biodegradable oil-based polymers; yet, these works are included as our primary focus is on the processing technique rather than the polymer itself.

3.1. Grafting and chemical modification of natural polymers

Chemical modification of natural polymers, most commonly polysaccharides, is a clever way to obtain new materials from abundant renewable feedstocks. A very effective approach to this task is the design of graft copolymers in which the former natural polymer constitutes the backbone and is covalently bound to numerous branches of synthetic polymer. For example, in order to obtain a biodegradable amphoteric material, chitosan was first reacted with phtalic anhydride to protect the amino groups along the chain and then used as an oxydrylic initiator for the stannous octoate catalyzed ROP of ε-caprolactone: after deprotection a chitosan-*g*-PCL copolymer with grafting percentage above 100% by weight was obtained and characterized (Scheme 4). The polymerization step was carried out smoothly under microwave irradiation using a domestic oven and best results in terms of increased reaction rate were obtained with high power levels of 450 W and 600 W (L. Liu et al., 2005). The β-cyclodextrin functionalization of chitosan was also performed under microwave irradiation yielding graft copolymers with a superior lead adsorption capacity if compared to analogous products obtained by conventional heating. The authors report that a superior yield was achieved despite the fact that no redox initiator was required (Sharma & Mishra, 2010). It is worth noting that chitosan itself can be obtained from chitin via a microwave mediated deacetylation reaction (Sahu et al., 2009). With a similar approach, PCL side chains were grafted onto pea starch (35% amylose, 65% amylopectin) in a microwave assisted (255 W for 3 min) one pot synthesis yielding a thermoformable hydrophobic copolymer with promising properties as eco-friendly or biomedical material. Regrettably, not much about the utilized microwave heating setup is reported by the authors (Chang et al., 2009). An efficient superabsorbent material capable of a high swelling ratio in distilled water was synthesized grafting sodium acrylate onto cornstarch under microwave irradiation and using potassium persulfate as initiator. Microwave power is reported to be the most influent factor to obtain best performing copolymers (i.e. with higher swelling ratio and lower solubility): 10 min of irradiation at 85-90 W proved to be the most suitable heating conditions (Tong & al., 2004). The microwave assisted preparation of a poly(acrylamide)-*g*-starch copolymer with promising properties as a flocculant for water treatment was recently described using ceric ammonium nitrate as initiator. The authors report high grafting rates and superior yields compared to the conventional procedure. Despite the use of a domestic 800 W microwave oven, and the manual switching off of irradiation whenever the reaction mixture was about to boil, reproducibility is claimed to be one of the advantages over ordinary methods (S. Mishra et al., 2011). Cellulose, being the most abundant natural polysaccharide, is also attracting a huge attention and indeed several examples of its chemical modification can be found in literature. Microwave heating was employed for the synthesis of common cellulose derivatives like esters (Satgé et al., 2002) and carbammates (Guo et al., 2009) as well as for the graft copolymerization of methyl methacrylate initiated with ceric amonium nitrate (Wan et al., 2011). Guar Gum is an edible galactomannan isolated from a legume native of the northwestern India and finds wide application as water thickener due to the ability of forming highly viscous colloidal dispersions. In order to tune its physicochemical and degradation properties a number of microwave assisted derivatizations were successfully carried out by the same group within short irradiation times. We report, among these, a poly(ethyl acrylate) graft copolymer for efficient removal of cadmium(II) from aqueous phases (Singh et al., 2009); a poly(acrylonitrile) graft copolymer with high water absorbing capacity (Singh et al., 2004) and the first example of poly(acrylamide) graft copolymer (Singh et al., 2004). It is worth noting how the authors achieve these results in the absence of redox initiators or catalysts: this is due, in their view, to non thermal effects and to radical formation as a consequence of molecular friction under microwave heating. Salient papers regarding microwave assisted synthesis deal also with grafting of methyl methacrylate onto carrageenan in which backbone decomposition is prevented by short reaction times (Prasad et al., 2006); poly(acrylonitrile)-*g*-xyloglucan copolymers by an initiator-free preparation method (A. Mishra et al., 2008) and poly(acrylic acid) grafting of *Cassia Javanica* seed gum to obtain a sorbent material for Hg(II) removal (Singh et al., 2010).

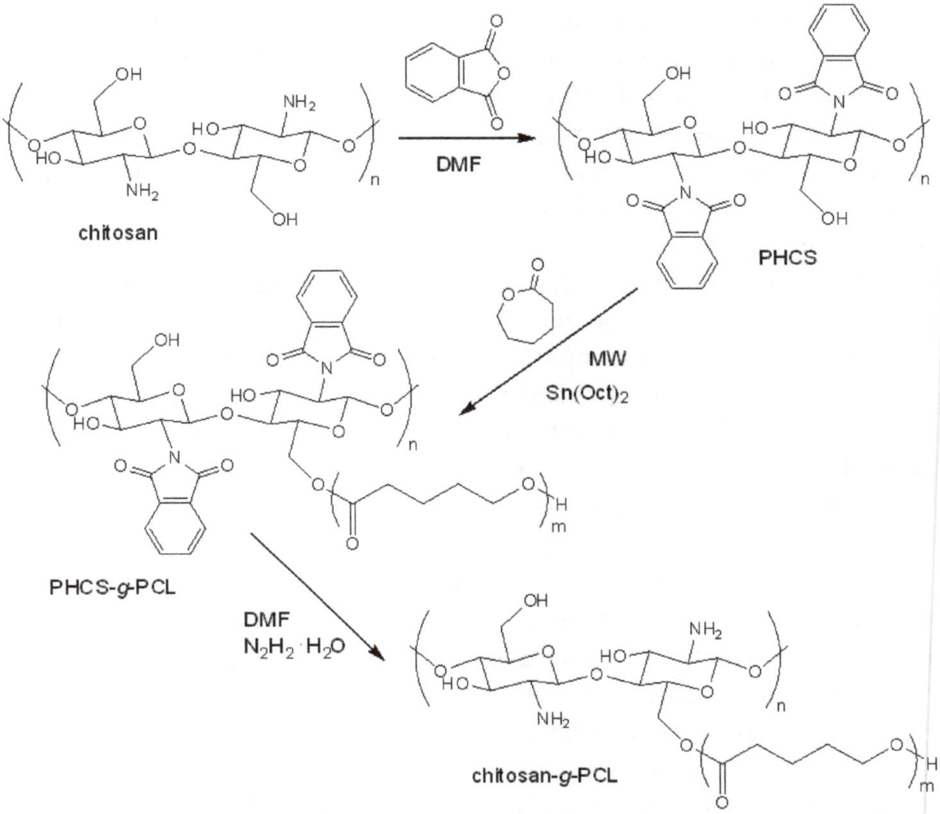

Scheme 4. Synthesis of a chitosan-g-PCL copolymer.

3.2. Polymer processing

With the development of industrial scale dedicated devices, microwave polymer processing techniques are nowadays sensible alternatives to conventional heating methods, setting new benchmarks in terms of time and energy efficiency. Comprehensive reviews about this chief research field can be found in literature (Clark & Sutton, 1996; Das et al., 2008; Leonelli & Mason, 2010; Thostenson & Chou, 1999); our aim here is to introduce some representative examples dealing with a variety of applications ranging from polymer crosslinking to welding of thermoplastics. Short irradiation times and a relatively low temperature (100 °C) turned out to be sufficient to carry out the one-pot crosslinking reaction of poly(vinyl alcohol) in the presence of metal salts, yielding polymer-inorganic nanocomposites with several different morphologies. On account of the good control over nanostructure's shape and size distribution and considering that similar products were previously prepared via hydrothermal treatment at 200 °C in 72 h, the efficiency of microwave heating applied to these syntheses results clear (Polshettiwar et al., 2009; Nadagouda & Varma, 2007). Polymer surface modification has been performed by a microwave induced plasma treatment (Leonelli & Mason, 2010): to cite an example, poly(dimethylsiloxane) (PDMS) slabs were irreversibly sealed within 30-60s by means of a 30 W plasma generated in glass bottles containing 2-3 Torr of oxygen (Hui et al., 2005). Similar procedures are reported in papers dealing with modification of PDMS surface hydrophilicity (Ginn & Steinbock, 2003) and polyaniline-aided welding of high density poly(ethylene) (Wu & Benatar, 1997). Welding of thermoplastics such as ultra high molecular weight polyethylene (UHMW PE), polycarbonate and ABS could be achieved by focused microwave irradiation of the specimen joint interface (Yarlagadda & Chai, 1998). An interesting comparison between thermal curing and microwave curing of the E44 epoxy resin showed a significant reduction of processing time and temperature; a lower quantity of curing agent (i.e. maleic anhydride) was needed under microwave irradiation and products with improved compressive strength and bending strength are obtained by the authors (Zhou et al., 2003). In a recent work epoxies of linseed oil and dehydrated castor oil were blended with PVA within 2 min of microwave irradiation (75 °C; 350 W) with the aim to develop new bio-based materials with desirable properties for packaging and mulching: as usual, the chief advantage over conventional blending procedures is reported to be the extremely short processing time (Riaz et al., 2011). According to several authors, microwave foaming under vacuum (Jaja & Durance, 2008) or at atmospheric pressure (Torres et al., 2007) proved to be an effective route to obtain porous materials for tissue engineering applications from polymeric gels. It is worth reminding that microwave transparent polymers, such as HDPE, may easily be processed employing microwaves as energy source in the presence of highly absorbing fillers and additives like carbon black (F. Liu et al., 2010). Theoretical models have been established to examine the ideal conditions for efficient heating; a recent study analyzed the application of microwaves on natural rubber and Nylon 66 slabs, emphasizing the significance of irradiation pulsing parameters and ceramic support plates (Durairaj & Basak, 2009).

4. CONCISE SUMMARY

Over the past decade, the proliferation of dependable research instruments has transformed microwave-assisted polymer chemistry from an experimental curiosity into a viable alternative heating strategy, while enhanced knowledge and engineering have optimized traditional polymer processing methods. Numerous studies in the relevant literature demonstrate enhanced reaction rates under microwave settings, with time efficiency acknowledged as the most apparent benefit compared to traditional heating. Despite the contentious nature of the theories for this advantageous impact, many contemporary authors prefer to exclude non-thermal phenomena and concentrate instead on accelerated energy transfer, elevated local temperatures, or the warming of solvents. Increased product yields are frequently, though not universally, cited as a significant characteristic of microwave reactors, along with modified selectivities, which may be attributed to differential radiation absorption in certain instances. Certain investigations demonstrate that product purity can be enhanced under microwave settings, reducing adverse side reactions through more uniform heating and decreased reaction durations. Notwithstanding these notable aspects, evaluating the energy efficiency of microwaves in terms of their environmental impact appears to be complex; numerous studies indicate efficiencies that are comparable to traditional oil baths and mantles (Devine & Leadbeater, 2011; Gronnow et al., 2005; Moseley & Kappe, 2011; Razzaq & Kappe, 2008). The selection of a particular experimental configuration significantly affects the overall equilibrium, with critical factors including the reaction scale, the employment of monomodal or multimodal devices, and the use of open or pressurized vials; however, the comparatively low efficiency of magnetrons (40% to 65%) typically mitigates the advantages of the considerably reduced reaction times. The energy efficiency of microwave-assisted chemistry should be meticulously assessed on a case-by-case basis rather than assumed.

5. REFERENCES

Adams N. Schubert U. S. 2007 Poly(2-oxazolines) in Biological and Biomedical Applications. Advanced Drug Delivery Review, 59 15 December 2007), 1504 1520 , 0016-9409X

Barbier-Baudry D. Brachais L. Cretu A. Gattin R. Loupy A. Stuerga D. 2003 Synthesis of Polycaprolactone by Microwave Irradiation- an Interesting Route to Synthesize this Polymer via Green Chemistry. Environmental Chemical Letters, 1 1 (January 2003), 19 23 , 1610-3661

Bezdushna E. Ritter H. Troev K. 2005 Poly(alkylene hydrogen phosphonate)s by Transesterification of Dimethyl Hydrogen Phosphonate with Poly(ethylene glycol). Macromolecular Rapid Communications, 26 6 March 2005), 471 476 , 1521-3927

Bogdal D. Penczek P. Pielichowsky J. Prociack A. 2003 Microwave Assisted Synthesis, crosslinking and Processing of Polymeric Materials. Advances in Polymer Science, 163 2003), 193 263 , 0065-3195

Borriello A. Nicolais L. Fang X. Huang S. J. Scola D. A. 2007 Synthesis of Poly(amide-ester)s by Microwave Methods. Journal of Applied Polymer Science, 103 3 February 2007), 1952 1958 , 0021-8995

Chang P. R. Zhou Z. Xu P. Chen Y. Zhou S. Huang J. 2009 Thermoforming Starch-graft-polycaprolactone biocomposites via One-pot Microwave Assisted Ring

Opening Polymerization. Journal of Applied Polymer Science, 113 5 September 2009), 2973 2979 , 0021-8995

Chatti S. Bortolussi M. Loupy A. Blais J. C. Bogdal D. Majdoub M. 2002 Efficient Synthesis of Polyethers from Isosorbide by Microwave-Assisted Phase Transfer Catalysis. European Polymer Journal, 38 9 September 2002), 1851 1861 , 0014-3057

Chatti S. Bortolussi M. Bogdal D. Blais J. C. Loupy A. 2006 Synthesis and Properties of New Poly(ether-ester)s Containing Aliphatic Diols Based on Isosorbide. Effects of the Microwave-Assisted Polycondensation. European Polymer Journal, 42 2 February 2006), 410 424 , 0014-3057

Chen R. Hao J. Y. Xiong C. D. Deng X. M. 2010 Rapid Synthesis of Biodegradable Poly(epsilon-caprolactone-co-p-dioxanone) Random Copolymers under Microwave Irradiation. Chinese Chemical Letters, 21 2 February 2010), 249 252 , 1001-8417

Clark D. E. Sutton W. H. 1996 Microwave Processing of Materials. Annual Review of Material Science, 26 299 331 , 0084-6600

Das S. Mukhopadhyay A. K. Datta S. Basu P. 2008 Prospects of Microwave Processing: an Overview. Bulletin of Material Science, 31 7 December 2008), 943 956 , 0973-7669

Devine W. G. Leadbeater N. E. 2011 Probing the Efficiency of Microwave Heating and Continuous-flow Conventional Heating as Tools for Organic Chemistry. Arkivoc, 2011 Part (v), 127 143 , 1424-6376

Dharman M. M. Ahn J. Y. Lee M. K. Shim H. L. Kim K. H. Kim I. Park D. W. 2008 Moderate Route for the Utilization of CO_{2-} Microwave Induced Coploymerization with Cyclohexene Oxide Using Higly Efficient Double Metal Cyanide Complex Catalysts Based on $Zn_3[Co(CO)_6]$. Green Chemistry, 10 6 2008), 678 684 , 1463-9270

Deetlef M. Seddon K. R. 2003 Improved Preparation of Ionic Liquids Using Microwave Irradiation. Green Chemistry, 5 2 2003) 181 186 , 1463-9270

Durairaj S. Basak T. 2009 Analysis of Pulsed Microwave Processing of Polymer Slabs Supported With Ceramic Plates. Chemical Engineering Science, 64 7 April 2009), 1488 1502 , 0009-2509

Frediani M. Sémeril D. Matt D. Rizzolo F. Papini A. M. Frediani P. Rosi L. Santella M. Giachi G. 2010 L-Lactide Polymerization by Calix[4]arene-titanium (IV) Complex Using Conventional Heating and Microwave Irradiation. e-Polymers, 019 February 2010) 1 8 , 1618-7229

Ginn B. T. Steinbock O. 2003 Polymer Surface Modification Using Microwave-Oven-Generated Plasma. Langmuir, 19 19 September 2003), 8117 8118 , 0743-7463

Glasnow T. N. Kappe C. O. 2007 Microwave-Assisted Synthesis Under Continuous Flow Conditions. Macromolecular Rapid Communications, 28 4 February 2007), 395 410 , 1521-3927

Gronnow M. J. White R. J. Clark J. H. Macquarrie D. J. 2005 Energy Efficiency in Chemical Reactions: a Comparative Study of Different Reaction Techniques. Organic Process Research & Engineering, 9 4 July 2005), 516 518 , 1083-6160

Guerrero-Sanchez C. Lobert M. Hoogenboom R. Schubert U. S. 2007 Microwave Assisted Homogeneous Polymerizations in Water-Soluble Ionic Liquids: an Alternative and Green Approach for Polymer Synthesis. Macromolecular Rapid Communications, 28 4 February 2007), 456 464 , 1521-3927

Guo Y. Zhou J. Song Y. Zhang L. 2009 An Efficient and Environmentally Friendly Method for the Synthesis of Cellulose Carbammate by Microwave Heating. Macromolecular Rapid Communications, 30 17 September 2009), 1504 1508 , 1521-3927

Hui A. Y. Wang G. Lin B. Chan W. T. 2005 Microwave Plasma Treatment of Polymer Surface for Irreversible Sealing of Microfluidic Devices. Lab on a Chip, 5 10 October 2005), 1173 1177 , 1473-0189

Jaja S. Durance T. D. 2008 Tailor-made Biopolymers Porous Scaffold Fabrication for Tissue Engineering: Application of Radiant Energy in the Form of Microwave under Vacuum. Bio-medical Materials and Engineering, 18 6 357 366 , 0959-2989

Jing S. Peng W. Tong Z. Baoxiu Z. 2006 Microwave-Irradiated Ring-Opening Polymerization of D,L-Lactide Under Atmosphere. Journal of Applied Polymer Science, 100 3 may 2006), 2244 2247 , 0021-8995

Keki S. Bodnar I. Borda J. Deak G. Zsuga M. 2001 Fast Microwave-Mediated Bulk Polycondensation of D,L-Lactic Acid. Macromolecular Rapid Communications, 22 13 September 2001), 1063 1065 , 1521-3927

Koroskenyi B. Mc Carthy S. 2002 Microwave-Assisted Solvent-Free or Aqueous-Based Synthesis of Biodegradable Polymers. Journal of Polymers and the Environment, 10 3 July 2002), 93 104 , 1566-2543

Kremsner J. M. Kappe C. O. 2006 Silicon Carbide Passive Heating Elements in Microwave-Assisted Organic Synthesis. Journal of Organic Chemistry, 71 12 June 2006), 4651 4658 , 0022-3263

Lei H. C. Wang P. Yuan W. B. 2009 Microwave-Assisted Synthesis of Poly(L-Lactic Acid) via Direct Melt Polycondensation using Solid Super Acids. Macromolecular Chemistry and Physics, 210 23 December 2009), 2058 2062 , 1521-3935

Leonelli C. Veronesi P. Grisoni F. 2007 Numerical Simulation of an Industrial Microwave Assisted Filter Dryer: Criticality Assessment and Optimization. Journal of Microwave Power and Electromagnetic Energy, 41 3 October 2007), 5 13 , 0832-7823

Leonelli C. Mason T. J. 2010 Microwave and Ultrasonic Processing: Now a Realistic Option for Industry. Chemical Engineering and Processing, 49 9 September 2010), 885 900 , 0255-2701

Li G. Zhao N. Bai W. Chen D. L. Xiong C. D. 2010 Microwave-Assisted Ring-Opening Polymerization of Poly(glycolic acid-co-lactic acid) Copolymers. E-Polymers, 051 May 2010), 1 6 , 1618-7229

Li Y. Wang X. L. Yang K. K. Wang Y. Z. 2006 A Rapid Synthesis of Poly(p-dioxanone) by Ring Opening Polymerization under Microwave Irradiation. Polymer Bulletin, 57 6 October 2006), 873 880 , 0170-0839

Liao L. Zhang C. Gong S. 2008 Preparation of Poly(trimethylene carbonate)-block-poly(ethylene glycol)-block-poly(trimethylen carbonate) Triblock Copolymers under Microwave Irradiation. Reactive and Functional Polymers, 68 3 March 2008), 751 758 , 1381-5148

Lidström P. Tierney J. Wathey B. Westman J. 2001 Microwave Organic Synthesis- a Review. Tetrahedron, 57 45 November 2001), 9225 9283 , 0040-4020

Liu F. Qian X. Wu X. Guo C. Ley Y. Zhang J. 2010 The Response of Carbon Black Filled High-density Polyethylene to Microwave Processing. Journal of Materials Processing Technology, 210 14 November 2010), 1991 1996 , 0924-0136

Liu L. Liao L. Q. Zhang C. He F. Zhuo R. X. 2001 Microwave-Assisted Polycondensation of L-2-hydroxy-3-phenylpropanoic Acid. Chinese Chemical Letters, 12 9 September 2001), 761 762 , 1001-8417

Liu P. Su Z. 2005 Thermal Stabilities of Polystyrene/Silica Hybrid Nanocomposites via Microwave-assisted In-situ Polymerization. Materials Chemistry and Physics, 94 3 December 2005), 412 415 , 0254-0584

Makino A. Kobayashi S. 2010 Chemistry of 2-Oxazolines: a Crossing of Cationic Ring-Opening Polymerization and Enzymatic Ring-Opening Polyaddition. Journal of Polymer Science- Part A: Polymer Chemistry, 48 6 March 2010), 1251 1270 , 1099-0518

Mishra S. Mukul A. Sen G. Jha U. 2011 Microwave Assisted Synthesis of Polyacrylamide Grafted Starch (St-g-PAM) and its Applicability as Flocculant for Water Treatment. International Journal of Biological Macromolecules, 48 1 January 2011), 106 111 , 0141-8130

Mishra A. Clark J. H. Vij A. Daswal S. 2008 Synthesis of Graft Copolymers of Xyloglucan and Acrylonitrile. Polymers for Advanced Technologies, 19 2 February 2008), 99 104 , 1042-7147

Moseley J. D. Woodman E. K. 2008 Scaling-Out Pharmaceutical Reactions in an Automated Stop-Flow Microwave Reactor. Organic Processes Research & Development, 12 5 September 2008), 967 981 , 1083-6160

Moseley J. D. Kappe C. O. 2011 A Critical Assessment of the Greenness and Energy Efficiency of Microwave-assisted Organic Synthesis. Green Chemistry, Advance Article, 1463-9270 1463 9270

Nikolic L. Ristic I. Adnadjevich B. Nikolic V. Jovanovich J. Stankovich M. 2010 Novel Microwave-Assisted Synthesis of Poly(D,L-Lactide): the Influence of Monomer/Initiatior Ratio on the Product Properties. Sensors, 10 5 May 2010), 5063 5073 , 1424-8220

Nadagouda M. N. Varma R. S. 2007 Preparation of Novel Metallic and Bimetallic Cross-Linked Poly(vinyl alcohol) Nanocomposites under Microwave Irradiation. Macromolecular Rapid Communications, 28 4 February 2007), 465 472 , 1521-3927

Nagahata R. Sano D. Suzuki H. Takeuchi K. 2007 Microwave-Assisted Single-Step Synthesis of Poly(lactic acid) by Direct Polycondensation of Lactic Acid. Macromolecular Rapid Communications, 28 4 February 2007), 437 442 , 1521-3927

Nuchter M. Ondruschka **B.** Bonrath W. Gum A. 2004 Microwave Assisted Synthesis-a Critical Technology Overview. Green Chemistry, 6 3 2004), 128 141 , 1463-9270

Perreux L. Loupy A. 2001 A Tentative Rationalization of Microwave Effects in Organic Synthesis According to the Reaction Medium, and Mechanistic Considerations. Tetrahedron, 57 45 November 2001), 9199 9223 , 0040-4020

Polshettiwar V. Nadagouda M. N. Varma R. S. 2009 Microwave-Assisted Chemistry: a Rapid and Sustainable Route to Synthesis of Organics and Nanomaterials. Australian Journal of Chemistry, 61 1 January 2009), 16 26 , 0004-9425

Razzaq T. Kappe C. O. 2008 On the Energy Efficiency of Microwave-Assisted Organic Reactions. ChemSusChem, 1 1-2 (February 2008), 123 132 , 0186-4564X

Riaz U. Ashraft S. M. Sharma H. O. 2011 Mechanical, Morphological and Biodegradation Studies of Microwave Processed Nanostructured Blends of

Some Bio-based Oil Epoxies with Poly(vinyl alcohol). Polymer Degradation and Stability, 96 1 January 2011), 33 42 , 0141-3910

Satgé C. Verneuil B. Branland P. Granet R. Krausz P. Rozier P. Petit C. 2002 Rapid Homogeneous Esterification of Cellulose Induced by Microwave Irradiation. Carbohydrate Polymers, 49 3 August 2002), 373 376 , 0144-8617

Sahu A. Goswami P. Bora U. 2009 Microwave Mediated Rapid Synthesis of Chitosan. Journal of Material Science: Materials in Medicine, 20 1 January 2009), 171 175 , 1573-4838

Sharma A. K. Mishra A. K. 2010 Microwave Induced β-cyclodextrin Modification of Chitosan for Lead Sorption. International Journal of Biological Macromolecules, 47 3 October 2010), 410 419 , 0141-8130

Singh V. Singh S. K. Maurya S. 2010 Microwave Induced Poly(acrylic acid) Modification of Cassia Javanica Seed Gum for Efficient Hg(II) Removal from Solution. Chemical Engineering Journal, 160 1 May 2010), 129 137 , 1385-8947

Singh V. Sharma A. K. Maurya S. 2009 Efficient Cadmium (II) Removal from Aqueous Solution Using Microwave Synthesized Guar Gum-Graft-Poly(ethylacrylate). Industrial & Engineering Chemistry Research, 48 10 May 2009), 4688 4696 , 0888-5885

Singh V. Tiwari A. Tripathi D. N. Sanghi R. 2004 Microwave Assisted Synthesis of Guar-g-polyacrylamide. Carbohydrate Polymers, 58 1 October 2004), 1 6 , 0144-8617

Singh V. Tiwari A. Tripathi D. N. Sanghi R. 2004 Grafting of Polyacrylonitrile onto Guar Gum under Microwave Irradiation. Journal of Applied Polymer Science, 92 3 May 2004), 1569 1575 , 0021-8995

Strauss C. R. Trainor R. W. 1995 Developments in Microwave-assisted Organic Chemistry. Australian Journal of Chemistry, 48 10 October 1995), 1665 1692 , 0004-9425

Thostenson E. T. Chou T. W. 1999 Microwave Processing: Fundamentals and Applications. Composites- Part A: Applied Science and Manifacturing, 30 9 September 1999), 1055 1071 , 0135-9835X

Tong Z. Peng W. Zhiqian Z. Baoxiu Z. 2004 Microwave Irradiation Copolymerization of Superabsorbents from Cornstarch and Sodium Acrylate. Journal of Applied Polymer Science, 95 2 January 2004), 264 269 , 0021-8995

Torres F. G. Boccaccini A. R. Troncoso O. P. 2007 Microwave Processing of Starch-based Porous Structures for Tissue Engineering Scaffolds. Journal of Applied Polymer Science, 103 2 January 2007), 1332 1339 , 0021-8995

Velmathi S. Nagahata R. Sugiyama J. Takeuchi K. 2005 A Rapid, Eco-Friendly Synthesis of Poly(butylene succinate) by a Direct Polyesterification under Microwave Irradiation. Macromolecular Rapid Communications, 26 14 July 2005), 1163 1167 , 1521-3927

Vogel B. M. Mallapragada S. K. Narasimhan B. 2003 Rapid Synthesis of Polyanhydrides by Microwave Polymerization. Macromolecular Rapid Communications, 25 1 January 2004), 330 333 , 1521-3927

Wan Z. Xiong Z. Ren H. Huang Y. Liu H. Xiong H. Wu Y. Han J. 2011 Graft Copolymerization of Methyl Methacrylate onto Bamboo Cellulose under Microwave Irradiation. Carbohydrate Polymers, 83 1 January 2011), 264 269 , 0144-8617

Waugh M. R. Parkin I. P. 2010 Enhanced Solubility and Covalent Solubilization of Single Walled Carbon Nanotubes via Atmospheric Pressure Microwave Reflux and the Subsequent Spray Coating of Transparent Conducting Thin Films. Current Nanoscience, 6 3 June 2010), 232 242 , 1573-4137

Wiesbrock F. Hoogenboom R. Abeln C. H. Schubert U. S. 2004 Single-Mode Microwave Ovens as New Reaction Devices: Accelerating the Living Polymerization of 2-Ethyl-2-oxazoline. Macromolecular Rapid Communications, 25 22 November 2004), 1895 1899 , 1521-3927

Wu C. Y. Benatar A. 1997 Microwave Welding of High Density Polyethylene Using Intrinsically Conductive Polyaniline. Polymer Engineering and Science, 37 4 April 2007), 738 743 , 1548-2634

Yarlagadda P. K. D. V. Chai T. C. 1998 An Investigation into Welding of Engineering Thermoplastics Using Focused Microwave Energy. Journal of Materials Processing Technology, 74 1-3 (February 1998), 199 212 , 0924-0136

Zhang C. Liao L. Liu L. 2004 Rapid Ring-Opening Polymerization of D,L-Lactide by microwaves. Macromolecular Rapid Communications, 25 15 August 2004), 1402 1405 , 1521-3927

Zhang C. Liao L. Gong S. 2007 Microwave-Assisted Synthesis of PLLA-PEG-PLLA Triblock Copolymers. Macromolecular Rapid Communications, 28 4 February 2007), 422 427 , 1521-3927

Zhang Y. Wang P. Han N. Lei H. 2009 Synthesis of Poly(D,L-Lactide) under Continuous Microwave Irradiation. Shiyou Huagong/Petrochemical Technology, 38 8 August 2009), 861 865 , 1000-8144

Zhou J. Shi C. Mei B. Yuan R. Fu Z. 2003 Research on the Technology and Mechanical Properties of the Microwave Processing of Polymers. Journal of Materials Processing Technology, 137 1-3 (June 2003), 156 158 , 0924-0136

Zhu X. Zhou N. He X. Cheng Z. Lu J. 2003 Atom Transfert Radical Bulk Polymerization of Methyl Methacrylate under Microwave Irradiation. Journal of Applied Polymer Science, 88 7 May 2003), 1787 1793 , 0021-8995

10 Chapter

POLYMERIC MATERIALS: MICROWAVE PYROLYSIS

1. INTRODUCTION: AN OVERVIEW

Microwave pyrolysis The pyrolysis of polymeric materials or plastic-containing wastes including scrap tires is a possible answer to the problem of their disposal because it let recover of gas, oil and solid able to be employed as a source of products and energy. Therefore the relevance of the pyrolysis processes of plastic waste has been growing.

A plethora of studies over the thermal degradation of polymeric materials are carried out using conventional heating method with internal or external heating source, under inert or oxidizing atmosphere. Generally the thermal decomposition needs operating temperature above 450°C (Kaminsky et al., 2004; Mastral et al., 2002; Whesterhout et al., 1998).

In this area also the interest to microwave heating technologies has become remarkable since they could represent a charming alternative to current technologies based on other conventional heating processes. Microwave heating provides a number of advantages with respect conventional heating methods, mainly referred to the rate and distribution of the heating. It is commonly thought that many limitations should occur nevertheless when switching microwave technology from a laboratory or pilot-scale process to a large scale plant. These limitations mainly concern the lack of data on raw materials amenability with microwave heating and therefore a proper set up of the experimental parameters of the process.

The main differences occurring between conventional and microwave heating are shown in Table 1.

Table 1. Comparison between conventional and microwave heating technologies for pyrolysis of plastics.

Conventional Heating	Microwave heating
Long reaction time (hours)	Short reaction time (minutes)
Hard transfer of heat, thermal conductivity of polymer is low	The low thermal conductivity of polymers may be overwhelmed, easy heating of polymers.

Conventional Heating	Microwave heating
Heating efficiency is hardly obtained	High heating efficiency
Every fuel source may be employed	Electrical power is required
Additive are not required	Microwave absorber are required

This chapter presents the current level of knowledge derived from scientific publications and patents, along with our experiences and findings, regarding the pyrolysis process utilizing microwave heating for polymers and complex plastics. Attention will be directed towards the impact of primary process variables on the resulting products: apparatus configuration, temperature, heating rate, microwave absorbent, among others. Significant focus will be directed towards the pyrolysis of used tires. Tires, possessing a high carbon content of up to 30%, are appropriate for direct microwave pyrolysis. They can absorb microwaves and rapidly convert them into heat (Meredith, 1998).

The chapter comprises two principal components. The initial section is only focused on the microwave pyrolysis of tires. We present a literature and patent review, analyze the findings based on our experience, and integrate them with our recent advancements in this domain. The liquid component of the goods is the most appealing due to its substantial concentration of valuable hydrocarbons, including benzene, toluene, xylenes, and limonene. We meticulously examine the relationship among layout technologies, operational characteristics, and product composition. We thoroughly examine the issue of optimizing the liquid fraction and precious hydrocarbons.The second section is focused on a review of microwave heating pyrolysis of plastic materials. These polymers don't absorb microwave and in order to carry out the pyrolysis is necessary to mix them with a microwave absorbent. Typical absorbent are: coal or carbon-containing materials (for instance tires), metals (Hussain et al., 2010) or metal-oxide (Horikoshi et al., 2009). We report about microwave pyrolysis of the main waste polymeric materials avoiding to send them to landfill (Kaminsky, 1992): polyethylene (PE), polypropylene (PP), polystyrene (PS), poly(ethylene terephthalate) (PET) and polyvinylchloride (PVC). We present the processes proposed and the operative parameters employed in order to obtain renewed raw materials, refinery feedstock or energy source.

2. TIRES PYROLYSIS

2.1. Why tires pyrolysis?

The recycling policy typically aims to optimize material recovery while minimizing the loss of materials and energy contained in garbage (Natural Resources Defense Council [NRDC], 2008). End-of-life tires can be repurposed for reconstruction methods that enhance material and energy recovery. This technique is constrained by the quality of waste tires, their degradation, and can be executed no more than one or two times. In instances where reuse and remanufacturing are unfeasible, scrap tires, whether intact or shredded, can be utilized in engineering projects for various applications, including: road construction (asphalt, where granules enhance mechanical strength, diminish noise, and prevent aquaplaning), urban furniture (curb beds, bollards, bicycle paths, parking areas, recreational zones), and sports facilities (synthetic grass football fields and athletic track surfaces). Sure enough not the whole production of scrap tires can be reprocessed or reused indeed 19% of them is disposed in Italy with heat treatment processes mainly for the recovery of energy content (FISE UNIRE, 2009). The main technological processes available for heat treatments are listed below:

- incinerator (including municipal solid waste incinerators) (Sharma et al., 2000);
- fuel for rotary kilns or furnaces to produce cement or steam (Giugliano et al.; 1999);
- pyrolysis.

Just the last process mentioned, the pyrolysis, has the characteristics to transform waste polymeric materials in products suitable for energy production or petrochemical feedstock.

2.2. Microwave versus non-microwave heating in pyrolysis processes: two distinct methodologies

Pyrolysis is a thermal cracking process performed in inert atmosphere, such as nitrogen, helium or CO_2, and let to the total recovery of mass as a solid (non-volatile material), liquid (condensable fraction) and gaseous (non-condensable fraction) products (Kaminsky, 2006). These three products are always obtained regardless the apparatus, heating source, operating temperature and heating speed. The aim in all research is usually to enhance yields in liquid product and to control products characteristics. The pyrolysis is an endothermic process, typical energy consumption 4.0-5.7 MJ Kg^{-1} (Piskorz et al., 1999), and the thermal energy required could be provided in several ways resembled in two main categories: conventional and non-conventional.

Conventional heating included heating sources internal (partial combustion of the load for instance) or external to the reactor: electrical resistance (Burrueco et al., 2005), flame (Williams et al., 1998) or microwave (Ludlow-Palafox & Chase, 2001). Among non-conventional heating may be included also plasma (Tang & Huang, 2004) and supercritical fluids (Chen et al., 1995). Up to now a wide number of apparatus and reaction conditions were experimented for conventional, non-microwave, heating pyrolysis. Alternative to this technology is microwave pyrolysis.

Only the pyrolysis process with conventional (external or internal) heating are up to now for large scale applications the only one employed.

Any apparatus, besides autoclave, is composed of few main parts: reactor, heat production and exchanger, collecting system.

The fundamental part of any apparatus is the reactor, where the heat is transferred from the source to the material. Many type of reactors are available, here are listed only the main classes: autoclave (de Marco Rodriguez et al., 2001), rotary kiln (Li et al., 2004), static bed reactor (Cunliffe & Williams, 1998; Burrueco et al., 2005) and fluidized bed reactor (Kaminsky & Mennerich, 2001, Aylòn et al., 2008; Aylòn et al., 2010). Their feature and main operating parameters are reported and compared in table 2.

Table 2. Reactor employed in non-microwave pyrolysis: main features, operating temperature and heating rate.

Reactor Type	Temperature range (°C)	Heating Rate (°C min^{-1})	Main Feature
Autoclave	300-700	15	Products don't vent, poor thermal transfer
Rotary kiln	450-650	-	Fixed speed feeding, heat transfer improved by lowering tire size
Static bed reactor	400-700	5-12	Long reaction time, products vent from reactor
Fluidized bed reactor	500-600	-	Continuous process, optimal thermal transfer

In table 2 are also underlined the temperature range and heating rate tested. The temperature and the heating speed are parameters of paramount importance and together with apparatus affect the products composition. In order to simplify the approach we focus just on these three process variables. Other variables such as load size, apparatus set-up, pyrolysis running time and material flux inside the reactor are discussed afterwards.

The choice of one of these reactors is linked with two aims: effectiveness and manageable transfer of heat to the polymeric material and volatilize products from the reactor towards collecting system. It is essential an efficient heat transfer due to the poor thermal conductivity of tires and polymeric material. The possibility to control the heat transfer is fundamental to obtain uniform products.

Without taking care of the others variables just the fluidized bed reactor fulfills all the above requirements. The heating medium is usually quartz sand preheated and kept at the prefixed temperature while the polymeric material is continuously fed (Kaminsky & Mennerich, 2001). Nevertheless the process is sensitive to fibers and high amounts of metals and fillers (Kaminsky et al., 2004). It requires high operating cost for heating and complex feedstock preparation making this kind of reactor relatively expensive (Juma et al., 2006). For instance Kaminsky use 9 kg of quartz sand to treat up to 4 kg of chopped polymers (Kaminsky & Mennerich, 2001).

On the other side a simpler approach is possible when working with static bed reactor or autoclave in a batch process. The reactor is just heated together with the polymer to the required temperature. It is kept to this temperature until the end of the experiment (Cunliffe & Williams, 1998; Mastral et al., 2000; Gonzàlez et al., 2001; Burrueco et al., 2005). Anyway the heat transfer is not effective in this type of reactors and it can be improved by upgrading the reactor with a continuous mixing of the load or with a spinning movement that is a rotary kiln (Li et al., 2004).

The heating rate and final temperature affect the properties and yields of the three products. It is deeply described in paragraph 2.3. Anyway different reactors may give different products even if the final or operating temperature is the same in each reactor.

An improving of the reactor technologies to effectively transfer the heat requires a control over the load size. A static bed reactor, autoclave, or rotary kiln doesn't need any particular load size, just small enough to be filled during the experiments. In the fluidized bed reactor usually a size of 1-2 mm (Kaminsky & Mennerich, 2001), 2 mm (Aylón et al., 2008) or no more than 5 mm (Aylón et al., 2010) is required.

The limitations displayed above are held back controlling the other experiment's variables: pyrolysis running time, apparatus set-up and material flux. Where a good heat transfer is achieved shorter reaction time is required. Kamisky reports 230 minutes to pyrolyze 4 kg with its fluidized bed reactor (Kaminsky & Mennerich, 2001) while Berrueco's process requires up to 300 minutes to pyrolyze 0.3 kg with its static-bed reactor (Berrueco et al., 2005).

Another way to overwhelm the not optimal heat transfer is the control of the flux of material and consequently the apparatus set up. Except for autoclave, where all the products are forced inside the reactor, all the others described reactors require a gas circulating system to remove the pyrolyzed products at the desired time.

Again a longer residence time of the volatilized products is used when the heat transfer is not optimal and it lets a more efficient cracking. Static-bed reactor requires up to 120 s of residence time of the gas in the reactor and this is achieved by a controlled nitrogen flux (Berrueco et al., 2005) or apparatus set-up (Cunliffe & Williams, 1998). The residence time of the gas in the fluidized-bed reactor is less than 3 s and it is achieved with strong flux of nitrogen or other inert gas (Kaminsky & Mennerich, 2001).

This description of main reactor technologies available is far away to be exhaustive; it wants only to give an idea of the possible approaches to the pyrolysis of tires and plastics.

When we switch to a microwave heating system the apparatus set up is almost the same: efficiently and manageable transfer of heat to obtain products with desired proprieties. Only the reactor is the feature which must be changed.

The scientific literature is extremely poor with respect to the patent literature. Unfortunately for scientific spread the patents don't give extensive information concerning the products achievable by the described apparatus. Anyway it is possible to understand the approach to the microwave pyrolysis of tire and plastics.

The goal in each patent is to provide a simple and economically convenient approach to the pyrolysis of both tires and plastics. It is achieved by treating the polymeric substrate in a static bed-type reactor. Holland (Holland, 1995) uses a static-bed reactor filled with carbonaceous materials (tires or coal) between 400 °C and 800°C to pyrolyze any kind of polymers. More recent patents showing similar approach, which always resemble a static bed reactor, are reported by many authors (Holland, 1992; Parker, 1992; Johnson et al., 1996; Pringle, 2006; Pringle, 2007; Kasin, 2009). Just the feeding system and the reactor design diversify the several patents. They are planned to be employed for industrial application. Strong attention is paid to productivity and energy efficiency.

Recently a more complete microwave pyrolysis plant was proposed from Scandinavian Biofuel Company (Scandinavian Biofuel Company [SBC], 2011). All the products were evaluated as source of electric energy and after subtraction of the energy required for the process give a production of 1.55 MWh/ton from tires or 3.98 MWh/1 ton from plastics. The hypothetical working capacity for SBC plant is reported in table 3.

Table 3. Project examples for electricity production from SBC plant

Feedstock	Plastics	Tires
Capacity/yr	25,000 tons	25,000 tons
Water Content	5%	1%
Net capacity/yr	25,000 tons	20,000 tons
Output oil	12,500 tons	5,600 tons
Output gas	10,000 tons	3,000 tons
Output carbon	1,200 tons	11,000 tons
Electricity output	103.4 GWh	44.5 GWh
Electricity use	3.79 GWh	4.57 GWh
Electricity sales	99.65 GWh	39.89 GWh

From the few scientific articles available up to now for microwave pyrolysis of polymeric materials two kind of apparatus came out: static bed batch reactor (Hussain et al., 2010) and fluidized bed reactor (Ludlow-Palafox & Chase, 2001).

The static bed reactor used by Hussain et al. is as simple as possible. The microwave absorbent is an iron mesh mixed with the polymer (just polystyrene in this article). The products are collected by two cold traps.

Ludlow-Palafox & Chase use a reactor resembling the fluidized bed reactor of Kaminsky, where the heat carrier is carbon instead of quartz sand (Kaminisky & Mennerich, 2001). The temperature is risen up and kept to the prefixed temperature before adding the polymers. Small amount of polymers is dropped at time intervals directly into the reactor, 50g of polymers for 1000g of heated carbon, to prevent large temperature fluctuation. It is required a small amount of time (120 s) to complete the pyrolysis in these conditions. Anyway with this productivity to pyrolyze 4 kg, as reported by Kaminsky & Mannerich with their fluidized bed reactor, is possible to extrapolate a reaction time of 120 minutes instead of 230 minutes for fluidized bed reactor. There's a huge difference. Furthermore the full potentiality of microwave oven is not used.

In our experiments we use a static bed reactor approach in a batch process to simplify most of the apparatus and to minimize the energy consumption.

The system is composed by a microwave oven, a heat exchanging pipes and collecting flasks. The volatilized material is moved towards the condensing traps by the products formed in the course of the pyrolysis and the partial condensation of gases. In this way the gases vented out are not diluted by transport gas. Up to now the experiments were carried out by heating the chopped tires and plastics from room temperature to the complete pyrolysis of material. The pyrolysis starts usually 20 seconds after switching on microwave. The experiments are conducted using an increasing microwave power. In this way only the energy necessary for the pyrolysis itself is supplied inside the oven.

In the best conditions tested until now we are able to pyrolyze 0.4 Kg of tires in 14 minutes. Here is possible to extrapolate 140 minutes for 4 kg of tires. It is a big improvement from the static bed reactor and slightly better than fluidized bed reactor from Ludlow-Palafox & Chase. Also in this case the full potential of the microwave oven is not used.

2.2.1. The problem of high temperature measurement when switching on microwave

When the microwave is not employed the temperature is easily measured by thermocouples. When microwave is turned on a metal thermocouple can't be used anymore. It absorbs the microwave radiation and provides a skewed temperature.

Up to now are available two technological solutions to investigate the high temperature during the microwave pyrolysis process: infrared and fiber optic thermometer. The infrared thermometer is a contactless thermometer which can work properly in a wide range of temperature. It measures the radiation emitted in the infrared region of the spectrum which is linked to the temperature. It consists in a lens to focus the infrared energy onto the detector which converts the energy to an electric signal that may be displayed in units of temperature after calibration. The main limitations of this instrument are its calibration and the medium value of the temperature of objects present. It is available at a relatively low price. The common fiber optic thermometer commercially available is not suited for high temperature detection. They can be used in a temperature range up to 400°C, over this temperature they melt. A quite new technology in optic fiber is now accessible based on sapphire crystalline fiber (Djeu, 2000). It can work up to temperature of 950°C, and it is perfectly suitable for temperature detection in a microwave pyrolysis experiment (Micromaterials Inc., 2010). The main advantage of this fiber optic probe is the detection of temperature in a single position where the probe is placed. The infrared thermometer on the other side averages the temperature over a large area and this could lead significant errors. We have to remember that the heating is not homogeneously distributed but localized on the microwave absorbent that are mixed with the polymeric materials.

2.3. Conventional tires pyrolysis: products characteristics

The three products, solid, liquid and gas, have been fully characterized and the results reported in many papers on non-microwave pyrolysis. Due to this reason it is really interesting to review them for a better understanding of microwave influence on products characteristics.

Tires have a variable composition, depending on brand, dimensions and use. In tables 4 and 5 are reported the average composition of truck and car tires (Rubber Manufacturers Association [RMA], 2011).

Table 4. Average composition of truck and car tires. Synthetic polymers include: polybutadiene (BR, butadien rubber), styrene-butadiene copolymers (SBR, styrene-butadiene rubber) and EPDM rubber (ethylene propylene diene Monomer (M-class) rubber).

Tires for	Polisoprene 1,4-cis (wt%)	Synthetic Polymers (wt%)	Carbon Black (wt%)	Steel (wt%)	Sulfur (wt%)	Textile, additive (wt%)
Car	14	27	28	14-15	1.5	16-17
Truck	27	14	28	14-15	2.5	16-17

Even if tire composition is variable, yields obtained from the pyrolysis are not strongly affected by it (Kyari et al., 2005). Variation in products characteristics are appreciated only when significant differences between tires occurred. Their characteristics are mainly affected by temperature, heating speed and feeding size as reported in paragraph 2.2.

Table 5. Average elemental composition (CHNSO) of tires.

Element	%
C	80.1
H	7.0
N	0.4
S	1.5
O	3.0
Ash	8.0

The main gas components are: H_2, H_2S, CO, CO_2, CH_4, C_2H_4, C_2H_6, C_3H_6, C_3H_8 and other organic substances in lower amount. It has a high gross calorific value of 68-84 MJ m^{-3} (de Marco Rodriguez et al., 2001) when it is rich in hydrocarbons and a lower one, 30-40 MJ m^{-3} (Williams et al., 1990), when depleted in organic compounds. It may be used as an energy source for sustaining the pyrolysis process itself (Williams et al. 1998; Kyari et al., 2005).

The liquid, or pyrolysis oil, is the most scientific and economical interesting product due to its high gross calorific value and content of valuable hydrocarbons such as: benzene, toluene, xylenes, limonene and so on. It can be employed as energy source, refinery feedstock or reservoir for aromatics and limonene (Laresgoiti et al., 2004).

The solid is mainly made of carbon and non volatilizable material: steel and inorganic additives (de Marco Rodriguez et al., 2001). The steel can be easily removed with electro-magnets. The carbonaceous fraction is employed up to now as fillers in tires manufacturing (de Marco Rodriguez et al., 2001), activated carbon production (Mui et al., 2004) and smokeless fuel (Dìez et al., 2004).

The yields for the three products are strictly connected with the process variables as introduced in paragraph 2.2.

In table 6 the yields obtained by various authors in different reaction conditions are reported.

A comparative study is not easy to carry out. The data reported in table 5 are not congruent among them due to the different reactors and conditions employed. Anyway a trend in all of them may be easily obtained. There's a critical temperature above which the pyrolysis is completed and the solid yield doesn't significative change. In any study it is around 500°C.

When the temperature is rising over 500°C, while the solid product remain almost constant, the yield of liquid product goes down due a stronger cracking which improve the yield of non-condensable fraction.

The yield of liquid is lower when the residence time of liquid in the oven increase. The autoclave, where the pyrolysis products don't vent out, the yield of non-condensable fraction is three times higher than the one achieved in a static bed reactor. In a fluidized bed reactor with a residence time of 3 seconds gives similar gas yield than a static bed reactor where the residence time is 120 seconds. When the reaction time increases the cracking of liquid products remarkable rises up to 3 times.

Table 6. Yields and parameters of non-microwave tires pyrolysis reported by various authors. S: solid; L: liquid; G: gas. *Estimated. NA: not available.

Author	Max Temp (°C)	Heating speed (°C/min)	Running time (min)	Yields			Comments
				S	L	G	
Burreco et al., 2005	400	NA	240	64.0	30.0	2.4	Static-bed batch reactor
	500	NA	240	52.7	39.9	3.6	
	550	NA	240	52.5	39.1	3.6	
	700	NA	240	51.3	42.8	4.4	
de Marco Rodiguez et al., 2001	400	15	55*	55.9	24.8	19.3	Autoclave
	500	15	62*	44.8	38.0	17.2	
	600	15	68*	44.2	38.2	17.5	
	700	15	75*	43.7	38.5	17.8	
Gonzàlez et al., 2001	400	-	30	55.1	42.9	2.0	Static-bed batch reactor, sample 4g.
	500	-	30	38.6	55.4	6.0	
	600	-	30	37.0	52.2	10.8	
	700	-	30	36.7	36.6	26.7	
Aylòn et al., 2010	600	-	NA	40.6	41.5	17.9	Moving bed reactor, without steel
	700	-	NA	39.0	31.3	29.7	
	800	-	NA	41.0	27.5	31.5	
Kaminsky & Mannerich, 2001	500	-	230	30	65	5	Moving bed reactor, tires composition simulate
	550	-	220	34	57	9	
	600	-	240	40	51	9	

Author	Max Temp (°C)	Heating speed (°C/ min)	Running time (min)	Yields			Comments
				S	L	G	
Cunliffe & Williams, 1998	450	5	150	37.4	58.1	4.5	Static-bed batch reactor, load pre-heated at 150°C
	500	5	160	38.3	56.2	5.5	
	560	5	172	38.1	55.4	6.5	
	600	5	180	38.0	53.1	8.9	
Li et al., 2004	450	-	200-250	43.9	43.0	13.1	Rotary kiln, 50 kg processed
	500	-	200-250	41.3	45.1	13.6	
	550	-	200-250	39.9	44.6	15.5	
	600	-	200-250	39.3	42.7	18.0	

2.3.1. Non-condensable fraction (gas)

Each apparatus is provided with a condensation system, in order to collect the liquid products. In order to maximize the yield of the liquid products different heat exchanging systems were tested and this is an extra variable affecting the gas composition in each report.

However the non-condensable fraction contains permanent gas (such as: H_2, H_2S, CO, CO_2 **and** CH_4) and organic substances with high vapor pressure difficult to condensate.

In table 7 **is summarized the amount of the main compounds identified in each gaseous mixture reported by various authors.**

Table 7. Main compounds present in the gas. * Values are extrapolated (Pressure of 25 atm) where they are reported as mol kg^{-1} of scrap tire. ** Values are calculated where are expressed as mass percentage.

Author	Max Temp (°C)	Compound (vol. %)									
		H_2	CO_x	CH_4	C_2H_6	C_2H_4	C_3	C_4	C_5	C_6	H_2S
Burreco et al., 2005	400	2.6	2.5	1.0	0.4	0.3	0.5	3.2	0.0	0.3	-
	500	14.2	1.3	4.3	1.5	0.7	1.3	2.3	0.0	0.4	-
	550	17.9	1.1	5.6	1.8	0.8	1.5	2.0	0.0	0.3	-
	700	10.1	1.6	5.4	1.6	1.2	1.6	2.5	0.0	0.5	-
de Marco Rodiguez et al., 2001	400	-	14.9	4.4	4.5	4.3	8.5	36.9	16.5	7.3	2.6
	500	-	14.2	19.8	9.1	9.4	10.8	21.3	7.6	2.8	5.1
	600	-	15.3	20.0	9.0	9.7	10.6	21.9	7.4	2.5	3.6
	700	-	21.8	20.6	8.1	8.9	7.7	19.8	6.7	2.5	3.9

Author	Max Temp (°C)	Compound (vol. %)									
		H_2	CO_x	CH_4	C_2H_6	C_2H_4	C_3	C_4	C_5	C_6	H_2S
Gonzàlez et al., 2001*	400	36.7	50.6	12.7	0.0	0.0	-	-	-	-	-
	500	19.8	34.4	22.9	15.3	7.6	-	-	-	-	-
	600	16.5	20.4	28.5	28.5	6.1	-	-	-	-	-
	700	18.7	7.1	34.7	32.5	7.1	-	-	-	-	-
Kaminsky & Mannerich, 2001 **	500	0.1	12.5	3.0	4.6	3.1	10.5	63.2	-	-	3.0
	550	0.2	7.5	3.8	5.4	5.1	13.7	62.5	-	-	1.8
	600	0.2	8.7	4.1	5.1	8.0	16.4	55.2	-	-	2.2
Dìez et al., 2004	350	24	3.3	20	29		12	5.7	3	1	-
	450	30	3.0	24	26		9	4.3	2	0.2	-
	550	40	3.0	26	20		6	2.8	1.1	0.1	-

The inorganic substances, H_2S, CO_x and H_2, are connected to three tire components. The sulfur compounds come from the vulcanization process of the polymeric tire constituents. A variable amount of sulfur is used to improve the rubber elasticity. In the gases only hydrogen sulfide is found as sulfur containing compound, any organic sulfur containing products are not found. Carbon dioxide and carbon monoxide find their origin from two sources: metal carbonates and oxygen containing inorganic substances. Both of them are additives in tire formulation. Hydrogen is a common cracking product. What happen in a pyrolysis reactor is close to a thermal cracking in a petroleum refinery. The organic substances find their origin from the random cracking of polymers. The growing presence of olefin instead of their paraffinic equivalents is linked to a working temperature in the range 400-700°C where alkenes are thermodynamically more stable than alkane.

2.3.2. Condensable fraction (liquid)

The condensable fraction, also known as liquid fraction or pyrolysis oil, is the most interesting and studied product. The maximum yield is obtained working with a pyrolysis temperature between 450 and 550 °C as shown in table 5, it varies between 30% and 58% depending if steel cords are considered in calculation. Its color is dull dark-brown and it has a bad-smelling flavor. It has a complex composition which is usually studied using a gas chromatograph coupled with a mass detector. An enormous amount of substances are identified, more than 300, but only few compounds are present in an amount greater than 1% (Laresgoiti et al., 2004).

In table 8 are reported the top 12 substances identified inside the condensable fraction from some authors.

Table 8. The 12 main substances present in the liquid fraction. The data reported are normalized to toluene. * It is the percentage over total liquid fraction.

Author	Max T (°C)	Compound											
		Benzene	Toluene	Ethylbenzene	Xylenes	Styrene	2-Ethyltoluene	α-methyl styrene	Limonene	1H-Indene	Naphthalene	Benzothiazole	1,6-dimethyl naphthalene
Laresgaoiti et al., 2004	400	0.65	1.00	1.07	0.69	0.36	0.15	3.22	0.17	0.18	0.53	0.25	0.65
	500	0.94	1.00	1.25	0.88	0.40	0.11	1.84	0.32	0.26	0.33	0.30	0.94
	600	2.34	1.00	1.09	1.03	0.16	0.11	1.70	0.34	0.34	0.39	0.36	2.34
	700	1.46	1.00	1.33	1.01	0.50	1.26	2.30	0.42	0.41	0.53	0.50	1.46
de Marco Rodiguez et al., 2001*	400	-	-	-	-	-	-	-	-	-	-	2.10	-
	500	-	-	-	-	-	-	-	-	-	-	1.80	-
	600	-	-	-	-	-	-	-	-	-	-	2.00	-
	700	-	-	-	-	-	-	-	-	-	-	1.80	-
Kyari et al., 2005*	500	0.85	1.00	0.11	0.52	1.91	0.27	3.34	0.16	0.05	0.17	0.28	0.85
Kaminsky & Mannerich, 2001	500	0.73	1.00	0.21	1.60	-	0.57	-	0.08	0.07	0.60	-	0.73
	550	1.00	1.00	0.38	0.87	-	0.37	-	0.09	0.08	0.40	-	1.00
	600	1.24	1.00	0.32	1.82	-	0.44	-	0.18	0.15	0.21	-	1.24

The liquid fraction is a mixture of organic compounds from C_4 **to** C_{24}, mainly olefins, single ring aromatics and limonene. Within the minor substances strong attention should be devoted to three classes of compounds always present: polycyclic aromatics hydrocarbons (PAHs), sulfur containing products and other heteroatoms containing substances.

In figures 1-4 are reported the mechanism proposed for the degradation of polymeric materials (Cunliffe & Williams, 1998; Mastral et al., 2000).

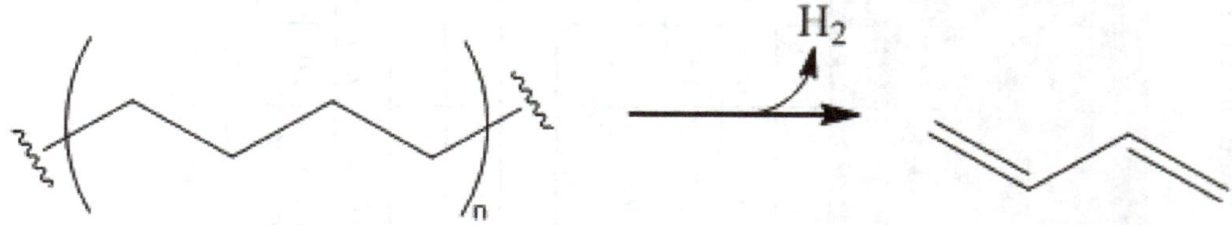

Figure 1. 1 Cracking of linear polyolefins.

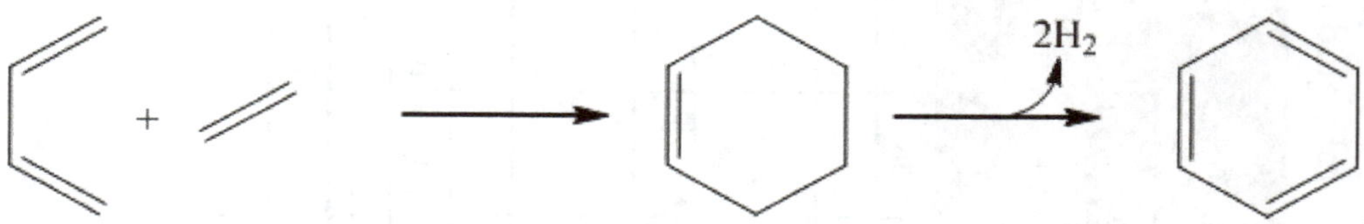

Figure 2. 2. Diels-Alder reaction followed by dehydrogenation reaction that yield benzene.

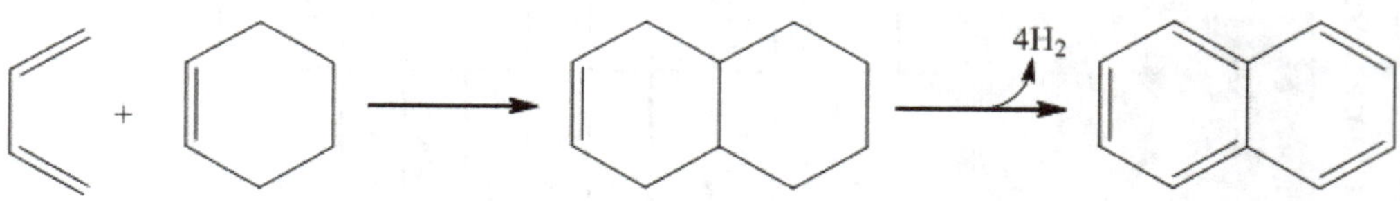

Figure 3. 3. PAHs via Diels-Alder and dehydrogenation reactions.

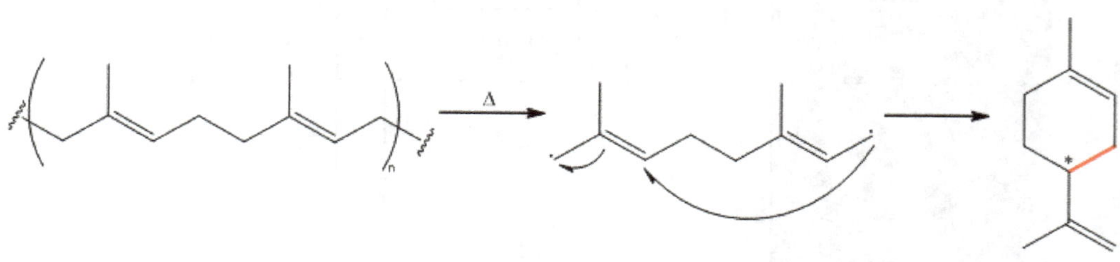

Figure 4. 4. Radical synthesis of rac-limonene from polyisoprene

These reactions explain the presence of huge amount of H_2, aromatics and limonene. The decrease of limonene when rising the reaction temperature can be connected with cracking and especially dehydrogenation reaction. Other reactive compounds may be involved in analogous reactions.

Benzothiazole derivates are present in tires and unsubstitued benzothiazole is widely found in condensable products. Other hetero compounds are also formed due to the high content of sulfur compounds in tire.

Anyway the sulfur content in the liquid is too high to let its use directly as fuel. This oil need refinery treatments or must be blended with refinery oil before any use as fuel.

The gas chromatographic analysis is just one way to characterize the liquid fraction. Other analysis becomes notable when the liquid is compared to petrochemical fuel. The simplest way to compare the liquid fraction with a petrochemical product is distillation. It let to split the liquid fraction in two common petroleum fractions: gasoline and diesel oil. Gasoline distillation temperature is in the range between 30 and 210°C and only 20% of liquid products are within this range. The typical interval of distillation temperature for diesel oil is between 150 and 370°C and 60% of pyrolysis oil distillate in this range. This product must be characterized as a fuel by the gross calorific value. It is between 43 and 44 MJ kg^{-1}, higher than commercial oil (de Marco Rodriguez et al., 2001; Laresgoiti et al., 2004).

2.3.3. Non-volatile fraction (solid)

The non-volatile fraction, or simply solid, contains steel cords, carbon, sulfur and other metals containing compounds. In table 9 is reported the average composition of a non-volatile fraction (Zabariotou et al., 2004; Pantea et al., 2003).

We have also identified by ICP-MS analysis the presence of others metal in minor amounts such as, in order of abundance: Co, Pb, Cu, Mn, Ni, Sn, Cr, Cd, V, As and V.

Table 9. Elemental composition of a non-volatile fraction (solid).

Element	Abundance (wt%)
C	71.5
O	13.3
Fe	5.4
S	2.8
Zn	2.3
Ca	1.3
Al	0.3

The steel cords are not scratched at pyrolysis temperature and they may be separated by an electromagnet. The remaining solid product is investigated and may be employed as active carbon in grain or pellets (Mui et al., 2004; Pantea et al., 2003). It is obtained by treating the raw carbonaceous non-volatile product at temperature between 800°C and 1000°C in the presence of an activating agent such as H_2O or CO_2. The active carbon obtained in this way has high active surface similar to the commercial product.

The carbon may be used as sorbent for organics (phenols and colorants), metals, chlorinated compounds and gases (butane for instance). Anyway the polluting metals present inside the active carbon (Cr, Cd) may be released and any application must be accurately monitored. Carbon may be also employed as a source of energy by combustion process.

2.4. Microwave tires pyrolysis

Up to now the microwave pyrolysis are exploited in laboratory and pilot plant scale. Anyway the greatest difference, if correlated to thermal heating, dwells in uniform and efficient heating.

We carry out experiments using a microwave oven that assure uniform distribution of the microwaves inside the chamber and equipped with the possibility to control the power of microwave supply. In this way it is possible to modulate the speed of thermal degradation and the yield of products together with their characteristics.

Yields of the liquids are in the range between 37-40%, similar to those reported for non-microwave pyrolysis (steel cords are counted) when the optimal conditions are achieved, otherwise fewer yields are obtained.

In table 10 are listed the yield of five representative experiments.

Table 10. Microwave pyrolysis: correlation among power, time and yields.

MW power (kW)	Time (min)	Solid (wt%)	Liquid (wt%)	Gas (wt%)
6	13	50.10	26.72	23.18
3	39	50.45	37.21	12.34
1.5	100	69.38	18.09	12.53
1.5 to 4.5	76	49.65	38.76	11.59
1.5 to 6	59	48.36	36.35	15.29

The cracking is already effective when half of the maximum power is employed. If the power is lower than 1.5 kW, in our oven, it is not possible to complete the pyrolysis itself. Anyway the properties of collected products vary amazingly.

The properties of non-volatile fraction anyway are slightly affected by the microwave power and the product has close characteristics to that one obtained from a non-microwave oven.

The composition of non-condensable organic fraction changes as function of microwave power and the results are summarized in table 11.

Table 11. Main substances identified in non-condensable product.

Substance	MW power (kW)				
	6	3	1.5	1.5 to 4.5	1.5 to 6
Hydrocarbons C1 + C2	46.54	76.79	70.29	27.39	29.22
Hydrocarbons C4	25.29	11.60	17.30	40.44	38.26
Hydrocarbons C5	1.20	0.75	0.94	1.79	2.07
2-methylbutadiene	10.23	3.60	5.18	-	10.88

High microwave power increases the production of volatile materials which flow quickly out of the reactor and they are not subsequently cracked. A slow heating is not good enough to break the gaseous molecules while a mid-power heating is suited to crack the polymers macromolecules and let them to stay for a long time in the reactor to be degraded to simple molecules such as methane, ethane and ethene. Minor amount of organic sulfur containing substances like ethylene sulfide are present in gases.

Among inorganic substances remarkable is the presence of H_2 which rises up to 36% when the microwave power is 3 kW. This result is not easy comparable with non-microwave results: just in few experiments where the residence time of gases inside the reactor is long similar results are achieved.

The condensable fraction (liquid) is deeply affected by microwave power. Its appearance switches from a dark dull liquid when the full power is used to a yellow clear solution when 1.5 kW is used. Decreasing the power also the viscosity is lower.

Anyway the GC/MS analysis gives a better idea how much the liquid product changes its properties.

The main substances identified and their relative abundances are reported in table 12.

Table 12. Main substances in liquid product, obtained using different microwave power

Compound	Abundance %				
	6 (kW)	3 (kW)	1.5 (kW)	1.5to4.5 (kW)	1.5to6 (kW)
2-butene	1.63	1.49	1.05	1.21	0.81
1,3-pentadiene	3.18	1.63	1.49	1.70	0.31
3-methyl-1-esene	2.09	1.57	1.40	1.63	0.64
Benzene	3.91	1.71	2.78	2.26	1.66
Toluene	4.55	2.87	3.52	3.11	2.51
Xylenes	6.92	5.22	7.48	7.08	3.43
Styrene	3.06	1.39	1.45	1.85	0.76
Trimethylbenzenes	3.23	3.03	4.57	4.54	2.53
α-methylstyrene	0.62	0.39	-	1.33	0.24
1-ethyl-3-methylbenzene	0.99	1.01	1.27	1.41	0.78
1,2,4,5-tetramethylbenzene	0.76	1.40	2.10	2.41	1.19
Limonene	3.72	3.43	2.49	3.49	0.82
1H-Indene	0.87	0.38	0.28	0.50	0.46
Naphthalene	0.97	0.43	-	0.20	0.77
Benzothiazole	0.59	0.57	0.25	0.32	0.58

A fast pyrolysis with a power of 6 kW yields a huge amount of aromatics (up to 20%). Decreasing the microwave power the aromatic yield decreases. Probably the microwave energy is not enough to perform dehydrogenation reactions.

The reduction of microwave power leads a change in other properties of liquid such as: viscosity, density and fraction of distilled product.

Viscosity and density are directly connected with microwave power. The density changes among 0.800 g cm^{-3} **and 0.900 g cm**$^{-3}$, while the viscosity is in the range 0.7 – 2.6 cPs.

The fraction of product distilling at temperature lower than 210°C changes deeply when the power is lowered as shown in table 13.

Table 13. The distillation of condensable fraction. Only the amount of distillate under 210°C is reported

MW Power (kW)	Distillate under 210°C (%)
6	22.93
3	37.61
1.5	74.61

The trend is almost linear and interesting results are achieved. In non-microwave pyrolysis only 20% of liquid fraction is distillable under 210°C.

Using a microwave heating is possible to manage properly the yields and proprieties of condensable and non-condensable products.

3. MICROWAVE PYROLYSIS OF POLYMERIC MATERIALS

As well as tires pyrolysis the microwave pyrolysis of single polymeric materials has been investigated by many authors (Kaminsky & Zorriqueta, 2007; Arandes et al., 2008; Kim, 2001; Yoshioka et al., 2004; de la Puente et al., 1998). Five main categories of polymers were tested: polyethylene (PE), polypropylene (PP), polystyrene (PS), poly(ethylene terephthalate) (PET) and polyvinylchloride (PVC). All of these polymers are not able to absorb microwave and turn it into heat. So a microwave absorbent must be employed to carry out the pyrolysis, as reported in a previous paragraph. Carbon powder (to avoid unwanted interference) or chopped tires has been employed as microwave absorbent.

The microwave influence on the products yields is poor but it is strong on the properties of liquid fraction and this influence is reported in the following paragraphs.

The non-microwave pyrolysis of the above mentioned polymers has been deeply studied and characterized. In the first instance the microwave pyrolysis seemed to be more suited for multi materials pyrolysis such as: Tetra Pak® or laminates (Ludlow-Palafox & Chase, 2001).

Usually the pyrolysis of a pure plastic material yields just a liquid and a gas product. In few experiments is reported the growing up of a solid product and usually low attention has been paid to this aspect. In our apparatus a low amount of non-volatilized fraction is always found when the polymers are pyrolyzed in the presence of coal, presumably due to a strong localized heating. In table 14 are listed some experiments with relative yields for the pyrolysis of single plastics.

Table 14. Microwave pyrolysis: correlation between microwave power and yield. * Tires instead of coal as microwave absorber.

MW power range (kW)	Polymer	Time (min)	Absorbent-polymer ratio	Solid (wt%)	Liquid (wt%)	Gas (wt%)
1.8 to 3*	HDPE	33	2:1	42.49	44.98	12.09
3	HDPE	75	1:2	0.40	83.92	15.68
1.2 to 6*	HDPE	265	2:1	35.52	45.83	18.65
1.8 to 3*	PP	39	2:1	35.29	54.00	10.72
3	PP	68	1:2	15.89	70.82	13.29

MW power range (kW)	Polymer	Time (min)	Absorbent-polymer ratio	Solid (wt%)	Liquid (wt%)	Gas (wt%)
1.8 to 3*	PVC	71	2:1	53.16	17.70	29.14
1.8 to 3	PVC	47	1:1	13.92	10.25	75.83
3	PVC	21	1:2	14.69	3.44	81.87
1.8 to 3*	PET	40	2.5:1	38.20	35.32	26.48
1.2 to 3*	PET	207	4.2:1	48.29	26.34	31.99
3 to 6	PS	59	1:2	6.83	89.25	8.92

3.1. PE microwave pyrolysis: products characteristics

The polyethylene (PE) is the most important and used plastic. It is a thermoplastic polymer, chemically inert and is used in an enormous range of applications. It is produced as a linear High Density (HDPE) or a branched Low Density (LDPE) polyethylene.

The thermal degradation of HDPE starts over 380°C (Bockhorn et al., 1998). When HDPE is heated to a temperature below 730°C it forms mainly a wax (a semi-solid product) and a liquid. Together with the increasing of the temperature the amount of wax decreases in advantages to liquid and gas products. At 730°C there is no further formation of wax, just liquid and gas (Mastral et al., 2002). Over 800°C the conversion of the polymer in gases is almost total: 91.2% at 850°C (Westerhout et al., 1998).

The cracking of the polymers chain leads to a large distribution of numerous linear hydrocarbons having different length. This distribution is a function of microwave power and working temperature. At high cracking temperature the product is liquid at room temperature otherwise it is solid.

The gas is composed of H_2, CH_4, C_2H_6, C_2H_4, C_3H_8, and C_3H_6. The liquid and wax are made of alkanes, from C_{11} to C_{57}, and their equivalent 1-alkene and 1,3-dialkene (Williams & Williams, 1999a).

Anyway too high temperature causes the formation of aromatics with the reaction path proposed in figure 3. Together with high temperature also long residence time promotes the formation of aromatic compounds.

The pyrolysis of HDPE using microwave heating requires HDPE flakes mixed with a microwave absorbent such as coal or chopped tires. Different microwave power gives completely different products. In table 14 are listed the yields in solid, liquid and gas using different microwave power and absorber.

When microwave power of 3 kW is used a wax is largely produced and the amount of a liquid product inside the wax for analysis is hardly recovered. Anyway when milder reaction conditions are used, lowering the microwave power, a liquid product is collected. It has been obtained using 1.8 - 1.2 kW, but in this condition it is not possible to complete the pyrolysis without the erogation of a higher power of microwave in a subsequent step. By the simple modulation of the microwave power is not possible to obtain, without other treatments or apparatus, only a liquid product.

The liquid fraction, analyzed via GC/MS, contains linear saturated hydrocarbons and their terminal unsaturated equivalent from C_6 to C_{32}.

The only way to obtain a liquid product instead of wax is the cracking of the wax in a second reactor, refluxing them again in the pyrolysis reactor or using a catalytic system which resemble a refinery unit (Arandes et al., 2008).

3.2. PP microwave pyrolysis: products characteristics

The polypropylene (PP) is the second most used polyolefin in the world due to the low-cost of monomer production together with flexible properties.

The degradation mechanism yields a chain cracking to a wax and a liquid. They are composed of hydrocarbons from C_{11} **to** $\mathbf{C_{57}}$ **(**Westerhout et al., 1998). Rising the pyrolysis temperature the wax yield decreases. Anyway these conditions lead to the formation of aromatics due to further reactions of hydrocarbons obtained in the first instance (see scheme reported in figure 3).

The PP is more unstable than HDPE to thermal degradation due the presence of a large extent of tertiary carbons giving more stable radicals with respect to primary or secondary radicals formed in the pyrolysis of PE. The pyrolysis of PP may be carried out at 500°C instead of 700°C as required for HDPE (Kaminsky & Zorriqueta, 2007)

An analogous effect is shown when the pyrolysis is performed using a microwave oven. The products are liquid at room temperature independent from the microwave power and the microwave absorbent: coal or chopped tires. In table 14 are reported two experiments performed at different microwave power.

The high reactivity of PP leads often to the formation of a solid product.

The liquid fraction contains organic compounds from C_6 to C_{30}. In table 15 are reported the main components identified by GC/MS analysis in the low boiling fraction (below 145°C) of the liquid product obtained using 3 kW as microwave power source.

Table 15. Main compounds present in the low boiling fraction of the microwave pyrolysis of PP using a power source of 3 kW.

Compounds	Composition (%)
4-methylhepthane	3.78
2,6-dimethyl-2-octene	3.69
1,3,5-trimethylcyclohexane	4.88
3,7-dimethyl-1-octene	57.3
1,2,4-trimethylcyclohexane	5.17
3,5,5-trimethylcyclohexane	3.45
1-nonene	18.93
Total	97.20

All the substances formed by PP degradation are methyl derivates. For instance 3,7-dimethyl-1-octene is very close to the repeating unit of PP as shown in figure 5.

Anyway few aromatic substances are present in liquid products when pure PP is pyrolyzed using a microwave oven with respect to the compounds present in a thermal pyrolysis of PP (Kaminsky, 2006). This is connected with a different way of heat transfer. The heat provided by microwave oven is quickly employed to break the polymers avoiding subsequent reactions.

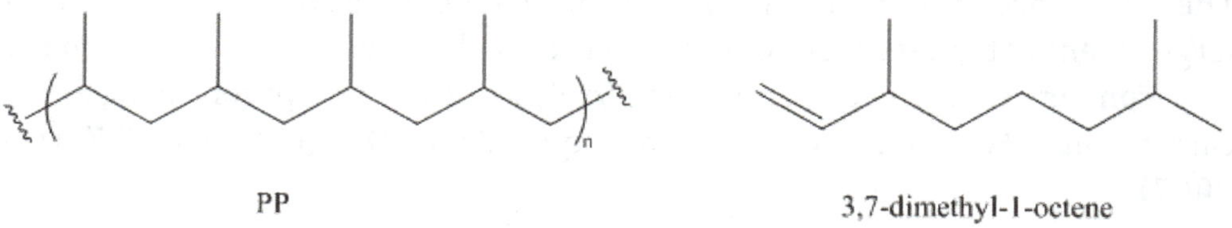

Figure 5. 5. A molecule formed in the liquid from pyrolysis of PP.

3.3. PVC microwave pyrolysis: products characteristics

The polyvinylchloride (PVC) holds a unique position among commercial polymers. It is cheap and used in a wide range of applications due to its high versatility.

Usually it is added with stabilizers, lubricating, plasticizers, fillers and so on. To avoid interference due to large amount of additives present, the microwave pyrolysis of PVC has been carried out on pure PVC.

Thethermaldegradation ofPVCtakesplaceintwo-stepsprocess:dehydrochlorination followed by residual chain cracking. The elimination of hydrogen chloride from the chain back bone is fast and complete. It happens at temperature among 300°C and 330°C (Bockhorn et al., 1998; Kim, 2001). The reaction scheme is reported in figure 6.

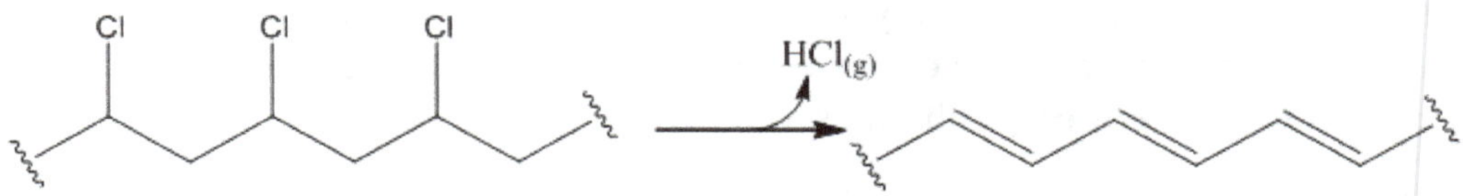

Figure 6. 6. Dehydrochlorination of PVC.

The gases are mainly composed of HCl, ethylene and propylene. The liquid fraction contains aromatic compounds such as benzene, toluene and styrene (Saeed et al., 2004).

The microwave pyrolysis of PVC follows the same two steps of thermal pyrolysis. The polymer quickly eliminates HCl that has been absorbed in NaOH traps after the collecting flasks. Up to 97% of chlorine in PVC chain is found in the traps. Few chlorine organic compounds are present in liquid products. In table 14 **are reported the PVC pyrolysis at different microwave power.**

As reported in the literature for thermal pyrolysis also in the microwave pyrolysis the main products present in liquid fraction are aromatic compounds, the main components of the liquid are listed in table 16.

Table 16. Main aromatic compounds formed in PVC/coal pyrolysis.

Compounds	Composition (%)
Benzene	18.56
Toluene	8.73
1,3-dimethyl-benzene	4.25
1,4-dimethyl-benzene	2.49
Styrene	0.46
1,2-dimethyl-benzene	3.44
Naphthalene	4.01
Total	41.94

The formation of aromatics in high yield may be attributed to Diels-Alder type reactions involving the poly-unsaturated chain formed after dehydrochlorination of PVC. This liquid may be a source of interesting compounds such benzene, toluene and xylenes.

3.4. PET microwave pyrolysis: products characteristics

The poly(ethylene terephthalate) (PET) is one of the most important polyester (together with polybutylenetherephatlate, PBT) and it is used to obtain: fiber textiles, carpets, medical accessories, belts for cars, electronics and items for cars, photographic film, magnetic tapes for audio and video, packaging, bottles and so on.

The gases obtained in PET pyrolysis are mainly CO_2 **(up to 22.71%) and CO (up to 13.29%) due to decarboxylation reactions involving the esters group present in the polymeric chain (**Williams & Williams, 1999b).

Oxygenated compounds, like benzoic acid or other compounds soluble in aqueous NaOH, are largely found. Anyway while rising the temperature from 510°C to 610°C the concentration of oxygenated compounds decreases from 45 wt% to 29% (Yoshioka et al., 2004).

A lot of aromatics and oxygenated compounds are found in the liquid fraction, all of them coming from the therephtalic moiety present in the polymer. Their concentrations increase when rising the pyrolysis temperature.

The PET microwave pyrolysis shows very different results from the non-microwave pyrolysis. The most significant difference comes from high amount of acetaldehyde and benzene. In table 14 **are reported the yields of two PET pyrolysis where chopped tires where used as microwave absorbent.**

The high amount of gas is linked with the CO_2 **and CO formation. The main substances present in the liquid fraction are reported in** table 17; for the pyrolysis with increasing power from 1.2 to 3 kW two liquid fractions where collected when the power is changed.

Table 17. The main products present in the liquid from microwave pyrolysis of PET.

Compounds	Composition %		
	1.8 to 3.(kW)	1.2 to 3 (kW)	
		Fraction 1	Fraction 2
Benzene	32.03	5.80	14.26
Toluene	9.40	4.37	9.57
1,2-dimethylbenzene	4.29	3.13	6.04
1,3-dimethylbenzene	3.70	2.31	5.25
Styrene	3.43	-	1.07
Benzaldehyde	-	1.19	-
1,2,3-trimethylbenzene	4.71	2.54	4.06
Limonene	14.51	2.31	-
1H-indene	2.44	-	-
Pentylbenzene	2.10	-	-
Benzoic Acid	5.15	1.14	-
Biphenyl	3.66	-	3.02
Total	85.42	22.79	43.27

The acetaldehyde was identified by trapping it with (2,4-dinitrophenyl)hydrazine. The hydrazone was characterized via ^{1}H-NMR.

The formation of acetaldehyde is not reported for non-microwave pyrolysis. It can be accounted as a microwave heating prerogative. The heat is provided from microwave absorbent, coal or tires, in close contact with the polymer. This heat is straight shifted from the coal to the surrounding macromolecules.

In figure 7 is reported the steps proposed for acetaldehyde formation.

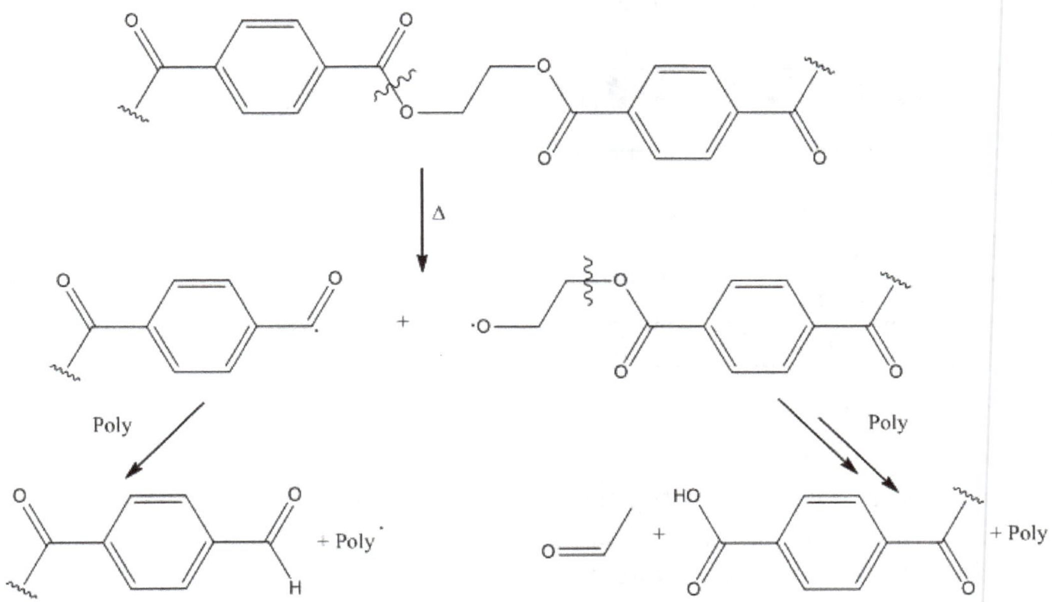

Figure 7. 7. Suggested reactions involved in microwave pyrolysis of PET.

3.5. PS microwave pyrolysis: products characteristics

The polystyrene (PS) is produced and sold as an amorphous polymer or as co-polymers with monomers such as acrylonitrile and butadiene.

The thermal pyrolysis of PS yields the formation of just a liquid product. A limited amount of gas is also collected (Williams & Williams, 1999b; Williams & Bagri, 2004). In the liquid product styrene is found up to 76.2 wt%. The remaining liquid fraction contains other aromatics such as: benzene, toluene, xylenes, ethylbenzene, cumene and α-methylstyrene.

The gas fraction contains: H_2, CH_4, C_2H_6, C_2H_4, C_3H_8, C_3H_6 and C_4H_8 (Williams & Williams, 1999b).

The PS is easy cracked also via microwave heating in the presence of a microwave absorber like with the thermal cracking. The pyrolysis leads to the complete depolymerization of PS to its monomer and other aromatics. The main products present in the liquid fraction are reported in table 18.

The yields are strongly different from a non-microwave pyrolysis (de la Puente & Sedran, 1998). The amount of styrene is lower than 50% but the amount of toluene and other aromatic compounds are higher with respect to non-microwave pyrolysis. These results may be explained only by a different way in which the heat is transferred from the absorber to the polymer.

Table 18. The main compounds present in liquid fraction from microwave pyrolysis of PS.

Compound	Composition (%)
Benzene	1.62
Toluene	8.99
1,2-dimethyl-benzene	8.70
Styrene	47.49
α-**methyl-styrene**	13.39
4-penthyl-benzene	5.06
Total	85.25

4. CONCISE SUMMARY

Microwaves are a valuable instrument that can be utilized in the pyrolysis process of plastics, including HDPE, PP, PVC, PET, and PS, as well as complicated polymeric materials like tires or mixed waste plastics. The utilization of a microwave heating source creates new avenues in waste disposal through pyrolysis, where the yields and properties of the products may be regulated solely by the microwave power employed. The issue of inadequate thermal conductivity in polymers can be addressed using simpler apparatuses, such as static bed reactors. With a straightforward equipment, the reaction time might be significantly decreased from hours to only minutes.

No pretreatments are necessary for the polymeric waste, save for the incorporation of a microwave absorbent. Another significant advantage of tire pyrolysis compared to thermal pyrolysis, for example, is the capability to heat unprocessed materials.

Microwave pyrolysis is an efficient method for converting both simple and complex plastic waste, as well as plastic-containing waste, into new products such as fuels, oils, and significant hydrocarbons like BTX. This process yields pyrolysis gas that is free from combustion gases and nitrogen from the air used in combustion, resulting in gases with a high calorific value.

5. REFERENCES

Arandes J. M. Torre I. Azkoiti M. J. Castaño P. Bilbao J. de Lasa H. 2008 Effect of catalyst properties on the cracking of polypropylene pyrolysis waxes under FCC conditions, Catalysis Today, 133-135 , April-June 2008, 413 419 , 0920-5861

Aylón E. Fernàndez-Colino A. Navarro M. V. Murillo R. Garcìa T. Mastral A. M. 2008 Waste Tire Pyrolysis: Comparison between Fixed Bed Reactor and Moving Bed Reactor, Industrial & Engineering Chemistry Research, 47 , 12 May 2008, 4029 4033 , 1520-5045

Aylón E. Fernàndez-Colino A. Murillo R. Navarro M. V. Garcìa T. Mastral A. M. 2010 Valorisation of waste tyre by pyrolysis in a moving bed reactor, Waste Management, 30 7 July 2010, 1220 1224 , 0000-0956-053X.

Berrueco C. Esperanza E. Mastral F. J. Ceamanos J. Garcìa-Bacaicoa P. 2005 Pyrolysis of waste tyres in an atmospheric static-bed batch reactor: Analysis of the gases obtained, Journal of Analytical and Applied Pyrolysis, 74 1-2 , August 2005, 245 253 , 0165-2370

Bockhorn H. Hornung A. Hornung U. 1998 Stepwise pyrolysis for raw material recovery from plastic waste, Journal of Analytical and Applied Pyrolysis, 46 1 June 1998, 1 13 , 0165-2370

Chen D. T. Perman C. A. Riechert M. E. Hoven J. 1995 Depolymerization of tire and natural rubber using supercritical fluids, Journal of Hazardous Materials, 41 1 November 1995, 53 60 , 0304-3894

Chen S. J. Sua H. B. Chang-En Juu. Leeb W. J. Huanga K. L. Hsieha L. T. Huang Y. C. Lind W. Y. Lin C. C. 2007 Emissions of polycyclic aromatic hydrocarbons (PAHs) from the pyrolysis of scrap tires, Atmospheric Environment, 41 6 February 2007, 1209 1220 , 1352-2310

Cunliffe A. M. Williams P. T. 1998 Composition of oils derived from the batch pyrolysis of tyres, Journal of Analytical and Applied Pyrolysis, 44 2 January 1998, 131 152 , 0165-2370

de la Puente G. Sedran U. 1998 Recycling polystyrene into fuels by means of FCC: performance of various acidic catalysts, Applied Catalysis B: Environmental, 19 3-4 , December 1998, 305 311 , 0926-3373

de Marco Rodriguez. I. Laresgoiti M. F. Cabrero M. A. Torres A. Chomòn M. J. Caballero B. 2001 Pyrolysis of scrap tyres, Fuel Processing Technology, 72 1 August 2001, 9 22 , 0378-3820

Dìez C. Martínez O. Calvo L. F. Cara J. Morán A. 2004 Pyrolysis of tyres. Influence of the final temperature of the process on emissions and the calorific value of the products recovered, Waste Management, 24 5 January 2004, 463 469 , 0095-6053X.

Djeu N. I. 2000 Fiber-optic high temperature sensor, United States Patent, April 2000, US6045259.

Federazione Imprese e Servizi, Unione Nazionale Imprese Recupero [FISE UNIRE]. 2009 L'Italia del recupero- Rapporto FISE UNIRE sul riciclaggio dei rifiuti, 10° edizione, Rimini, Italy.

Freedonia. 2010 World Rubber & Tire to 2013 - Demand and Sales Forecasts, Market Share, Market Size, Market Leaders, in: Freedonia Industry Research, 15.02.2011, Available from: http://www.freedoniagroup.com/DocumentDetails.aspx?ReferrerId=FG-01&studyid=2575.

Giugliano M. Cernuschi S. Ghezzi U. Grosso M. 1999 Experimental Evaluation of Waste Tires Utilization in Cement Kilns, Journal of the Air & Waste Management Association, 49 December 1999, 1405 1414 , 1047-3289

Holland K. M. 1992 Process of destructive distillation of organic material, United States Patent, January 1992, US005084141A.

Holland K. M. 1995 Apparatus for Waste Pyrolysis, United States Patent, February 1995, US005387321A.

Horikoshi S. Matsubara A. Takayama S. Sato M. Sakai F. Kajitani M. Abe M. Serpone N. 2009 Characterization of microwave effects on metal-oxide materials: Zinc oxide and titanium dioxide, Applied Catalysis B: Environmental, 91 1-2 , September 2009, 362 367 , 0926-3373

Hussain Z. Khan K. M. Hussain K. 2010 Microwave-metal interaction pyrolysis of polystyrene, Journal of Analytical and Applied Pyrolysis, 89 1 September 2010, 39 43 , 0165-2370

Johnson A. C. Lauf R. J. Bible D. W. Markunas R. J. 1996 Apparatus and method for microwave processing of materials, United States Patent, May 1996, US005521360A.

Juma M. Koreňovà Z. Markoš J. Annus J. Jelemenský L. 2006 Pyrolysis and combustion of scrap tire, Petroleum & Coal, 48 1 2006, 15 26 , 1335-3055

Kaminsky W. 1992 Plastics, Recycling, In: Ullmann's Encyclopedia of Industrial Chemistry, 57 73 , Wiley-VCH, 3-52730-385-5 Germany

Kaminsky W. Mennerich C. 2001 Pyrolysis of synthetic tire rubber in a fluidised-bed reactor to yield 1,3-butadiene, styrene and carbon black, Journal of Analytical and Applied Pyrolysis, 58-59 , 1 April 2001, 803 811 , 0165-2370

Kaminsky W. Predel M. Sadiki A. 2004 Feedstock recycling of polymers by pyrolysis in a fluidised bed, Polymer Degradation and Stability, 85 3 September 2004, 1045 1050 , 0141-3910

Kaminsky W. 2006 Feedstock Recycling and Pyrolysis of Waste Plastics, John Wiley & Sons, 0-47002-152-7

Kaminsky W. Zorriqueta I. J. N. 2007 Catalytical and thermal pyrolysis of polyolefins, Journal of Analytical and Applied Pyrolysis, 79 1-2 , May 2007, 368 374 , 0165-2370

Kasin K. I. 2009 Microwave gasification, pyrolysis and recycling of waste and other organic materials, United States Patent Application Publication, January 2009, US2009000938A1.

Kim S. 2001 Pyrolysis kinetics of waste PVC pipes, Waste Management, 21 7 2001, 609 616 , 0165-2370

Kyari M. Cunliffe A. Williams P. T. 2005 Characterization of Oils, Gases, and Char in Relation to the Pyrolysis of Different Brands of Scrap Automotive Tires, Energy & Fuels, 19 3 April 2005, 1165 1173 , 0000-1520-5029.

Laresgoiti M. F. Caballero B. M. de Marco I. Torres A. Cabrero M. A. Chomón M. J. 2004 Characterization of the liquid products obtained in tyre pyrolysis, Journal of Analytical and Applied Pyrolysis, 71 2 June 2004, 917 934 , 0165-2370

Li S. Q. Yao Q. Chi Y. Yan J. H. Cen K. F. 2004 Pilot-Scale Pyrolysis of Scrap Tires in a Continuous Rotary Kiln Reactor, Industrial & Engineering Chemistry Research, 43 , 17 June 2004, 5133 5145 , 1520-5045

Ludlow-Palafox C. Chase H. A. 2001 Microwave-Induced Pyrolysis of Plastic Waste, Industrial & Engineering Chemistry Research, 40 , 22 October 2001, 4749 4756 , 1520-5045

Mastral A. M. Murillo R. Callén M. S. Garcìa T. Snape C. E. 2000 Influence of Process Variables on Oils from Tire Pyrolysis and Hydropyrolysis in a Swept Fixed Bed Reactor, Energy & Fuels, 14 4 July/August 2000, 739 744 , 1520-5029

Mastral F. J. Esperanza E. Garcìa P. Juste M. 2002 Pyrolysis of high-density polyethylene in a fluidised bed reactor. Influence of the temperature and residence time, Journal of Analytical and Applied Pyrolysis, 63 1 March 2002, 1 15 , 0165-2370

Meredith R. 1998 Engineers' handbook of industrial microwave heating, The Institution of Electrical Engineers, 0-85296-916-3 United Kindom.

Micromaterials Inc. Tampa F. L. U. S. A. 2010 Fiber Optic Thermometer- Up to 950°C, In: Products, 25.02.2011, Available from: http://www.micromaterialsinc.com/specsSensor.html.

Mui E. L. K. Ko D. C. K. Mc Kay G. 2004 Production of active carbons from waste tyres-a review, Carbon, 42 14 2004, 2789 2805 , 0008-6223

Natural Resources Defense Council [NRDC]. (February 2008 The 3r's Still Rule, In: This Green Life, 20.02.2011, Available from: http://www.nrdc.org/thisgreenlife/0802.asp.

Pantea D. Darmstadt H. Kaliaguine S. Roy C. 2003 Heat-treatment of carbon blacks obtained by pyrolysis of used tires. Effect on the surface chemistry, porosity and electrical conductivity, Journal of Analytical and Applied Pyrolysis, 67 1 March 2003, 55 76 , 0165-2370

Parker T. H. 1992 Apparatus for pyrolysis of tires and waste. United States Patent, December 1992, US005167772A.

Piskorz J. Majerski P. Radlein D. Wik T. Scott D. S. 1999 Recovery of Carbon Black from Scrap Rubber, Energy & Fuels, 13 3 March 1999, 544 551 , 1520-5029

Plastics Europe. 2010 Plastics- the Facts 2010 An analysis of European plastics production, demand and recovery for 2009, In: Plastic Market, 09.03.2011, Available from: http://www.plasticseurope.org/documents/document/20101028135906 -final_plasticsthefacts_26102010_lr.pdf.

Pringle F. G. 2007 Microwave-based recovery of hydrocarbons and fossil fuels, United States Patent Application Publication, June 2007, US20070131591A1.

Pringle J. A. 2006 Microwave pyrolysis apparatus for waste tires, United States Patent, September 2006, US007101464B1.

Rubber Manufacturers Association [RMA]. 2011 Scrap Tire Characteristics, In: Scrap Tires and the Environment, 28.02.2011, Available from http://www.rma.org/scrap_tires/scrap_tire_markets/scrap_tire_characteristics/

Saeed L. Tohka A. Haapala M. Zevenhoven R. 2004 Pyrolysis and combustion of PVC, PVC-wood and PVC-coal mixtures in a two-stage fluidized bed process, Fuel Processing Technology, 85 14 September 2004, 1565 1583 , 0378-3820

Scandinavian Biofuel Company [SBC]. 2011 Microwave Assisted Pyrolysis, In: Microwave pyrolysis, 09.03.2011, Available from: http://www.sbiofuel.com/pyrolysis.html.

Sharma V. K. Mincarini M. Fortuna I. F. Cognini F. Cornacchia G. 2000 Disposal of waste tyres for energy recovery and safe environment-review, Energy Applied Energy, 65 1-4 , April 2000, 381 394 , 0306-2619

Tang L. Huang H. 2004 An investigation of sulfur distribution during thermal plasma pyrolysis of used tires, Journal of Analytical and Applied Pyrolysis, 72 1 August 2004, 35 40 , 0165-2370

Westerhout R. W. J. Waanders J. Kuipers J. A. M. van Swaaij W. P. M. 1998 Development of a Continuous Rotating Cone Reactor Pilot Plant for the Pyrolysis of Polyethene and Polypropene, Industrial & Engineering Chemistry Research, 37 6 May 1998, 2316 2322 , 1520-5045

Williams P. T. Besler S. Taylor D. T. 1990 The pyrolysis of scrap automotive tyres: The influence of temperature and heating rate on product composition, Fuel, 60 12 December 1990, 1474 1482 , 0016-2361

Williams P. T. Bottrill R. P. Cunliffe A. M. 1998 Combustion of Tyre Pyrolysis Oil, Process Safety and Environmental Protection, 76 4 November 1998, 291 301 , 0957-5820

Williams P. T. Williams E. A. 1999a Fluidised bed pyrolysis of low density polyethylene to produce petrochemical feedstock, Journal of Analytical and Applied Pyrolysis, 51 1-2 , July 1999, 107 126 , 0165-2370

Williams P. T. Williams E. A. 1999b Interaction of Plastics in Mixed-Plastics Pyrolysis, Energy & Fuels, 13 1 1999, 188 196 , 1520-5029

Williams P. T. Bagri R. 2004 Hydrocarbon gases and oils from the recycling of polystyrene waste by catalytic pyrolysis, International Journal of Energy Research, 28 1 January 2004, 31 44 , 0036-3907X.

Yoshioka T. Grause G. Eger C. Kaminsky W. Okuwaki A. 2004 Pyrolysis of poly(ethylene terephthalate) in a fluidised bed plant, Polymer Degradation and Stability, 86 3 December 2004, 499 504 , 0141-3910

Zabaniotou A. Madau P. Oudenne P. D. Jung C. G. Delplancke M. P. Fontana A. 2004 Active carbon production from used tire in two-stage procedure: industrial pyrolysis and bench scale activation with H_2O-CO_2 mixture, Journal of Analytical and Applied Pyrolysis, 72 2 November 2004, 289 297 , 0165-2370

PART 5
SYNTHESIS

11

Chapter

MICROWAVE HEATING FOR THE SYNTHESIS OF SIC POWDERS AND WHISKERS

1. INTRODUCTION: AN OVERVIEW

Silicon carbide (SiC) is extensively utilized in industry as both a structural and functional material owing to its exceptional hardness, strength, superior wear resistance, lightweight nature, outstanding oxidation and corrosion resistance, reduced thermal expansion coefficient, and high thermal conductivity. The main method for the production of SiC powders (SiC*p*) in industry is known as the Acheson process, in which SiC ingots are synthesized at about 2200 °C through the carbothermal reduction of quartz sand by coke lasting for a few days in general(Krstic, 1992), following with cracking, grinding, and meshing, etc.. It is energy- intensive and difficult to produce fine powders with a narrowed size distribution and controllable morphology. This process has been modified as so called rapid carbothermal reduction (RCR) to produce fine and pure SiC powders(Setiowati & Kimura, 1997). Unlike conventional carbothermal reduction process, the aerosol consisting of ultrafine carbon and silica precursors have been used as starting materials in the RCR process(Weimer et al., 1991; Weimer et al., 1993; Weimer et al., 1994), so that the carbothermal reduction can be completed in a few seconds through the reactor tube by coupling the thermal radiations and the fast kinetics. It indicates the probability that SiC could be synthesized in a very short time with appropriate methods.

Therefore, new methods, such as sol-gel(Hatakeyama & Kanzaki, 1990; Seog & Kim, 1993), plasma(Guo et al., 1995), CVD(Klein et al., 1998; Halamka et al., 2003), and laser(Li et al., 1994) were attempted during the last few decades to synthesize fine SiC powders with narrow particle size distribution and high purity. However, these processes have not satisfied the industry requirement of the quick production of ultra-fine SiC powders and whiskers due to its complex process, high cost and less productivity.

Compared with that of the other methods, the microwave heating has been demonstrated to be an attracting method in synthesis of ceramic materials(Rao et al., 1999) because of its rapid and bulk heating, selective coupling, improvement of heat transfer, enhanced reaction kinetics, acceleration of the reaction rate and these all results in quick synthesis of ultra-fine SiC powders(Satapathy et al., 2005). The basic idea behind microwave heating is that, it transfers heat from inner to surface of materials and as a result the temperature of the inner part will be higher than that of the other parts and the first the reaction will begin in the inner core. Thus a significant reduction in the reaction temperature and thermal processing time will be achieved, which is energy-effective and economical.

SiC whiskers (SiC*w*) are widely used in reinforcing metallic matrix and ceramic matrix composites. At present, solid phase synthesis method is used for the industrial production of SiC*w*, by using Fe, Ni and NaF etc. as catalyst via VLS mechanism. However this method will introduce metallic impurities which lead to lower the quality of SiC*w* by forming stacking fault, dislocation, twin boundary, rough surfaces and nodular structure. In addition, CVD, which use gaseous precursor or template (carbon nanotube) to prepare SiC*w*, is also a common method; however, low yield and complex equipment are still the major disadvantage. As the inner part of the reactants will be firstly heated to the reaction temperature under the microwave irradiation, the reactions begin at the inner will increase the concentration of gaseous phase reactants, and benefit the synthesis of SiC*w* by VS or VLS mechanisms.

In this chapter, the effect of the silicon sources and carbon sources, reaction temperature and holding time on the synthesis of SiC powders and whiskers have been discussed and reviewed mainly from our published papers and dissertations (Kuang et al, 2011; F. Wang, 2008a; F. Wang et al, 2008b, 2009; Q. Wang 2009;) and the cited works from the other groups..

2. SYNTHESIS OF SIC POWDERS

Various carbon sources were chosen to examine their impact on the particle size and morphologies of the generated SiC powders, as detailed below.

2.1 Experimental

Silicon powders and different carbons (their chemical and physical properties are listed in Table 1) were mixed as starting materials at the Si/C mol-ratio of 1/1. The mixtures were well homogenized by ball-milling in a polyethylene milling jar with alumina balls, ratio of balls to powders was 3/1, with the ball milling time of 4 h, and alcohol was used as the dispersant. The prepared mixtures were dehydrated at 110 °C for 24 h. The obtained powders were charged into alumina crucible, and then heated in a 2.45 GHz microwave filed in flowing Ar atmosphere with pressure of 0.1 MPa. The schematic diagram of heating system is shown in Fig. 1. The output microwave power is adjusted automatically ranged from 0~4 kW. The as-prepared products were decarbonized at 650°C for 2 h.

Table 1. Chemical composition and physical properties of silicon and carbon black

Powder	element (wt %)							O_2 (wt %)	BET (m^2/g)	grain size (μm)
	Si	Fe	Al	Ca	Sn	C	S			
Silicon	99.1	0.18	0.11	0.04	0.20	-	-	2.17	1.32	< 44 μm
activated carbon	-	-	-	-	-	>99.5	-	-	-	1~30 μm
Graphite	-	-	-	-	-	>99.5	-	-	-	1~20 μm
carbon black	-	-	-	-	-	96.2	3.8	6.28	76	μm

2.2 Synthesis of SiCp by silicon and activated carbon powders

Silicon powders and activated carbon are used as starting materials to synthesize SiC powders whose detail is given in the incoming lines.

The SEM images of the starting activated carbon, silicon powders and their mixtures after ball milling are shown in Fig. 2. Activated carbon particles are irregular in shape and wide in particle size distribution. The size was distributed from about 1 to 30 μm. While that of the silicon particles with sharp angle is about 50 μm. After ball milling, the shape of the mixture powders is still irregular with sharp angles and edges and its size has been reduced to less than 20 μm.

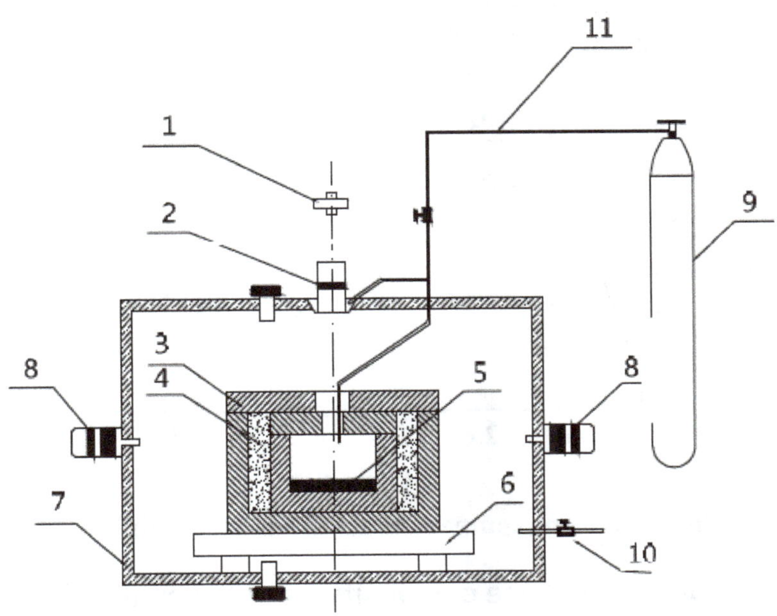

Fig. 1. Schematic diagram of the microwave furnace. 1- infrared radiation thermometer; 2- silica glass; 3- porous alumina crucible; 4- alumina fiber; 5- reactants; 6- alumina plate; 7- stainless steel chamber; 8- microwave generator; 9- argon gas tank; 10- outlet valve; 11- inlet valve.

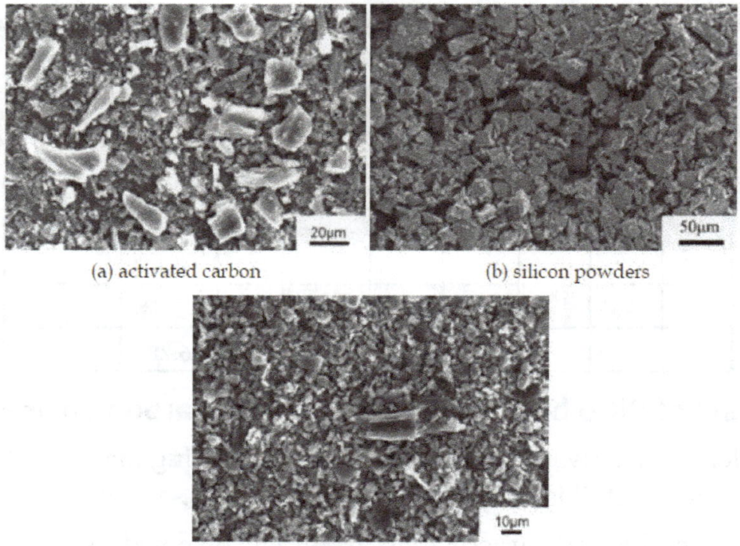

(a) activated carbon (b) silicon powders

Fig. 2. SEM images of the starting materials

The heating temperature has been varied from 900 to 1200°C (100°C interval) and holding for 5 to 30 min. The curve of the heating temperature vs. heating time is shown in Fig. 3.

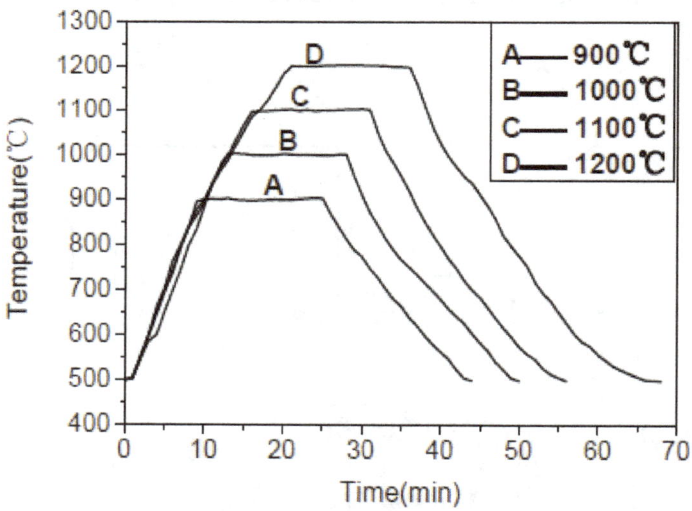

Fig. 3. Heating temperature (°C) vs. time (min)

2.2.1 Identification of the phase compositions of the synthesized products

Schematic composition structure of the synthesized product is illustrated in Fig. 4. After completion of the reaction, the color of the top and inner parts of the synthesized product was changed into dark green (part A in Fig. 4), however that of the parts closed to the wall and bottom of the crucible was still black (part B in Fig. 4).

The X-ray diffraction peaks reveal that the black parts are unreacted silicon and carbon black. Microwave heats from inside of the reactants and is driven to the surface. This is reverse of the conventional heating, where the surface of the reactants is heated first (Metaxas, 1991; Shawn et al., 2008). Heat loss will happen at the sidewall and the bottom of the crucible due to the heat transfer, which further decreases the heating temperature and leads to the uncompleted reaction. However, the heat will transfer upwards and thus the reaction temperature difference between that of the top part and the inner part is very small. Then the reaction on the top surface also can complete. Thus, only the inner part of the products was used for further characterization in the later experiment.

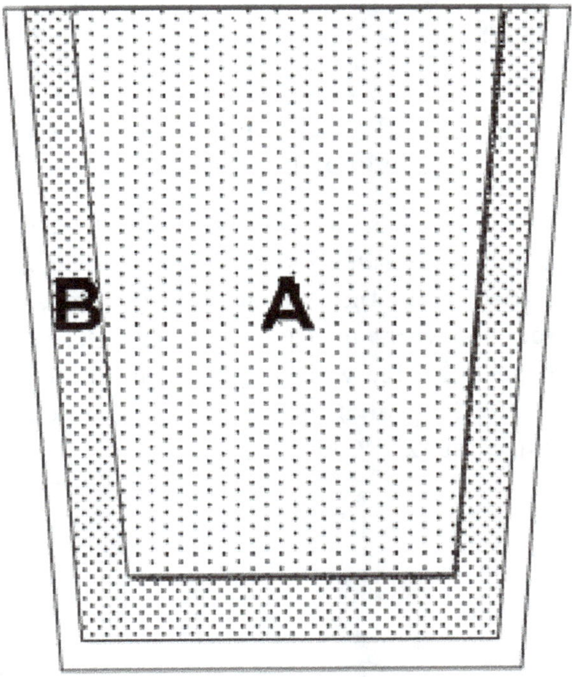

Fig. 4. Schematic composition structure of the product synthesized in the crucible

Figure 5 shows the XRD patterns of the products synthesized at different heating temperatures holding for 15 min. In the case of 900°C, β-SiC was formed but the reactant Si still could be identified in the products due to the uncompleted reaction. The content of β- SiC in the products was increased with the increase of the heating temperature while that of the unreacted Si decreased. When the heating temperature was increased to 1200°C, pure β- SiC powders were synthesized successfully.

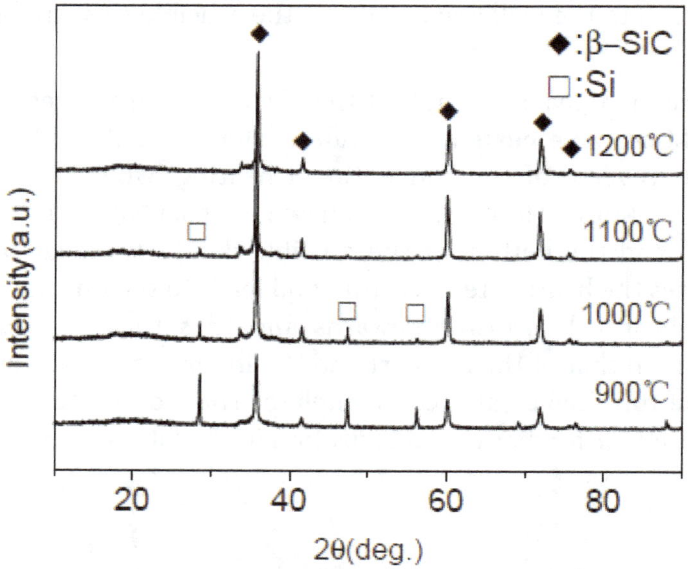

Fig. 5. XRD patterns of the products synthesized at different heating temperatures holding for 15 min

The effect of holding time on the synthesis of SiC at 1200 °C was investigated. The XRD patterns of the products synthesized at 1200°C holding for different time are shown in Fig. 6. When the holding time was set as only 5 min, no unreacted Si could be detected in the products, which means the reaction has proceeded completely. The crystallinity and particle size will be further increase with the increase of holding time according to the peak intensity. It can be concluded that pure phase β-SiC could be synthesized at 1200°C holding for only 5 min.

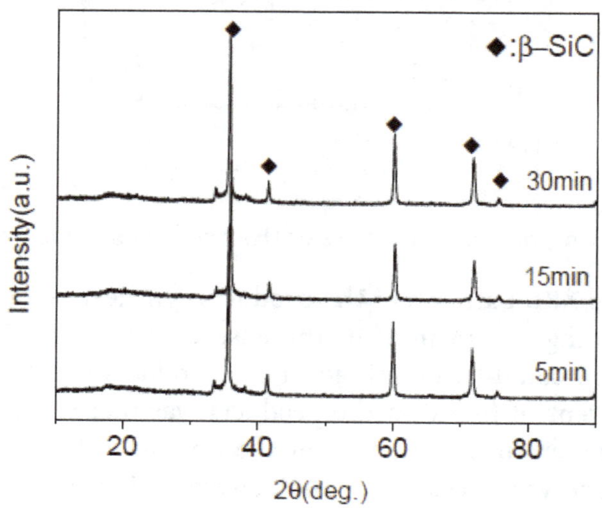

Fig. 6. XRD patterns of the products synthesized at 1200°C holding for different time

2.2.2 Effect of reaction temperature on the morphology of the synthesized products

The effect of reaction temperature on the morphology of the products has been investigated. The SEM images of the products prepared at different reaction temperatures holding for 15 min are shown in Fig. 7. In the case of 900 °C, similar morphology with that of the starting materials can be seen in the products due to the uncompleted reaction. There are still some unreacted silicon and carbon particles, which is consistent with that from the X-ray diffraction analysis. When the reaction temperature was increased to above 1000 °C, there were large number of agglomerated SiC particles in the products and a little unreacted silicon particles could be found. Agglomerated pure SiC particles with size of 1~10 μm have been synthesized when the reaction temperature was increased to 1200°C.

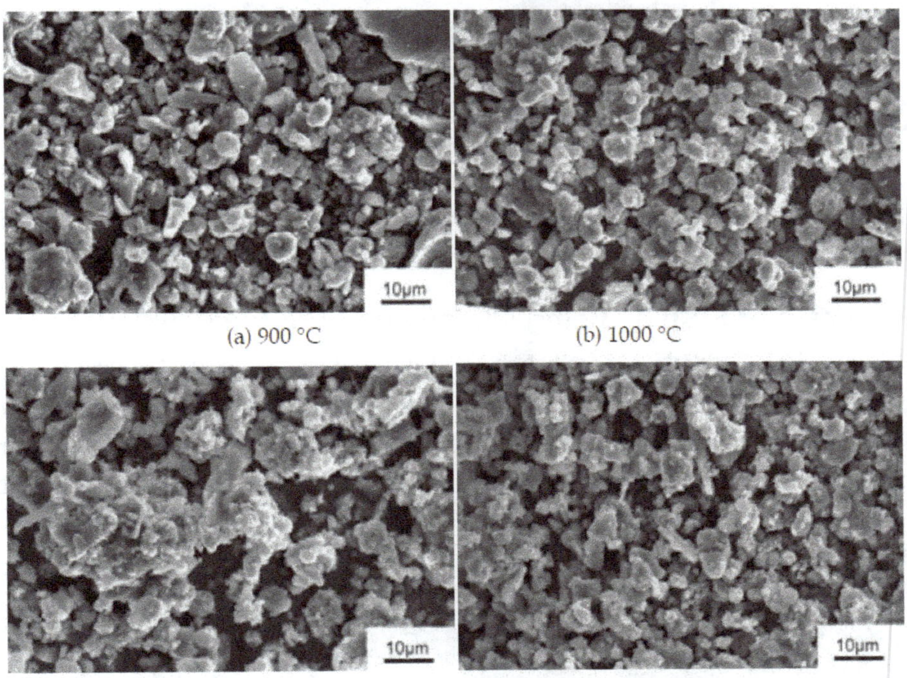

(a) 900 °C (b) 1000 °C (c) 1100 °C (d) 1200 °C

Fig. 7. SEM images of products synthesized at different heating temperatures holding for 15 min

2.2.3 Effect of heating time on the morphology of the synthesized products

The SEM images of the products prepared at 1200 °C holding for different time are shown in Fig. 8. The shape of the SiC particles tends to be equiaxial with the increase in the holding time. When holding for 30 min, equiaxial particles with size of about 1~10 μm were synthesized.

(a) 5 min (b) 15 min

(c) 30 min

Fig. 8. SEM images of the products synthesized at 1200°C holding for different time

2.3 Synthesis of SiCp by using silicon and graphite powders

Figure 9 shows the SEM images of the graphite and the as-milled mixture powders. The average size of the original graphite particles is 1~20 μm and that of the mixture was decreased to about ~10 μm after ball milling. Both the original and the ball-milled particles have sharp angles and edges due to their brittle properties.

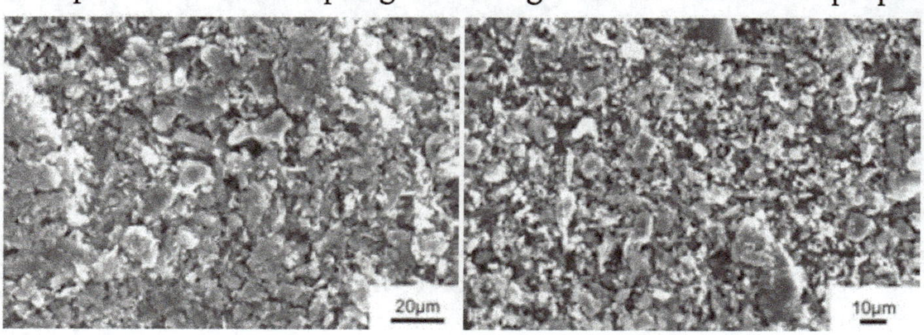

(a) (b)

Fig. 9. SEM images of (a) graphite and (b) mixtures of graphite and silicon powders

2.3.1 Identification of phase compositions of the synthesized products

The XRD patterns of the products prepared at various heating temperatures holding for 15 min are shown in Fig. 10. All the peaks are corresponding to those of the β-SiC. Pure phase β-SiC can be synthesized when the heating temperature was as low as 900°C. While with the further increase of the heating temperature, the peak intensity of the synthesized SiC increased, which was assigned to the increase in grain size and crystallinity. It was found that the heating temperature has positive impact on the reaction. Higher heating temperature benefits the progress of reaction. Figure 11 shows the XRD patterns of products reacted at 1200°C holding for different heating time. The results show that even the holding time was as less as 5 min, the reactants could be completely converted into pure β-SiC. And the further increase in the holding time gives the increase in the crystallinity and grain size.

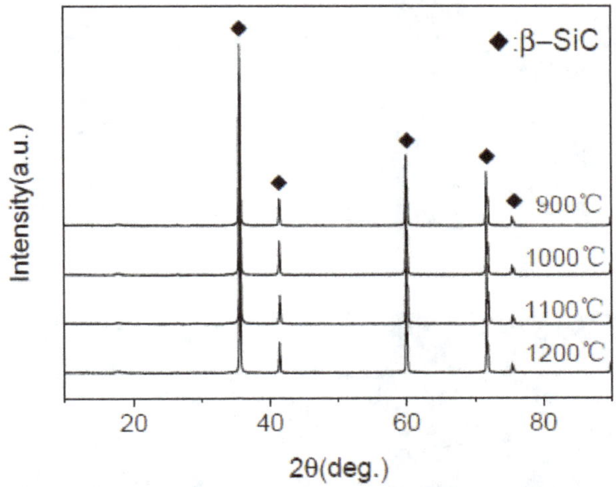

Fig. 10. XRD patterns of the products synthesized at different heating temperatures holding for 15 min

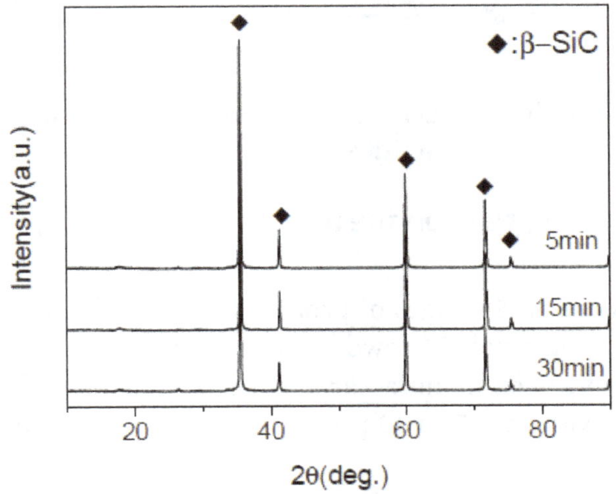

Fig. 11. XRD patterns of the products synthesized at 1200°C holding for different time

2.3.2 Effect of reaction temperature on the morphology of the synthesized products

The SEM images of the products prepared at different heating temperatures holding for 15 min are shown in Fig. 12. It is clear that the particle size of the synthesized SiC increased with the increase of the heating temperature. The synthesized particles tend to be equiaxial and have a narrower size distribution with the increase of the heating temperature. In the case of 1200°C, equiaxed SiC powders with averaged size of about 5 µm were prepared.

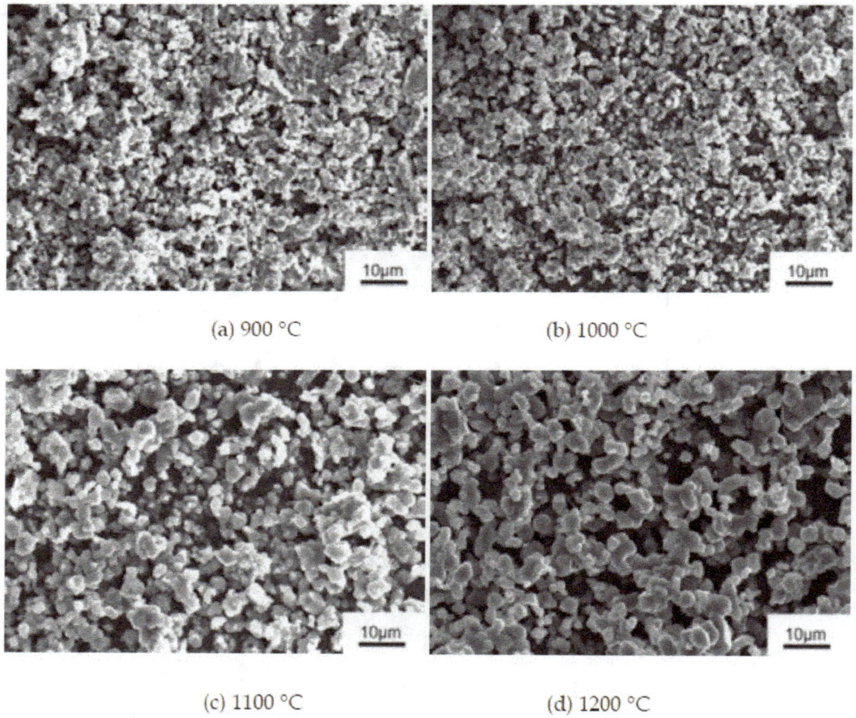

Fig. 12. SEM images of the products synthesized at different heating temperatures holding for 15 min

2.3.3 Effect of heating time on the morphology of the synthesized products

Figure 13 displays the SEM images of products synthesized at 1200 °C for varying durations. Uniform equiaxial SiC powders with enhanced particle size and reduced size distribution have been manufactured with prolonged holding time. Equiaxed SiC powders, approximately 5 to 10 µm in size, were produced after a 30-minute holding period..

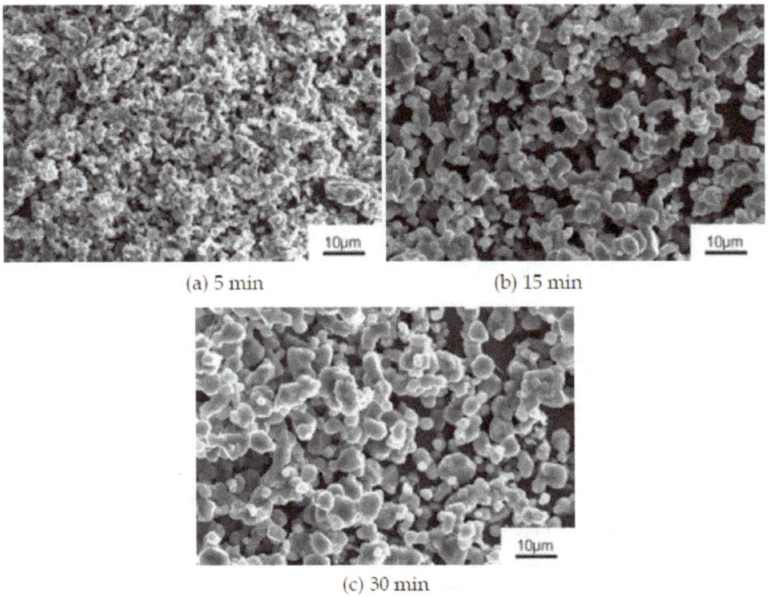

Fig. 13. SEM images of the products synthesized at 1200°C holding for different time

2.4 Synthesis of SiCp using silicon powders and carbon black

The SEM images of the carbon black and the mixtures with silicon after ball milling are indicated in Fig. 14. The averaged particle size of the carbon black is about 500 nm. After ball milling, the averaged particle size of the silicon in the mixtures has been decreased to about 10 μm. The silicon particles, which still have sharp angles, have been dispersed homogeneously in the carbon black after ball milling.

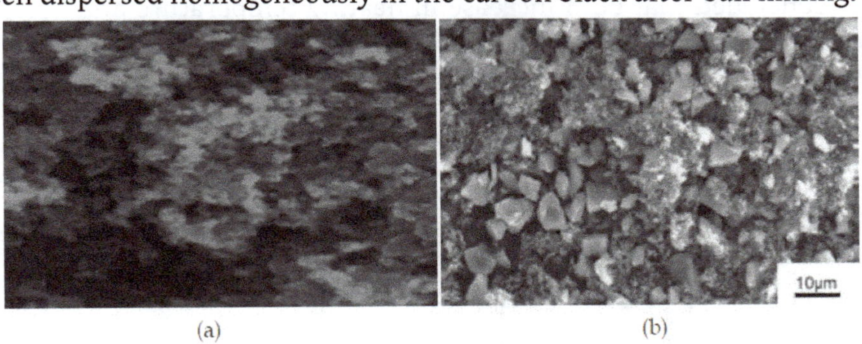

Fig. 14. SEM images of (a) carbon black and (b) mixtures of carbon black and silicon powders

2.4.1 Identification of phase compositions of the synthesized products

Figure 15 shows the XRD patterns of the products synthesized at various heating temperatures holding for 30 min. β-SiC starts to form at the heating temperature of

900°C. There are still some unreacted silicon can be detected in the products. When it was increased to 1100°C, all the diffraction peaks are indexed to β-SiC, which means that pure phase β-SiC was prepared at 1100°C for 30 min. Further increasing the heating temperature to 1200°C, the increased intensity of the diffraction peaks was attributed to the grain growth and improved crystallinity.

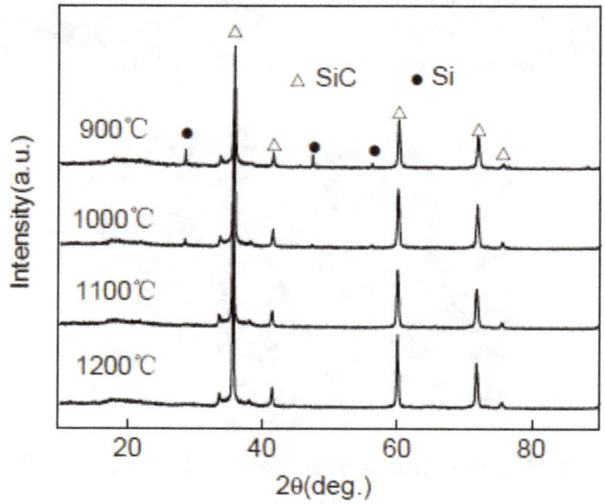

Fig. 15. XRD patterns of the products synthesized at various heating temperatures holding for 30 min

XRD patterns of the products synthesized at 1200°C holding for various time are shown in Fig. 16. The major composition in the product was β-SiC and a small amount of unreacted silicon was also detected when heated for 5 min. Pure phase SiC was obtained when the holding time was set as 15 min and 30 min.

Thus it is concluded that both increase in heating temperature and holding time can promote the formation of β-SiC.

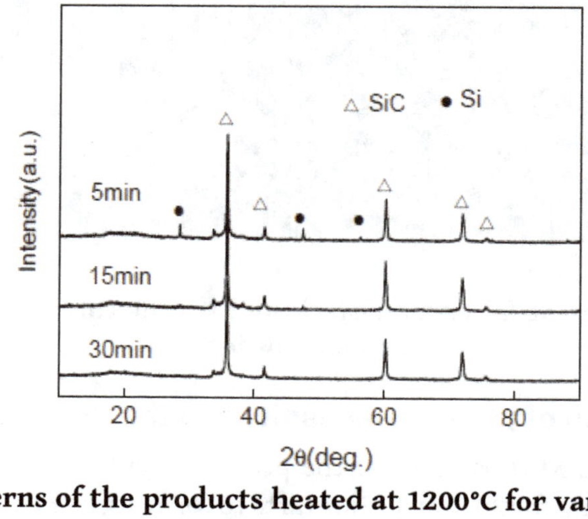

Fig. 16. XRD patterns of the products heated at 1200°C for various holding time

2.4.2 Effect of reaction temperature on the morphology of the synthesized products

Figure 17 shows that the SEM images of the products synthesized at various heating temperatures holding for 30 min. It is clear that all the products were composed of particles with irregular shapes. In the case of 900 and 1000 °C, the particle size was about several hundred nanometers. With the increase in heating temperature to 1100°C an obvious grain growth was found, further increase to 1200°C gives particles having size of 0.5~2 μm.

(a) 900 °C (b) 1000 °C

(c) 1100 °C (d) 1200 °C

Fig. 17. SEM images of the products synthesized at various heating temperatures holding for 15 min

2.4.3 Effect of the heating time on the morphology of the synthesized products

Figure 18 shows the SEM images of the products prepared at 1200°C for various holding time. It can be seen that when the holding time was increased from 5 min to 30 min the particle size of the products rapidly increased from 0.1~1 μm to 1~3 μm, and became narrower in particle size distribution. The above results indicate that increasing heating temperature and extending holding time can also promote the growth of SiC particles.

Under high-temperature SiC sublimates to form Si, Si_2C and SiC_2 gas, and the crystal growth depends on the coacervation of these gaseous phases. These gas phases transfer from high- temperature part to low-temperature part and then condense on the surface of SiC crystal; or directly condense on the nearby larger crystal and promote the growth of crystal. The relationship between saturated vapor pressure, surface tension and radius of curvature is shown in Eq. (1) (Young-Laplace equation). The smaller crystal has the larger saturated vapor pressure which means that under the same vapor pressure, it is saturated for the large crystal, but it must be under saturated for the smaller one. Based on this phenomenon, with the increase in the reaction time, the smaller crystal will evaporate continuously until it disappears, however the larger crystal will continue to grow to form large crystal(Tairov, 1995).

$$P = 2\gamma / R \tag{1}$$

(a) 5 min (b) 15 min

(c) 30 min

Fig. 18. SEM images of the products synthesized at 1200°C holding for various time

2.4.4 Effect of the silicon particle size on the morphology of the synthesized products

There are two mechanisms about the effect of particle size of the reactants on the morphology of the synthesized SiC. The first one is dissolution-precipitation mechanism which considers that the morphology of the synthesized SiC is determined by that of silicon(Yamada et al., 1989), and the second is so called diffusion mechanism which believes that the morphology of the SiC is controlled by carbon(Pampuch et al., 1987; Narayan et al., 1994).

Silicon powder with particle size of ~44 μm and 15 μm were tried with the same kind of carbon black to synthesize SiC powders at 1200°C for holding 15 min. SEM images of the silicon powders, carbon black (500 nm) and the corresponding products are shown in Fig. 19. It is clear that the particle size of the as-prepared SiC was about 1~2 μm in both cases, which suggests that the morphology of the SiC was dominated by that of carbon by diffusion mechanism as the size was much more close to that of the carbon black.

Fig. 19. SEM images of (a) ~44 μm and (c) ~ 15 μm silicon powders; products prepared by (b) ~44 μm and (d) ~15 μm silicon powders and acetylene carbon black at 1200°C for 15 min

Synthesis of SiC by quartzite and coke powders

In order to further reduce the cost of the synthesis of SiC, quartzite and coke were selected as raw materials whose chemical compositions are listed in Table 2 and Table 3, respectively. The proportion of reactant is strictly based on Eq. (2). There were two kinds of particle size of coke, less than 20 μm and 100~200 μm, each one mixed with ~1 μm quartzite, and then pressed under 15 MPa to form compacts. The heating temperature was varied from 1300 to 1600°C, holding time changed from

15 to 60 min and the heating rate was set as 50°C/min.

Table 2. Chemical composition of quartzite

Composition	SiO_2	Fe_2O_3	Al_2O_3	CaO	MgO	Burn reduces
Content (%)	>98.8	<0.3	0.24	<0.16	<0.2	<0.2

Table 3. Chemical composition of coke

Composition	Fixed carbon	Ash	Volatile	Water
Content (%)	82.5	15.7	1.1	0.8

$$SiO_2(s) + 3C(s) = SiC(s) + 2CO(g) \quad (2)$$

2.5.1 Identification of phase compositions

The coke with the particle size of less than 20 μm was selected to investigate the effect of reaction temperature and holding time on the phase compositions. Figure 20 shows the XRD patterns of products heated at various temperatures holding for 30 min. In the case of 1300°C, SiO_2 **is the major composition and very small amount of** β-SiC could be identified; with the increase in reaction temperature, the relative amount of β-SiC increased, and that of SiO2 decreased; when the reaction temperature was 1600°C holding for 30 min, pure phase β-SiC was synthesized.

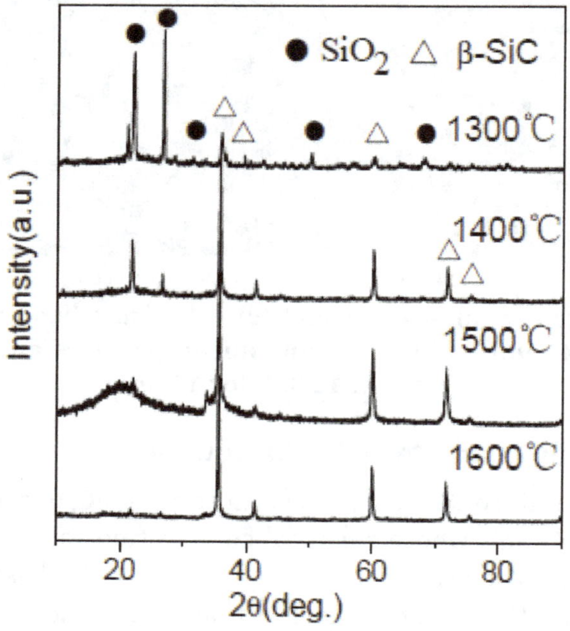

Fig. 20. XRD patterns of products synthesized at various temperatures holding for 30 min

Holding time is one of the most important factors which affect the conversion of the SiC. Different holding times have been tried to investigate their effect on the progress of reaction. XRD patterns of the products synthesized at 1500°C holding for various time are shown in Fig. 21.

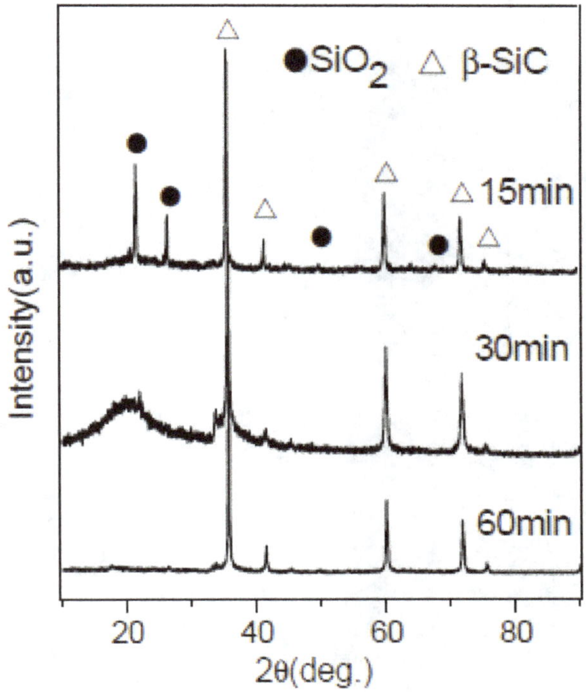

Fig. 21. XRD patterns of products synthesized at 1500°C holding for various time

It is clear that when the holding time was set as 15 or 30 min, there are still some unreacted SiO_2 **could be detected in the products. But the content of SiO_2 was decreased with the increase of the holding time, which suggests that the holding time should be further increased. At the case when the holding time was increased to 60 min, pure** β-SiC has been synthesized by XRD. This heating temperature and holding time are remarkably lower than that of the conventional Acheson process.

2.5.2 Effect of coke particle size on the morphology of SiC

The effect of the particle size of coke on the morphology of the synthesized products has been investigated. Figure 22 shows SEM images of coke particles with size of less than 20 μm and 100~200 μm, and the products prepared by using them with ~1 μm quartzite heated at 1600°C for holding 30 min. The synthesized SiC has almost the similar morphology and particle size as those of the coke. Size distribution of SiC powders prepared from coke with various particle size is shown in Table 4. The D50 value of SiC is very close to that of the corresponding coke, which was considered that the morphology and size of the SiC were controlled by those of coke through the diffusion mechanism(Pampuch et al., 1987).

Thus it can be concluded that in the microwave synthesis process the morphology and particle size of the synthesized SiC*p* could be easily controlled by adjusting the corresponding coke size.

Fig. 22. SEM images of (a) 20 μm and (c) 100~200 μm coke powders; products prepared by (b) 20 μm and (d) 100~ 200 μm silicon powders and carbon black at 1600°C for 30 min

Table 4. Size distribution of SiC prepared with various grain size of coke

No.	Temperature/ oC	Holding time/min	D50 of coke/μm	D50 of product/μm
1	1600	30	13.795	8.604
2	1600	30	154.235	156.907

Except our works, the data from the researches on the microwave synthesis of SiC*p* by using other silicon and carbon sources are listed in Table 5. These works also reveal that microwave heating can lower the synthesis temperature and reduce the reaction time.

Table 5. Synthesis of SiC*p* by microwave heating. 1-(Ramesh et al., 1994); 2-(Dai et al., 1997b); 3-(Satapathy et al., 2005); 4-(Ebadzadeh & Marzban-Rad, 2009); 5-(Jin et al., 2010).

Silicon source	Carbon source	Form temperature	Complete reaction	Products	Morphology
silicon powder[1]	activated charcoal	<977°C	-	β-SiC	-
precipitated silica[2]	char resin carbon black	1180°C for 20 min	-	β-SiC and a little α-SiC	3~80 nm
Silicon powder[3]	amorphous carbon powder	900°C for 30 min	1200°C for 15 min	agglomerated β-SiC particles	~20 μm
silica gel[4]	graphite and carbon black	1225°C	1454°C	equiaxed β-SiC particles	13.6 and 58.2 nm
silicon powder[5]	carbon Black	-	1200°C for 15 min	homogeneous β-SiC particles	1 μm

3. PREPARATION OF SIC WHISKERS

3.1 Preparation of SiCw by silica-sol and activated carbon

As shown in Fig. 23, the temperature of the inner part of the reactants is always higher than that of other parts under microwave irradiation, which means the reaction will be initiated firstly inside the reactants, results in the reduce of the loss of the produced gaseous intermediates. Then the gaseous intermediates could be maintained with a higher concentration in the inner part, which benefits the synthesis of SiCw by VS or VLS mechanisms and the increase of the SiCw productivity.

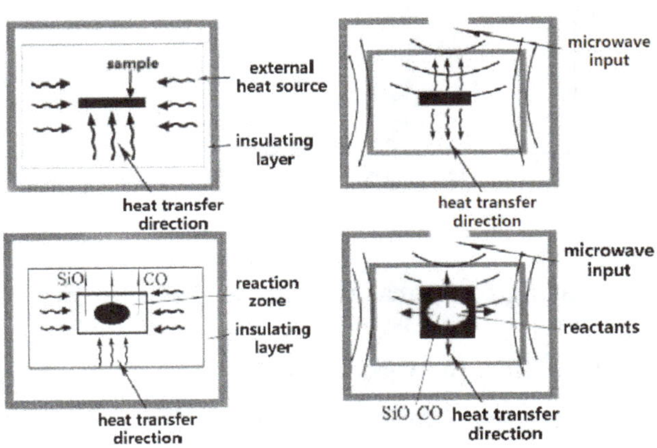

Fig. 23. Conventional heating vs. microwave heating

This section will discuss the synthesis of SiC*w* by using silica-sol as silicon source and activated carbon (or carbon black) as carbon source by microwave heating without the use of any catalysts, such as Fe, Ni, NaF, etc..

3.2 Experimental

Due to the ultrafine silica particles dispersed in silica sol could be coated well on the surface of the carbon particles, thus the silica and carbon could be uniformly mixed. So silica sol is considered as one of the excellent silica sources for the synthesis of SiC*w*. The content, particle size of SiO_2 in silica sol is about 30% and 10~20 nm, respectively. Certain amount of silica sol and activated carbon were mixed and stirred for 2 h as starting materials. The Si/C molar ratio was set as 1/4. 1 *wt*% sodium hexametaphosphate was used as the dispersant for better mixing. The prepared mixtures were dehydrated at 110 °C for 24h, and then crushed and screen separated into three kinds of powders with size distribution of 200-300 μm, 30-80 μm and 1-10 μm. The powders were charged into alumina crucibles and heated in a microwave furnace. The heating temperature varied from 1200 to 1600°C (100°C interval), while holding time changed from 5 to 60 min. The heating temperature vs. heating time is shown in Fig. 24. The as-prepared products were decarbonized in air at 650°C for 2 h.

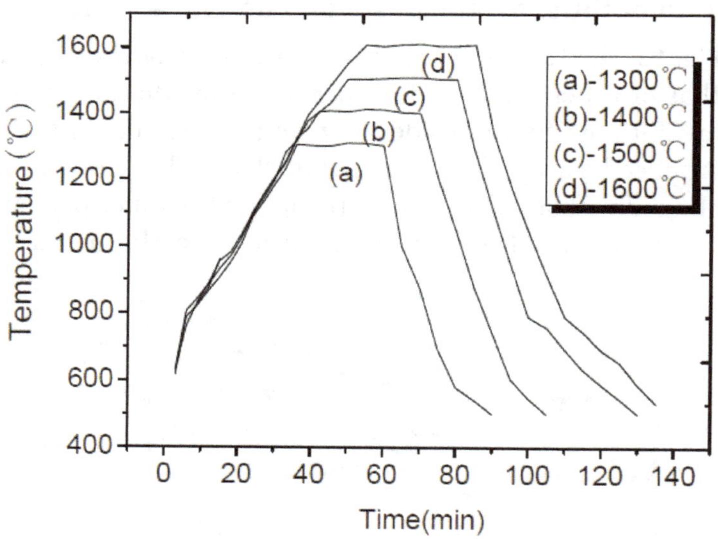

Fig. 24. Heating temperature (oC) vs. heating time (min)

3.3 Identification of phase compositions

The 1-10 μm mixtures were used to investigate the effect of processing parameters on the phase compositions of the synthesized products. The XRD patterns of different parts of the product in the crucible heated at 1500°C for 15 min were shown in Fig. 25. The patterns show that different parts of the products were composed with different compositions. The bottom part consists of unreacted SiO2 and very small amount of β-SiC; the inner part consists of pure phase β-SiC without any SiO_2; while the top part consists of β-SiC and very small amount of SiO_2. This phenomenon is similar with that of the synthesis of SiC*p* as mentioned before. Thus, only the inner parts of the products synthesized under different conditions were selected for further characterization in the followed investigates.

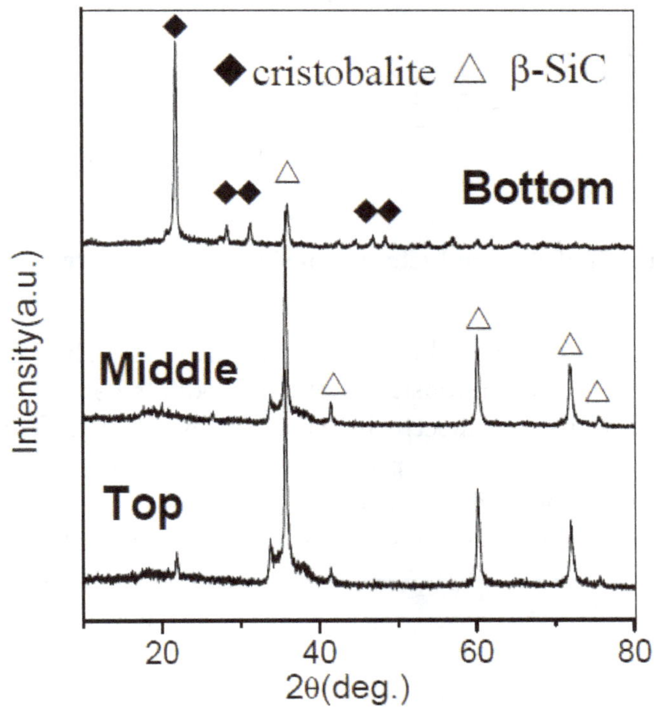

Fig. 25. X-ray diffraction patterns of the different parts of the products synthesized at 1500°C for 15 min

The X-ray diffraction patterns of the products heated at different temperatures for 15 min are shown in Fig. 26. In the case of 1200°C, SiO_2 is the main composition phase and very small amount of β-SiC could be detected, which shows that the reaction rate is slow and it is hard to synthesize SiC at this reaction temperature by microwave heating. The peak intensity of β-SiC increased and that of SiO2 decreased with the increase of reaction temperature. Pure β-SiC could be prepared at 1500°C holding for 15 min. While when the reaction temperature was increased to 1600°C, the crystallinity of β-SiC has been remarkably increased.

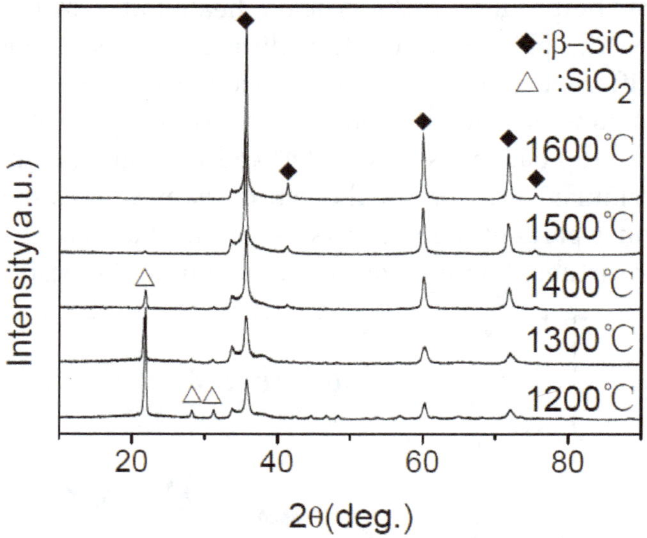

Fig. 26. XRD patterns of the products synthesized at various temperatures holding for 15 min

The effect of the holding time on the synthesis of SiC at 1500°C has been investigated. Figure 27 shows the phase compositions of the products heated at 1500°C holding for 5, 10, 15, and 30 min, respectively. The content of SiO2 decreased with the increase of the holding time. It is impossible to complete the reaction when holding for less than 15 min. The pure phase β-SiC could be synthesized when holding for at least 15 min. The further increase of the holding time will improve the crystallinity.

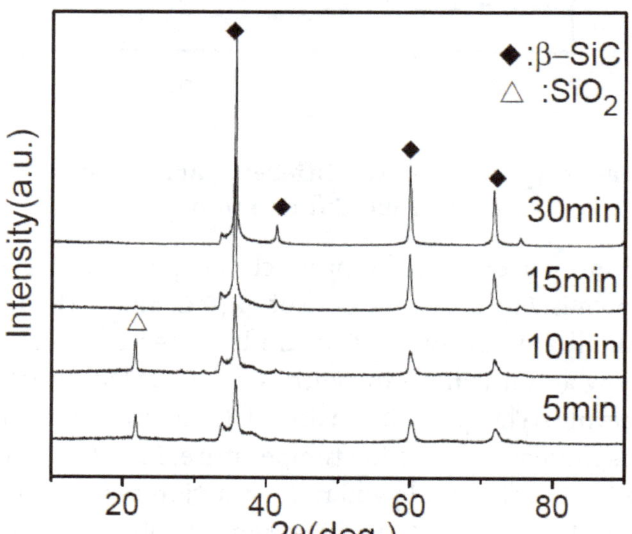

Fig. 27. XRD patterns of the products synthesized at 1500°C holding for different time

It can be concluded from the above results that carbothermal reduction could be promoted by increasing either the heating temperature or the holding time.

3.4 Effect of reaction temperature on the morphology of the products

Figure 28 shows the SEM images of the products synthesized at various temperatures holding for 15 min. It is clear that there was only very small amount of SiCw could be observed in the images of the product synthesized at 1200°C. When the heating temperature was increased from 1200°C to 1500°C, the amount of SiCw increased gradually. However at the case of 1600°C, the product is almost composed of SiC particles. This could be explained by the VS formation mechanism of SiCw. In the carbothermal reduction the formation of SiC mainly depends on Eq. (3)~(5). According to Eq. (3), SiO and CO will be formed firstly, and then the formed SiO will react with C or CO to form SiC. The products synthesized through Eq. (4) will mainly be SiC particles, and the products synthesized through the reaction Eq. (5) will primarily be composed of SiC whiskers. When the heating temperature is higher than 1300°C and the partial pressure of CO is higher than 0.027 MPa, the reaction Eq. (5) will dominate the process to synthesize SiCw. When the reaction temperature is too high, violent reaction rate will result in excessive production of SiO and CO, thus SiC will grow in several directions to form SiC particles. So, 1500°C is the advisable reaction temperature while the temperatures over that are disadvantage for the synthesis of SiCw via microwave heating (Sharma et al., 1984; Biernacki & Wotzak, 1989; Chrysanthou et al., 1991).

$$SiO_2(s) + C(s) = SiO(g) + CO(g) \tag{3}$$

$$SiO(g) + 2C(s) = SiC(s) + CO(g) \tag{4}$$

$$SiO(g) + 3CO(g) = SiC(s) + 2CO_2(g) \tag{5}$$

(a) 1200°C (b) 1300°C

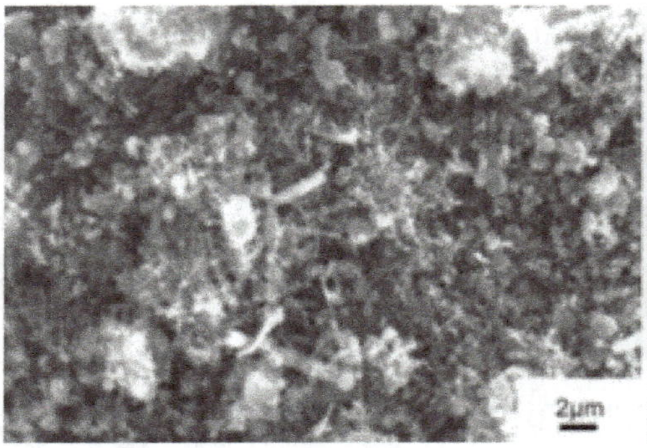

(c) 1400°C

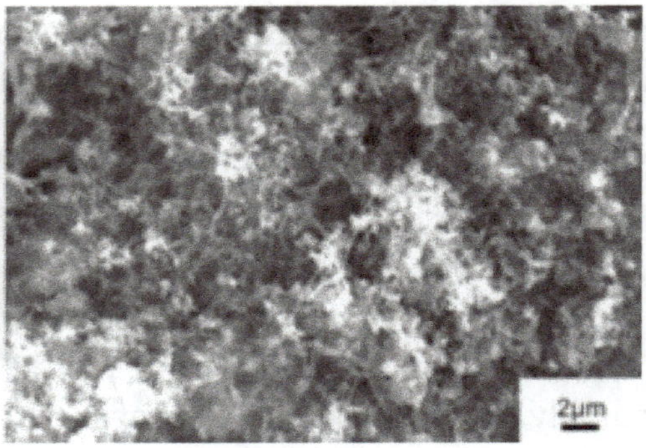

(d) 1500°C

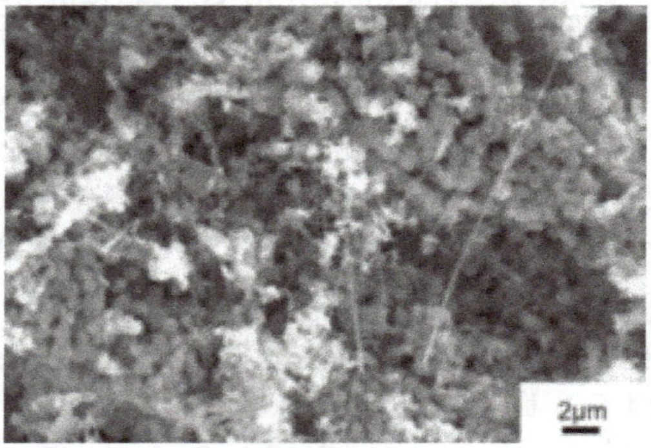

(e) 1600°C

Fig. 28. SEM images of products synthesized at different heating temperatures holding for 15 min

3.5 Effect of reaction time on the morphology of the products

Figure 29 shows the SEM images of the inner part of the products synthesized at 1500°C for various holding time. The images show that SiC*w* could be synthesized in all cases, and the amount of SiC*w* increased with the increase in holding time. When the holding time was set as 30 min, SiC*w* has the maximum aspect ratio of 100:1. The results reveal that increasing the holding time is benefit for the growth of SiC*w*.

3.6 Effect of reaction temperature and holding time on the yield of SiCw

The yield of SiC*w* synthesized at various temperatures and time were measured and listed in Table 6. With the same holding time of 30 min, the yield of SiC*w* prepared at 1300 and 1400°C are very low because lower heating temperature inhibits the progress of the reaction. The yield reached 56.9% when the temperature was increased to 1500°C and it will hit 60.1 wt% if the holding time was extended to 60 min. While when the heating temperature was further increased to 1600°C, the yield was decreased to 28.7wt% due to the formation of equiaxial SiC*p*. On the other hand, with the decrease in the particle size of raw materials, the yield of SiC*w* was increased which is shown in Table 6. The results indicate that, the product synthesized at 1500°C for 60 min has the maximum yield of SiC*w*, but holding for 30 min is more economic in our cases.

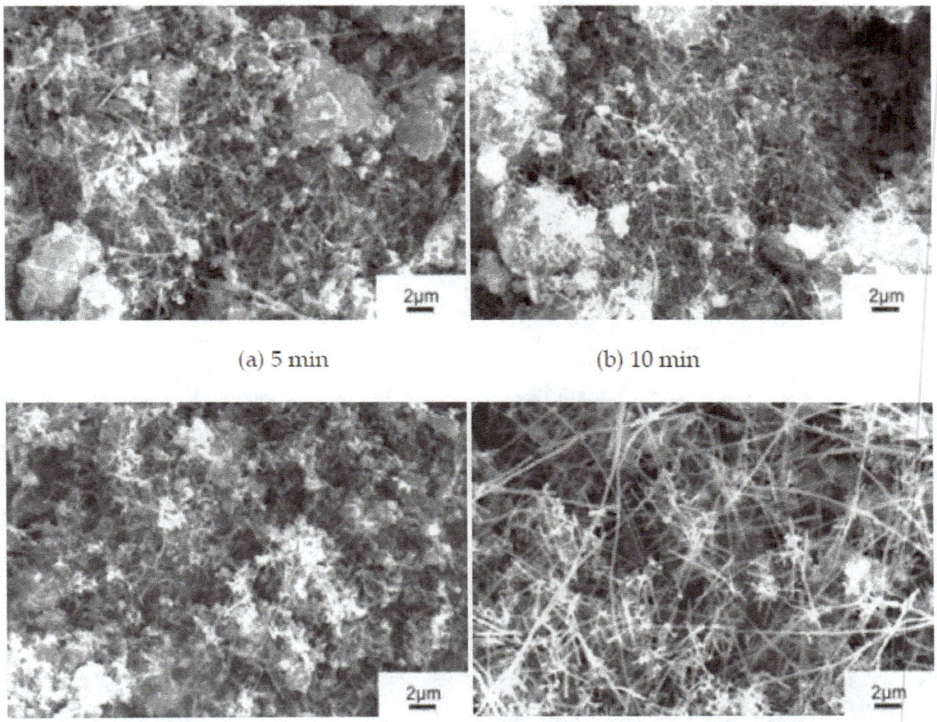

(a) 5 min (b) 10 min

(c) 15 min (d) 30 min

Fig. 29. SEM images of products heated at 1500°C holding for various time

Figure 30 shows the SEM images of the products prepared at 1500°C for 30 min, by using raw materials with particle size of 30~80 and 200~300 μm, respectively. It can be seen that there were many large grains of SiC have been synthesized in the products, and no SiC*w* could be observed on the surface of them. As the particle of raw materials is relatively large, it is hard to form SiO and CO and put the formation of SiC*w* at a disadvantage.

Table 6. The yield of SiC*w* prepared at various conditions

Heating temperature / oC	Holding time / min.	Particle size of raw materials / μm	Yield of SiC*w* / weight %
1300	30	1-10	4.8
1400	30	1-10	7.1

Heating temperature / oC	Holding time / min.	Particle size of raw materials / µm	Yield of SiCw / weight %
1500	30	1-10	56.9
1600	30	1-10	28.7
1500	15	1-10	30.3
1500	60	1-10	60.1
1500	30	30-80	34.6
1500	30	200-300	28.3

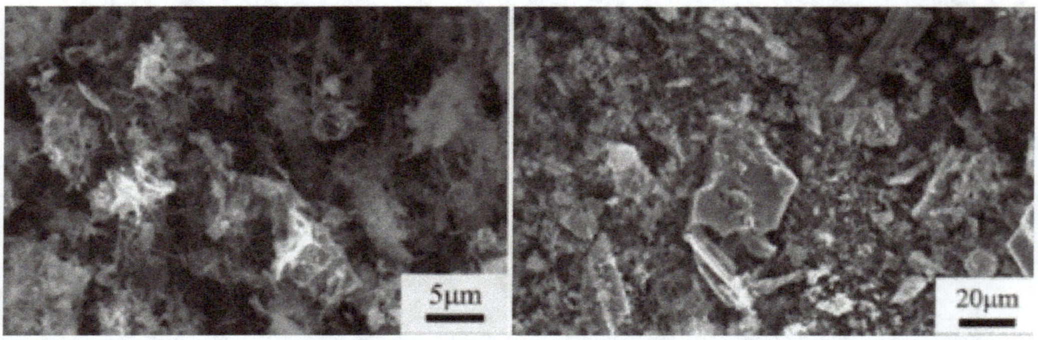

(a)30~80 µm (b) 200~300 µm

Fig. 30. SEM image of products prepared by using different grain size of raw materials

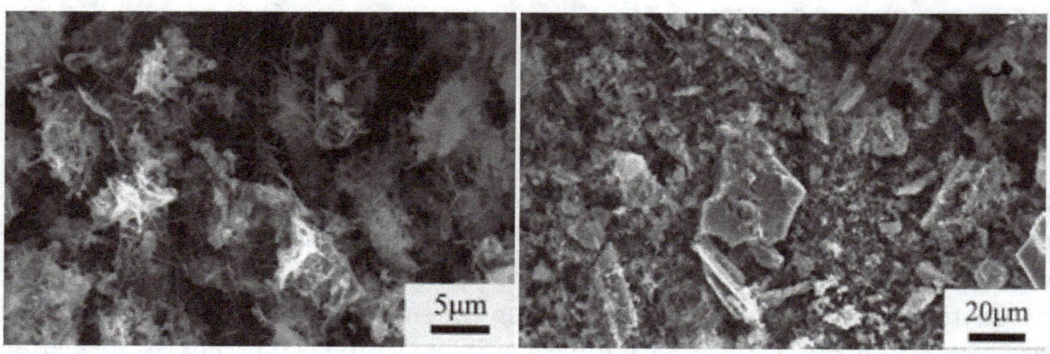

(a)30~80 µm (b) 200~300 µm

3.7 Effect of carbon particle size on the morphology of the products

In order to investigate the effect of particle size of carbon on the synthesis of SiCw, the carbon black (500 nm) was selected instead of activated carbon as carbon source, mixed with silica sol to prepare SiCw. The preparation process is the same with that when using activated carbon. The phase compositions and morphology of product synthesized at 1500°C for 30 min are shown in Fig. 31. The obtained product was β-SiC however its morphology was different from that prepared from the activated carbon-silica sol system. The product is composed of equiaxed particles of SiC with a narrowed particle size distribution, and SiCw could not be observed by SEM.

The carbothermal reduction process between the nano SiO2 and the carbon black could be illustrated by Fig. 32 according to the literature (Wang et al., 1992). Firstly, SiO and CO were formed in the carbothermal reduction, and the formation of SiC was mainly depended on the gas-solid reaction between SiO and C, because its Gibbs free energy was lower than that of the reaction between SiO and CO. SiC formed on the surface of the carbon particles and the formed SiC layer hindered the diffusion of carbon atom and then the reaction rate between SiO and C was decreased. When the carbon particle was ultrafine (500 nm) the formed SiC layer was also thin, thus the carbon atom could diffuse through this layer and react with SiO to form SiC, thus, the morphology of as-prepared SiC is similar to carbon particle. As the size of carbon particle was several microns, thicker SiC layer would form,

which hindered the diffusion of carbon atom and the reaction between SiO and C. In this case the formation of SiC was depending mainly on the gas-solid reaction between SiO and CO. Generally the product of this reaction was believed to synthesize one-dimension SiC, such as whisker, nanowire, and nanorod by VS mechanism.

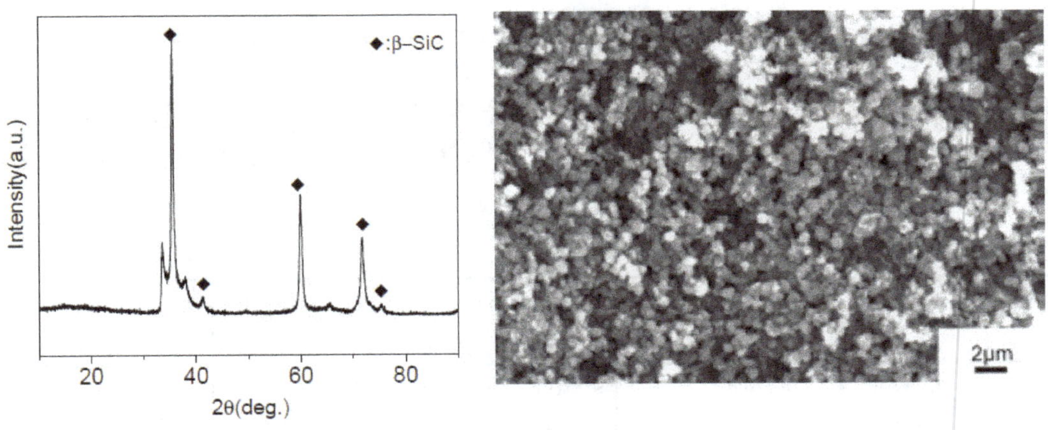

Fig. 31. XRD pattern (left) and SEM image (right) the product synthesized at 1500°C for 30 min by using carbon black as carbon source

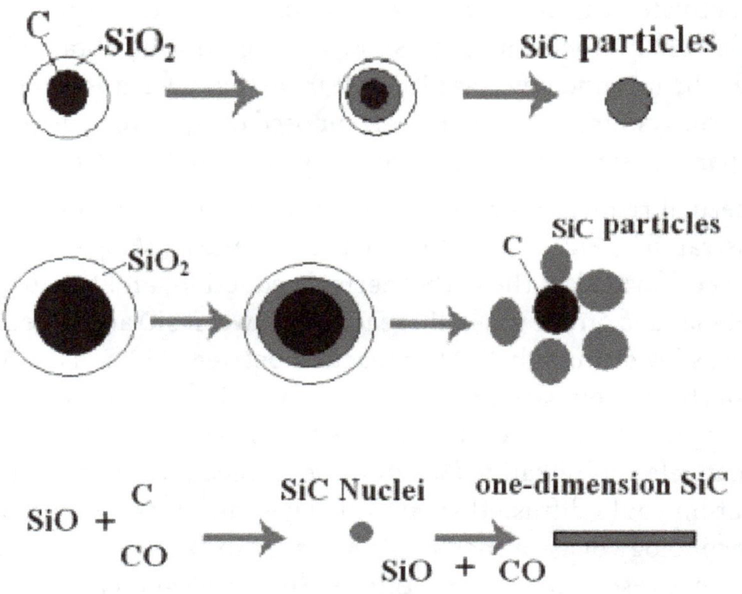

Fig. 32. Schematic diagram of carbothermal reduction

Except for silica sol-activated carbon and silica sol-carbon black systems we used, other works that using different silicon and carbon source to prepare SiC*w* by microwave heating have been published, and some of them are listed in Table 7. No data on yield of SiC*w* has been mentioned in their papers.

Table 7. Synthesis of SiC*w* by microwave heating. 1-(Dai et al., 1997a); 2-(Lu et al., 2007); 3- (Wei et al., 2008); 4-(Fu et al., 2009).

Silicon Source	Carbon Source	Preparation Method	Products	Morphology
precipitated silica[1]	char carbon black	1180, 1240 and 1350°C for 20 min	β-SiC and a little α-SiC	D: 20 ~ 50nm L: several microns
silicon powder[2]	phenolic resin	1300~1400°C for 0.5-2h	SiC	D: 20 ~ 100nm L: tens microns
tetraethylortho silicate (TEOS)[3]	sucrose	> 1300°C for 30 min	6H-SiC	D: 5~200 nm L: tens to hundreds of microns
quartz plate[4]	methane	1050~1200°C for 60 min	β-SiC	D: 100nm L: several tens of microns

3.8 Effect of packing density of the reactants on the morphology of the products

It is thought that the growth of SiCw by VLS or VS mechanism will firstly start in the free space in the reactants. So the porosity in the reactants will greatly affect the morphology and productivity of SiCw.

The mixture powders of silica sol and activated carbon were well homogenized by ball-milling for 15min, 30min, 1h, 2h, 4h and 8h, which resulting in six different powder mixtures with different particle size. Each powder mixtures (30 g) was loose packed in crucible and heated at 1500°C for 30 min in flowing Ar atmosphere with pressure of 0.1 MPa.

(a) (b)

Fig. 33. SEM images of (a) powder mixtures after 8 h-ball-milling; (b) nano-SiO2 coated on the surface of activated carbon (local magnified image of Fig. 33(a))

Figure 33 shows the SEM images of powder mixtures after 8 h-ball-milling and its local magnified image. It can be seen that the nano-SiO2 was coated uniformly on the surface of activated carbon. The relationship between the particle size of raw material powders and ball milling time is indicated in Fig 34(a). AC represents activated carbon. When the ball milling time was increased from 15 min to 8 h, the D10, D50 and D90 of powders decreased from 5.1, 14.1 and 30.2 **μm** to 2.1, 4.4 and 9.0 **μm** respectively. Figure 34(b) shows the loose

packing density and the corresponding porosity of the powder mixtures after different ball milling time, which were calculated based on the Eq. (6) and (7).

$$p = 1 - (V_P/V) \tag{6}$$

$$\rho = W_P/V \tag{7}$$

The symbols p, VP, V, ρ, WP represent the porosity, true volume, total volume, packing density, and weight of raw material powders respectively. The value of V and WP were measured in our experiment; VP was calculated by the density and weight of SiO_2 (2.33 g/cm^3) and activated carbon (2.2 g/cm^3). With the increase in ball milling time, porosity of powders increased from 80.2% to 89.2%, and loose packing density decreased from 0.44 g/cm3 to 0.24 g/cm^3.

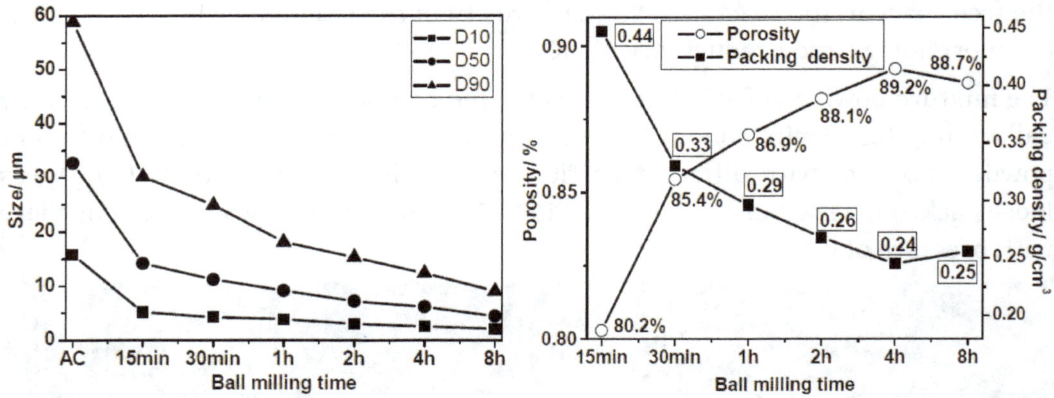

(a) (b)

Fig. 34. (a) particle size (b) porosity and loose packing density of powders vs. ball milling time

Figure 35 shows the SEM images of the synthesized SiC whiskers. When the loose packing density was 0.44 and 0.33 g/cm^3 (Fig. 35 a,b), two kinds of whiskers, flexural one and straight one, have been synthesized. The diameter of the whiskers was 100-300 nm, and the whiskers ranged in length from 1 µm to up to several microns. At the case of 0.29 g/cm3 (Fig. 35 c), flexural whiskers were almost disappeared, but still were about several microns in length. As the packing density was less than 0.26 g/cm^3 (Fig. 35 d,e,f), the morphologies of the SiC whiskers look quite different compared with that of the previously synthesized samples. SiC whiskers were straight and dispersal and the length of them were increased to more than 10 microns. But the diameter of the SiC whiskers was almost the same as that of the previous samples.

The variation in the morphologies of SiC whiskers from flexural to straight with the decrease of packing density of the raw materials could be explained by the different volume of the growing space. When the growing space was very small, the straight growth of whiskers was limited; whiskers would have a curving or circular movement during the growing process. Straight whiskers could be synthesized as a result having enough growing space. This tendency was consistent with the change of the loose packing density. The concentration of gaseous reactants should be also considered(Fu et al., 2005). When the packing density increases and the porosity decreases, the relative concentration of the,gaseous phase will increase and then the whiskers will grow in several directions results in the formation of flexural and shorter whiskers.

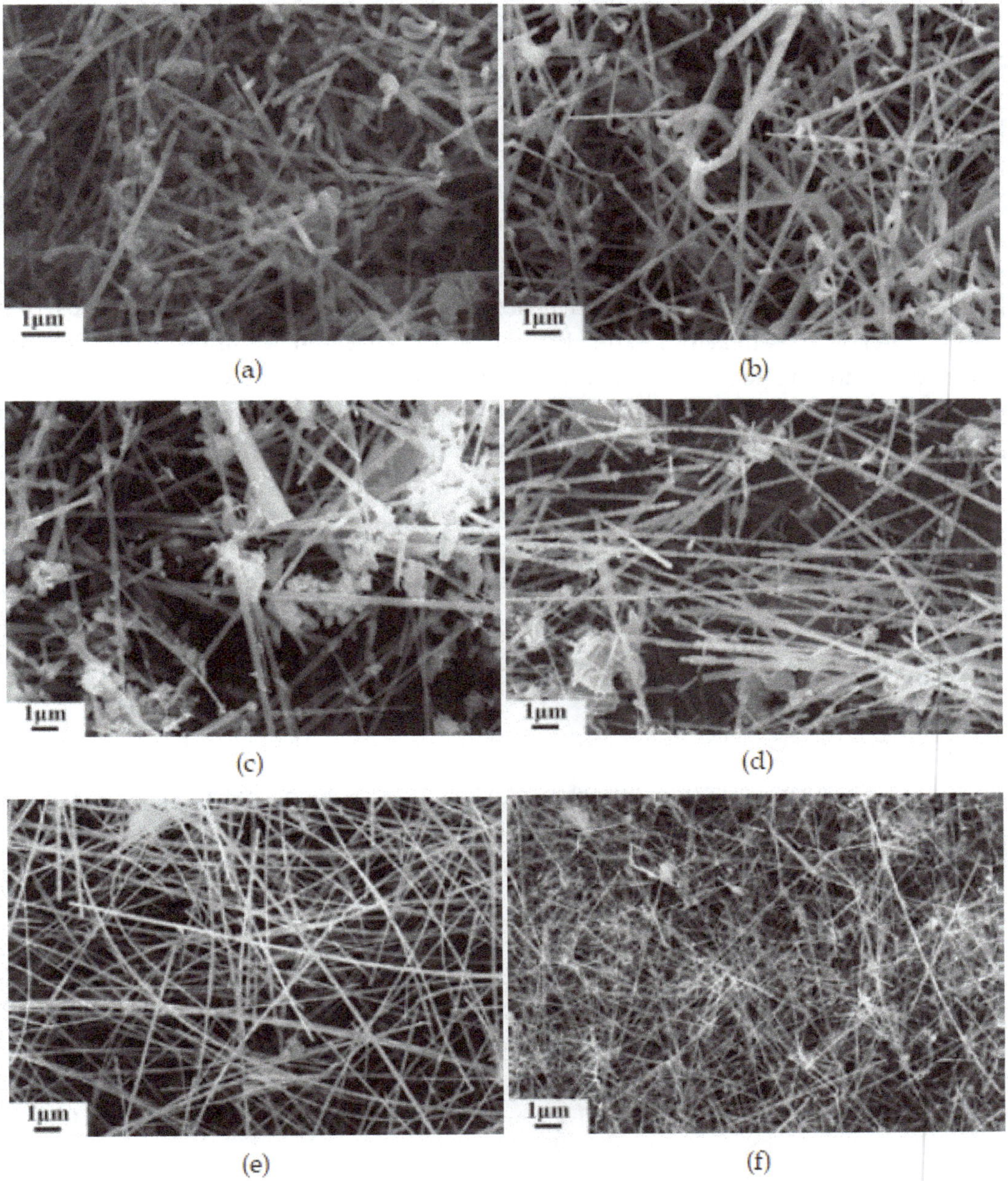

Fig. 35. SEM images of samples synthesized at 1500 °C for 30 min with the loose packing density(g/cm³) of (a) 0.44; (b) 0.33; (c) 0.29 (d) 0.26; (e) 0.24; (f) 0.25

4. CONCISE SUMMARY

Pure β-SiC powders were generated via microwave heating at temperatures ranging from 900°C to 1600°C, maintained for durations of 5 to 60 minutes, through the reaction of various silicon and carbon sources.

Aggregated SiC particles have been produced at 1200 °C using activated carbon and silicon. In the graphite-silicon system, β-SiC can be synthesized at 900 °C, partially addressing the issue of agglomeration of the produced products. Uniform equiaxial SiC particles with a limited particle size distribution were produced at 1200 °C utilizing carbon black and silicon. The complete conversion in the coke-silica system occurs at 1600 °C for a duration of 30 minutes, resulting in a SiC concentration of 98.1% in the reacted powders. The dimensions and morphology of the produced SiC particles are predominantly influenced by carbon sources.

Silicon carbide whiskers were synthesized from silica sol and activated carbon using microwave heating. The influence of reaction temperature, holding duration, and the loose compact density of the initial materials on the phase composition, morphology, and yield of the result has been examined. The yield of SiCw increases with elevated heating temperatures, reaching a maximum of 60.1 wt% when subjected to 1500 °C for 60 minutes. Subsequently, the yield will diminish with any additional rise in heating temperature. The morphology of the SiCw was seen to transition from flexural to linear when the packing density of the raw materials diminished, which was elucidated by the varying volume of the growth environment. A larger growing space will promote the development of straight whiskers, whereas a very little growing space will provide curved or circular whiskers.

5. REFERENCES

Biernacki, J. J. & Wotzak, G. P. (1989). Stoichiometry of the C + SiO2 Reaction. *Journal of the American Ceramic Society*, Vol. 72, No. 1, pp. 122-129, IS SN 1551-2916

Chrysanthou, A., Grieveson, P. & Jha, A. (1991). Formation of silicon carbide whiskers and their microstructure. *Journal of Materials Science*, Vol. 26, No. 13, pp. 3463-3476, IS SN 0022-2461

Dai, C., Zhang, X. & Zhang, J. (1997a). Microwave Synthesis of Ultrafine Silicon Carbide Whiskers. *Journal of the American Ceramic Society*, Vol. 80, No. 5, pp. 1274-1276, IS SN 1551-2916

Dai, C., Zhang, X. & Zhang, J. (1997b). The synthesis of ultrafine SiC powder by the microwave heating technique. *Journal of Materials Science*, Vol. 32, No. 9, pp. 2469- 2472, IS SN 0022-2461

Ebadzadeh, T. & Marzban-Rad, E. (2009). Microwave hybrid synthesis of silicon carbide nanopowders. *Materials Characterization*, Vol. 60, No. 1, pp. 69-72, IS SN 1044-5803

Fu, D., Zeng, X. & Zou, J. (2009). In situ synthesis and photoluminescence of SiC nanowires by microwave-assisted pyrolysis of methane. *Journal of Alloys and Compounds*, Vol. 486, No. 1-2, pp. 406-409, IS SN 0925-8388

Fu, Q., Li, H. & Shi, X. (2005). Microstructure and growth mechanism of SiC whiskers on carbon/carbon composites prepared by CVD. *Materials Letters*, Vol. 59, No. 19-20, pp. 2593-2597, IS SN 0167-577X

Guo, J. Y., Gitzhofer, F. & Boulos, M. I. (1995). Induction plasma synthesis of ultrafine SiC powders from silicon and CH4. *Journal of Materials Science*, Vol. 30, No. 22, pp. 5589- 5599, IS SN 0022-2461

Halamka, M., Kavecky, S. & Docekal, B. (2003). Synthesis of high purity Si3N4 and SiC powders by CVD method. *CERAMICS - SILIKATY*, Vol. 47, No. 3, pp. 88-93, IS SN 0862-5468

Hatakeyama, F. & Kanzaki, S. (1990). Synthesis of Monodispersed Spherical β-Silicon Carbide Powder by a Sol-Gel Process. *Journal of The American Ceramic Society*, Vol. 73, No. 7, pp. 2107-2110, IS SN 1551-2916

Jin, H. B., Cao, M. S. & Zhou, W. (2010). Microwave synthesis of Al-doped SiC powders and study of their dielectric properties. *Materials Research Bulletin*, Vol. 45, No. 2, pp. 247-250, IS SN 0025-5408

Klein, S., Winterer, M. & Hahn, H. (1998). Reduced-Pressure Chemical Vapor Synthesis of Nanocrystalline Silicon Carbide Powders. *Chemical Vapor Deposition*, Vol. 4, No. 4, pp. 143-149, IS SN 1521-3862

Krstic, V. D. (1992). Production of Fine, High-Purity Beta Silicon Carbide Powders. *Journal of The American Ceramic Society*, Vol. 75, No. 1, pp. 170-174, IS SN 1551-2916

Kuang, J. L., Li, L. & Cao, W. B. (2011). Preparation of SiC Whiskers by Microwave Heating. *Materials Science Forum*, In press, ISSN 0255-5476

Li, Y. L., Liang, Y. & Zheng, F. (1994). Carbon Dioxide laser Synthesis of Ultrafine Silicon Carbide Powders from Diethoxydimethylsilane. *Journal of The American Ceramic Society*, Vol. 77, No. 6, pp. 1662-1664, IS SN 1551-2916

Lu, B., Liu, J., Zhu, H. (2007). SiC Nanowires Synthesized by Microwave Heating. *Materials Science Forum*, Vol. 561-565, No. pp. 1413-1416, IS SN 1662-9752

Metaxas, A. C. (1991). Microwave heating. *Power Engineering Journal*, Vol. 5, No. 5, pp. 237- 247, IS SN 0950-3366

Narayan, J., Raghunathan, R. & Chowdhury, R. (1994). Mechanism of combustion synthesis of silicon carbide. *Journal of Applied Physics*, Vol. 75, No. 11, pp. 7252-7257, IS SN 0021-8979

Pampuch, R., Stobierski, L. & Lis, J. (1987). Solid combustion synthesis of β SiC powders. *Materials Research Bulletin*, Vol. 22, No. 9, pp. 1225-1231, IS SN 0025-5408

Ramesh, P. D., Vaidhyanathan, B. & Ganguli, M. (1994). Synthesis of β-SiC powder by use of microwave radiation. *Journal of Materials Research*, Vol. 9, No. 12, pp. 3025-3027, IS SN 2044-5326

Rao, K. J., Vaidhyanathan, B. & Ganguli, M. (1999). Synthesis of Inorganic Solids Using Microwaves. *Chemistry of Materials*, Vol. 11, No. 4, pp. 882-895, IS SN 0897-4756

Satapathy, L. N., Ramesh, P. D. & Agrawal, D. (2005). Microwave synthesis of phase-pure, fine silicon carbide powder. *Materials Research Bulletin*, Vol. 40, No. 10, pp. 1871- 1882, IS SN 0025-5408

Seog, I. & Kim, C. (1993). Preparation of monodispersed spherical silicon carbide by the sol- gel method. *Journal of Materials Science*, Vol. 28, No. 12, pp. 3277-3282, IS SN 0022- 2461

Setiowati, U. & Kimura, S. (1997). Silicon Carbide Powder Synthesis from Silicon Monoxide and Methane. *Journal of The American Ceramic Society*, Vol. 80, No. 3, pp. 757-760, IS SN 1551-2916

Sharma, N. K., Williams, W. S. & Zangvil, A. (1984). Formation and Structure of Silicon Carbide Whiskers from Rice Hulls. *Journal of the American Ceramic Society*, Vol. 67, No. 11, pp. 715-720, IS SN 1551-2916

Shawn, A., Holly, S. & Morgana, F. (2008). Microwave Heating Technologies. *Heat Treating Progress*, Vol. 8, No. 3, pp. 39-42, IS SN 1536-2558

Tairov, Y. M. (1995). Growth of bulk SiC. *Materials Science and Engineering B*, Vol. 29, No. 1-3, pp. 83-89, IS SN 0921-5107

Wang, L., Wada, H. & Allard, L. F. (1992). Synthesis and characterization of SiC whiskers. *Journal of Materials Research*, Vol. 7, No. 1, pp. 148-163, IS SN 0884-2914

Wang, F., Wang, Q., Cao, W. B. (2009). Microwave Synthesis of SiC Powders with Coke and Quartzite. *Journal of Materials Engineering*, No. 7, pp. 32-35, ISSN 1001-4381

Wang, F. (2008a). Study on preparation of β-SiC powders. Beijing: University of Science & Technology Beijing

Wang, F., Sun, J. L., Cao, W. B. (2008b). Microwave synthesis of silicon carbide whisker. *Refractories*, Vol. 42, No. 5, pp. 357-361, ISSN 1001-1935

Wang, Q. (2009). Research on the synthesis of silicon carbide powders by microwave heating method. Beijing: University of Science & Technology Beijing

Wei, G., Qin, W. & Wang, G. (2008). The synthesis and ultraviolet photoluminescence of 6H- SiC nanowires by microwave method. *Journal of Physics D: Applied Physics*, Vol. 41, No. 23, pp. 235102, IS SN 0022-3727

Weimer, A. W., Moore, W. G. & Rafaniello, W. (1994). Process for Preparing Silicon Carbide, U.S. Patent 5,340,417

Weimer, A. W., Nilsen, K. J. & Cochran, G. A. (1993). Kinetics of carbothermal reduction synthesis of beta silicon carbide. *AIChE Journal*, Vol. 39, No. 3, pp. 493-503, IS SN 1547-5905

Weimer, A. W., Roach, R. P. & Haney, C. N. (1991). Rapid carbothermal reduction of boron oxide in a graphite transport reactor. *AIChE Journal*, Vol. 37, No. 5, pp. 759-768, IS SN 1547-5905

Yamada, O., Hirao, K. & Koizumi, M. (1989). Combustion Synthesis of Silicon Carbide in Nitrogen Atmosphere. *Journal of the American Ceramic Society*, Vol. 72, No. 9, pp. 1735-1738, IS SN 1551-2916

INDEX

O

P

Q

R

S

www.ingramcontent.com/pod-product-compliance
Lightning Source LLC
LaVergne TN
LVHW062220200926
848484LV00006B/1444